Senior **Physics**

FOLENS

GEORGE PORTER

FOLENS

Preface

This book is intended for students following the Leaving Certificate syllabus in Physics and it covers this course in full to Higher Level standard. This edition has been prepared in the light of experience gained from using the first edition and from the comments and suggestions of teachers. Material which is required for Higher Level only is indicated by a vertical line beside the text. This material may be omitted without any break in continuity.

Each chapter concludes with a set of questions, consisting of two sections. Section I are completion-type questions similar to those found in Section A of the Leaving Certificate Examination papers. Section II consists mostly of numerical questions which are designed to test understanding as well as to provide practice in manipulation of formulae. These questions are intended to be an integral part of the text. In addition there are five sets of revision exercises which may be used as felt appropriate. These include a section of questions based on practical work, similar to those in Section B of the Examination papers. All questions on Higher Level material are marked with an asterisk. Multiple choice questions are not included in the text but are supplied in a separate booklet. In this there is a set of multiple choice questions for each chapter in the book. SI units and symbols are used throughout.

Practical work is an essential part of Physics and receives due emphasis in the syllabus. Throughout the text, continual reference is made to experimental investigations and results. Many of these investigations could be carried out as student experiments and/or teacher demonstrations, depending on time and equipment available. More specifically, a total of 40 student experiments, including all of those listed as mandatory in the syllabus, are described, with step-by-step instructions given for carrying out the experimental procedure. A number of questions have been included with each experiment to encourage students to think about the principles involved. The experiments have been integrated into the text to emphasise the essential unity of theory and experiment. Since it is not always possible for student experiments to be carried out in conjunction with the appropriate theory, a list of student experiments is given on page 345.

In addition to being an experimental subject, Physics is also a practical subject in the sense that it is relevant to everyday experience. This relevance is stressed wherever appropriate, with many references to everyday applications of the various physical principles. Worked examples are based, as far as possible, on problems which are likely to be relevant to the student's experience.

I am indebted to all who helped in the preparation of the first edition of this book. In particular, I would like to thank Prof. M. Sexton, Dr. A. Somerfield, Messrs. D. Condren, J. Coughlan, R. Henly, R. Kerr, J. Lynch and I. Travers for their time and expertise, so freely given. I would also like to record my gratitude to all who offered comments and suggestions during the preparation of the present edition and I am especially grateful to the following for their invaluable assistance: Messrs. B. Campbell, D. Condren, H. Dorgan, D. Kennedy and E. O'Flaherty. My thanks go also to Mr. J. Blandford of The Slide File and the staff of Folens, who as always, have been most helpful and co-operative throughout. Finally, I would like to thank my wife and family for their continuing support and encouragement.

G. C. Porter,
Beneavin College,
Dublin 11
April 1987.

Artwork Michael Phillips
Design Coordination Philip Ryan
Editor Antoinette Higgins
Layout Unlimited Design Company and Design Factor
Printed at the Press of the Publishers.
ISBN 0 86121 217 7
©Folens Publishers, Airton Road, Tallaght, Dublin 24.

Photos: The Slide File; Photo Library International (Figs. 5.6, 7.4, 10.7, 23.6); Daily Telegraph (Fig. 2.10); and as indicated in text. Cover courtesy of the National Microelectronics Research Centre, U.C.C. Thanks to the Department of Education for questions reproduced.
While considerable effort has been made to locate all holders of copyright material used in this text, we have failed to contact some of these. Should they wish to contact Folens Publishers, we would be glad to come to some arrangement with them.

Contents

1 — General Introduction
1.1 What is Physics   5
1.2 Measurements and Units   6
1.3 Experiments   7

2 — Reflection of Light
2.1 Light Rays   9
2.2 Plane Mirrors   9
2.3 Concave Mirrors   12
2.4 Convex Mirrors   18

3 — Refraction of Light
3.1 Refractive Index   23
3.2 Total Internal Reflection   26
3.3 Converging Lenses   31
3.4 Diverging Lenses   35

4 — Optical Instruments
4.1 The Human Eye   39
4.2 The Microscope   41
4.3 The Astronomical Telescope   42
4.4 Prism Binoculars   45
4.5 The Spectrometer   45
Revision Exercises A   49

5 — Linear Motion
5.1 Scalar and Vector Quantities   51
5.2 Displacement, Velocity and Acceleration   51
5.3 Equations of Motion   52
5.4 Momentum   56

6 — Force I
6.1 Newton's Law   62
6.2 Gravity   66

7 — Force II
7.1 Friction   72
7.2 Moments   74
7.3 Pressure   77

8 — Energy
8.1 Work   87
8.2 Energy   87
8.3 Power   92

9 — Vectors
9.1 Addition of Vectors   95
9.2 Resolution of Vectors   97

10 — Circular Motion
10.1 Angular Speed   100
10.2 Centripetal Acceleration   101
10.3 Satellite Motion   103

11 — Simple Harmonic Motion
11.1 Simple Harmonic Motion   108
11.2 The Simple Pendulum   110
Revision Exercises B   113

12 — Wave Motion
12.1 Waves   117
12.2 Diffraction   119
12.3 Interference   120
12.4 Reflection and Refraction   124
12.5 Resonance   125
12.6 Polarisation   125
12.7 The Doppler Effect   126

13 — Wave Nature of Light
13.1 The Corpuscular Theory   131
13.2 The Wave Theory   131
13.3 The Speed of Light   135
13.4 Dispersion   136
13.5 The Electromagnetic Spectrum   138
13.6 Spectra   139

14 — Sound
14.1 Wave Nature of Sound   143
14.2 Harmonics   145
14.3 Musical Notes   149
14.4 Speed of Sound   149
14.5 Ultrasonics   152
14.6 Intensity of Sound   153

15 — Temperature
15.1 Temperature Scales   158
15.2 Thermometers   159
15.3 Boyle's Law   161
15.4 The Ideal Gas Temperature Scale   164

16 — Kinetic Theory
16.1 Brownian Movement   169
16.2 The Kinetic Theory of Gases   170

17 — Heating and Internal Energy

17.1 Specific Heat Capacity	176
17.2 Specific Latent Heat	179
17.3 Energy Transfer	182
Revision Exercises C	187

18 — Electrostatics I

18.1 Electric Charges	190
18.2 Atomic Theory	193
18.3 Electric Fields	194

19 — Electrostatics II

19.1 Potential	199
19.2 Electric Field Intensity	202

20 — Capacitance

20.1 Capacitance	209
20.2 Practical Capacitors	212

21 — Electric Circuits

21.1 Electric Current	215
21.2 Resistance	217
21.3 Resistors	219
21.4 Measuring Resistance	225
21.5 Resistance and Temperature	228

22 — Electrical Energy

22.1 Joule's Law	234
22.2 Domestic Wiring	237

23 — Chemical Effect of a Current

23.1 Electrolysis	241
23.2 Cells	245

24 — Current and Magnetism

24.1 Current and Force	251
24.2 Magnetic Fields	251
24.3 Force on a Current in a Magnetic Field	256
24.4 The d.c. Motor	258
24.5 Meters	259

25 — Electromagnetic Induction

25.1 The Laws of Electromagnetic Induction	264
25.2 Generators	266
25.3 Alternating Current	268
25.4 Inductance	270
25.5 The Induction Motor	274
Revision Exercises D	280

26 — Electron Beams

26.1 The Electron	283
26.2 Thermionic Emission	284
26.3 X-rays	287
26.4 Photoelectric Emission	288
26.5 Spectra and Electron Energies	291
26.6 Conduction in Gases	292

27 — Semiconductors

27.1 Semiconductors	296
27.2 The P-N Junction	299

28 — Semiconductors II

28.1 Transistors	309
28.2 Transistor Circuits	312

29 — Radioactivity

29.1 Nuclear Structure	318
29.2 Detectors	319
29.3 Types of Radiation	321
29.4 The Law of Radioactive Decay	325
29.5 The Uses and Hazards of Radiation	327

30 — Nuclear Energy

30.1 Fission	331
30.2 Fusion	333
Revision Exercises E	336

Appendices

A. Equations in Mechanics Using Calculus	337
B.1. Torque on a Coil in a Magnetic Field	339
2. The Law of Radioactive Decay	340
C. Physical Quantities, Units and Symbols	341
D. Values of Physical Constants and Other Useful Data	342
E. Useful Equations	343
F. List of Experiments.	345

Answers	346
Index	353
Syllabus	356

1 General Introduction

1.1 What is Physics

Scientists study nature with a view to understanding how particular events occur. Human beings are, by nature, curious about what goes on around them and so the gaining of knowledge and understanding is, in itself, both exciting and fulfilling. In addition, the knowledge gained by scientists can very often be used to make life more comfortable and enjoyable for all of us.

Fig. 1.2

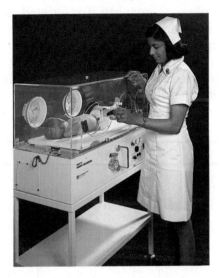

Fig. 1.1

The study of physics ranges from investigation of the very small to the very large. Physicists study atoms which are less than 10^{-10} metres (one ten millioneth of a millimetre) in diameter and galaxies which are millions of light years across (a light year is the distance light travels in one year — about 10^{13} kilometres). Time scales also vary enormously. Particles are known which exist for less than 10^{-20} seconds, while the lifetimes of stars are measured in thousands of millions of years.

The applications of physics are all around us. Telephone cables stretch around the world, both over land and under water. Radio and television programmes are broadcast, not only within countries, but also around the world via satellites which orbit 36 000 km above the earth's surface. Modern developments in telecommunications include the use of light, passing along very thin glass fibres, to carry information from place to place. The last twenty years have seen the beginning of the second industrial revolution with the development of microelectronics, whereby it is possible to construct thousands of circuits on a silicon chip only a few millimetres square. This has made possible the development of calculators, microcomputers, *etc.*, together with a vast range of automated machinery.

Fig. 1.3

The applications of physics are not confined to any particular area but are to be found in every aspect of modern engineering and medicine, from the construction of a bridge to the treatment of cancer. It must also be said, however, that not all the applications of scientific knowledge have been for the benefit of mankind — the mis-use of nuclear energy is but one example. Even beneficial applications can have unfortunate side effects, *e.g.* the pollution of the environment by traffic and industry.

However, despite the many problems, it is true to say that the work of scientists down through the years has resulted in a vastly improved standard of living for all.

1.2 Measurement and Units

Almost all scientific observations involve the measurement of physical quantities, *e.g.* length, weight, time, *etc.* When we make any measurement we are, in fact, comparing the size of something with an accepted standard. For example, when we measure the length of a room and say that it is 4 metres, we mean that it is four times as long as a particular standard length called the metre. There are many other standards we could have used, *e.g.* the foot, the yard, *etc.* All of these standards have one thing in common — there is general agreement on the exact size of each.

The SI System of Units

As we have seen, length may be measured in any one of a number of different units. The same is true of most physical quantities. Since scientists in different parts of the world often wish to compare results, *etc.*, it is obviously desirable that all should agree to use the same unit for any given physical quantity. This agreement has now been achieved with the International System of Units (SI). In this system there are seven basic, or fundamental, units. In this book we shall use six of these fundamental units. These are:

Unit	Physical Quantity
metre	length
kilogram	mass
second	time
ampere	electric current
kelvin	temperature
mole	amount of substance

The size of each of these fundamental units is defined very accurately. The units of all the other physical quantities which you will meet in this book are defined in terms of these fundamental units.

For convenience, multiples or fractions of the SI unit are sometimes used. For example, large distances may be measured in kilometres (10^3 metres) and small distances may be measured in millimetres (10^{-3} metres). However, these should always be converted to the standard SI unit before calculations are done or a final result given.

Symbols

Standard symbols have been agreed for most physical quantities. Thus, the letter m is used for mass, the letter t for time, *etc.* Likewise, units are represented by symbols, *e.g.* kg for kilogram, s for second, *etc.* Note that when a symbol represents a physical quantity it is printed in *italics*, while symbols for units are printed in ordinary roman type. A list of physical quantities, with their units and symbols, is given, for reference, in *Appendix C*.

Prefixes are used to indicate multiples and fractions of the standard unit, *e.g.* **kilo**metre, **milli**second, *etc.* The names of the most common prefixes, with their meanings and symbols, are:

Prefix	Symbol	Meaning
pico	p	10^{-12}
nano	n	10^{-9}
micro	μ	10^{-6}
milli	m	10^{-3}
kilo	k	10^3
mega	M	10^6
giga	G	10^9

Note that the indices are all multiples of 3. The prefix 'centi' (10^{-2}) is also used, but normally only with the metre.

The Scientific Method

In arriving at an understanding of a particular event, all scientists follow essentially the same method, which has therefore come to be known as the **scientific method**, although much the same procedure is followed by a doctor diagnosing an illness or by a detective investigating a crime.

The scientist starts by observing nature, by carrying out carefully controlled experiments and making an accurate, detailed record of the results. He or she then tries to summarise these results into simple statements. These statements are usually called **laws** or **principles**. The scientist then tries to develop a theory which explains the observations. The theory is tested by further experiments and, if found satisfactory, is generally accepted. An acceptable, and useful, theory is one which explains adequately a number of observations. A theory should also be capable of predicting further observations which may then be investigated experimentally. Further knowledge gained over the years may cause a theory to be modified or replaced entirely by a more comprehensive theory.

Energy

Physics, originally called Natural Philosophy, is the most basic of the sciences. It is concerned with energy and the properties of matter. Probably the most useful and common form of energy is electrical energy. However, energy can exist in a number of other forms, *e.g.* light, sound, chemical, nuclear, *etc.* During this course you will study each of the forms of energy in some detail.

1.3 Experiments

No measurement can give the true value of the measured quantity. Some uncertainty, depending largely on the instruments used, will always exist. This uncertainty is referred to as **experimental error**. While such errors can never be eliminated, their effect can be reduced by:

1. Taking several measurements of the same quantity.
2. Measuring the quantity under different conditions.
3. A suitable treatment of the results.

In quoting results you must be careful never to claim a greater accuracy than you are able to justify. As a general rule, no measurement can be more accurate than the smallest division on the instrument used. Thus, for example, if you measure the length of a line with a metre stick, your result can only be accurate to within 1 millimetre of the true value. It is therefore wrong to give the result as, say, 52.68 mm, since this implies that you measured the length to an accuracy of one hundredth of a millimetre, which is clearly impossible. The result should be given as 53 mm and you should be aware that the true value could lie anywhere between 52 mm and 54 mm, *i.e.* (53 ± 1) mm.

It should also be noted that the smaller the quantity being measured the more significant the error. For example, 1 mm in a length of 100 mm is an error of 1%, but 1 mm in a length of 2 mm is an error of 50% and makes the measurement practically worthless. Experiments should therefore be designed so that quantities to be measured should be as large as possible.

Significant Figures

Most results follow from calculations based on a series of measurements. The error in the final result will therefore reflect the errors in the individual measurements. A general rule is that the number of significant figures in the final result should not be greater than the number of significant figures in the *least* accurate of the measurements. In practice, in a school laboratory, this normally means that no result should contain more than three significant figures.

Tables and Graphs

Results may be presented in the form of a table or a graph or, frequently, both. A table gives a series of numerical values of a particular quantity or quantities. A graph shows the relationship between the numerical values of a pair of quantities.

A symbol for a physical quantity represents both a numerical value and a unit. For example, $l = 3$ m means that l represents a length which is three times as great as a standard length which we call a metre. Thus, the numerical value is equal to the symbol divided by the unit, *i.e.* $3 = l/m$. Since tables contain columns of numerical values each column should have a heading of the form (physical quantity divided by unit). For example, a table showing the distance, s travelled by a car at various times t might look like this:

s/m	t/s
20	1.6
30	2.6
40	3.4
50	4.1
60	4.9
70	5.8
80	6.6
90	7.5

Table 1.1

The axes of graphs should be labelled in the same way as the headings of a table *(see below)*.

In drawing a graph the following points should be noted.

1. A suitable scale should be chosen, *e.g.* 1 cm = 1, 2 or 5. The scale should also be such that the graph covers most of the page rather than being confined to a small corner of the page.
2. Axes should be clearly labelled as explained above.

3. Points should be plotted with a sharp pencil.
4. Points should be marked with a small dot surrounded by a circle.
5. The line should be drawn so that the points are equally distributed on either side of it. The origin should not be taken into account unless it represents measured values of the quantities in question.

With these points in mind the data shown in *Table 1.1* might be presented on a graph as shown in *Fig. 1.4*.

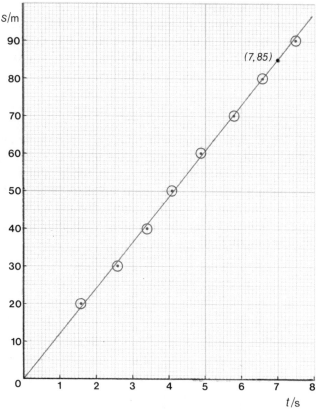

Fig. 1.4

The graph of *Fig. 1.4* is a straight line through the origin, *i.e.* it is of the form y = mx. The slope of the graph is distance divided by time, *i.e.* speed *(see Chapter 5)*. Thus, we may find the speed of the car by finding the slope of the graph. To do this we choose two points on the line which are as far apart as reasonably possible and apply the formula

$$m = \frac{y_2 - y_1}{x_2 - x_1}$$

where (x_1, y_1) ánd (x_2, y_2) are the coordinates of the chosen points. Since the graph passes through the origin we shall choose that as one point and as the second we shall take the point (7, 85). The slope is then

$$m = \frac{85 - 0}{7 - 0}$$

$$= 12.1 \text{ (unit)}$$

Since the slope is equal to the speed we now know that the speed of the car was 12 metres per second. Note that we say 12 metres per second and not 12.1 metres per second. Our data were given correct to two significant figures so our final result should also be given correct to two significant figures.

It should be clear that the unit of the slope is the metre per second. It is good practice to give the unit of the slope since it serves as a means of checking the relationship between the slope and the quantity which we wish to find.

2 Reflection of Light

2.1 Light Rays

Light is a form of energy. It is the form of energy which stimulates the nerve endings in our eyes, thus causing the sensation of 'seeing'.

The path taken by light as it travels between two points is called a **ray**. Light usually travels in straight lines — a fact sometimes referred to as the rectilinear propagation of light. However, there are some exceptions to this as we shall see later.

When a body, *e.g.* an electric lamp, emits light, it does so in all directions. We 'see' the body when one or more rays from it leads to our eye. A body which emits its own light, *e.g.* the sun or an electric lamp, is called a **luminous** body. However, most bodies are **non-luminous** and we can see such bodies only when they reflect light from a luminous body into our eye. For example, you can see this page because it is reflecting light from the sun, or perhaps from an electric lamp, into your eyes, *Fig. 2.1*. As shown in *Fig. 2.1* the light is reflected from the page in all directions. This is called **diffuse reflection**. Not all surfaces behave in this way. A highly polished surface reflects light in certain directions only, *Fig. 2.2*. This type of reflection is called **regular reflection**.

2.2 Plane Mirrors

A highly polished metal surface reflects 80% to 90% of the light falling on it. Most ordinary mirrors are made by depositing a thin layer of silver on the back of a piece of glass. If the glass is flat, the mirror is a plane mirror.

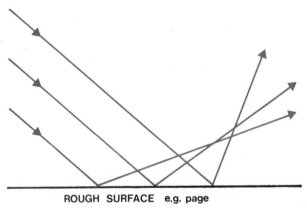

ROUGH SURFACE e.g. page

Fig. 2.1 Diffuse reflection

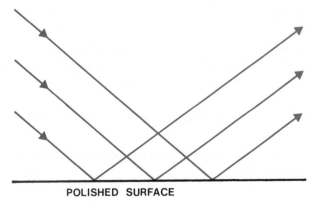

POLISHED SURFACE

Fig. 2.2 Regular reflection

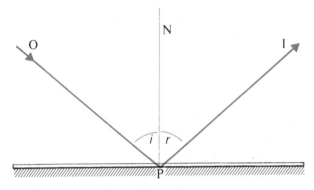

Fig. 2.3 Reflection at a plane mirror

Fig. 2.3 shows light being reflected by a plane mirror. The line, PN, drawn perpendicular to the surface of the mirror is called the **normal**. The line OP marks the path of the light coming from some object to the mirror and is called the **incident ray**. The path taken by the reflected light, PI, is called the **reflected ray**. Note that in each case there is an arrow on the line to indicate the direction in which the light is travelling. The angle between the incident ray and the normal is called the **angle of incidence**, i, and the angle between the reflected ray and the normal is called the **angle of reflection**, r.

Experiments with plane mirrors lead to two conclusions regarding the way in which light is reflected. These conclusions are known as the **Laws of Reflection**.

> 1. **The incident ray, the normal and the reflected ray are in the same plane.**
>
> 2. **The angle of incidence equals the angle of reflection.**

EXPERIMENT 2.1

To verify that the Angle of Incidence Equals the Angle of Reflection

Method 1:

Apparatus: Ray box, plane mirror and holder, protractor.

Procedure:

1. Place the mirror in the holder on a sheet of paper and mark the position of the back of the mirror.
2. Direct a ray of light at the mirror and mark the positions of the incident and reflected rays.
3. Remove the mirror and draw the incident and reflected rays and the normal at the point where they meet. Measure the angles of incidence and reflection.
4. Repeat the procedure for different values of the angle of incidence.
5. Prepare a table of results as shown.

RESULTS

$i/°$	$r/°$

Questions

1. Why should the position of the back of the mirror be marked rather than the front?
2. Why should the measurements be repeated a number of times for different values of the angle of incidence?

Method 2:

Apparatus: Four pins, plane mirror and holder, protractor.

Procedure:

1. Place the mirror in the holder on a sheet of paper and mark the position of the back of the mirror.
2. Draw a line at an angle to the mirror and place two of the pins on it.

3. Place the other two pins in front of the mirror so that they are in line with the images of the first two pins, *Fig. 2.4*. Mark the positions of the second pair of pins and remove them.
4. Remove the mirror and draw the incident and reflected rays and the normal at the point where they meet. Measure the angles of incidence and reflection.
5. Repeat the procedure for different values of the angle of incidence.
6. Prepare a table of results as shown in Method 1 and answer the questions at the end of Method 1.

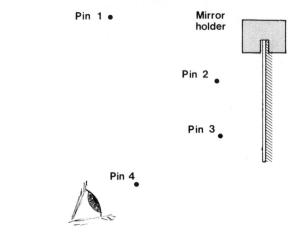

Fig. 2.4

How Images are Formed

When we look at a plane mirror we see an image of ourselves which appears to be behind the mirror. The ray diagram of *Fig. 2.5* shows how such an image is formed. O is, for convenience, a point object, *i.e.* a very small object. OA and OB are any two rays from O. To a person looking at the mirror the two rays entering his/her eye appear to come from a point behind the mirror. This point, I, is called the image of the point O. In this case, the light does not actually pass through the image, it only appears to do so. Such an image is called a **virtual** image.

It can be shown experimentally, or from the diagram of *Fig. 2.5*, that:

1. The line joining the image and the object cuts the mirror at right angles.

2. The image is the same distance behind the mirror as the object is in front of it.

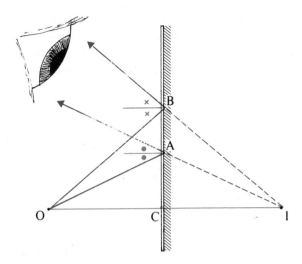

Fig. 2.5 Formation of an image in a plane mirror

EXPERIMENT 2.2

To Verify that the Image Distance Equals the Object Distance for a Plane Mirror.

Apparatus: Two pins, plane mirror and holder.

Procedure:

1. Place the mirror in the holder on a sheet of paper and mark the position of the back of the mirror.
2. Place one pin some distance in front of the mirror.
3. Move the second pin behind the mirror until it is seen to coincide with the image of the first pin. (They coincide if they appear together when viewed from different directions. If, when you move your head, the image seems to move in the same direction then the image is behind the pin; if the image seems to move in the opposite direction it is in front of the pin. This is known as the **no-parallax** technique.) Mark the positions of the two pins.
4. Measure the distance from the first pin to the back of the mirror. This is the object distance, u. Measure the distance from the second pin to the back of the mirror. This is the image distance, v.
5. Repeat the procedure for different positions of the object pin and prepare a table of results as shown.
6. Check that the line joining the image and object cuts the mirror at right angles.

RESULTS

u/cm	v/cm

Lateral Inversion

If you look at yourself in a plane mirror and close your right eye you will notice that your image seems to close its left eye; if you raise your left arm the image raises its right arm, and so on. This effect is known as **lateral inversion** and is most noticeable when print is viewed in a plane mirror, *Fig. 2.6.*

Fig. 2.6 Lateral inversion (Courtesy of the Inst. of Physics)

The Periscope

A simple periscope can be made by fixing two plane mirrors in a long tube, *Fig. 2.7*. The mirrors should be parallel to each other and at an angle of 45° to the axis of the tube. The light from an object then follows the path shown in the diagram. A periscope like this can be used to see over a crowd or over the top of an obstacle.

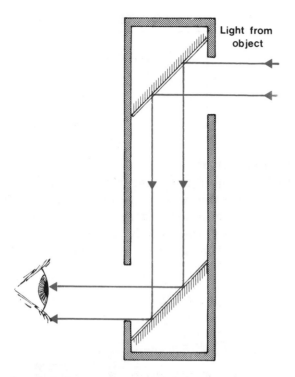

Fig. 2.7 *Simple periscope*

2.3 Concave Mirrors

A concave mirror is one in which the reflecting surface is curved inwards — like a cave! Most concave mirrors are spherical, *i.e.* are part of the surface of a sphere.

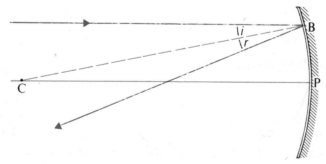

Fig. 2.8 *Reflection at a concave mirror*

Terms used in connection with concave mirrors are shown in *Fig. 2.8*. P, the centre of the mirror, is called the **pole** of the mirror. C, the centre of the sphere of which the mirror is a part, is called the **centre of curvature** of the mirror. Since

C is the centre of a sphere any line drawn through C, *e.g.* CB, is perpendicular to the surface of the sphere, *i.e.* is a **normal** to the mirror. The distance PC is the **radius of curvature**, R, and the line passing through P and C is called the **principal**

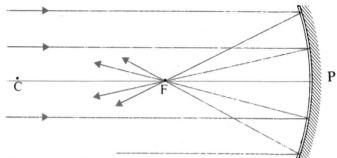

Fig. 2.9 *Rays parallel to the principal axis are reflected through the principal focus*

axis. Experimentally it is found that all rays parallel to the principal axis are reflected through the same point, *Fig. 2.9*. (This is strictly true only if the rays are near the axis.) The point through which the rays are reflected is called the **principal focus**, F. The distance PF is called the **focal length**, f.

Fig. 2.10 *Solar furnace: A large concave mirror is used to focus the sun's rays*

Focal Length and Radius of Curvature

We can show that the principal focus of a concave mirror is approximately half way between the centre of curvature and the pole of the mirror, *i.e.* that the radius of curvature is approximately equal to twice the focal length. Referring to the diagram of *Fig. 2.11* which shows an incident ray parallel to the principal axis being reflected through the principal focus:

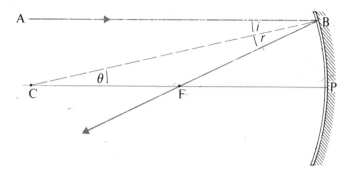

Fig. 2.11

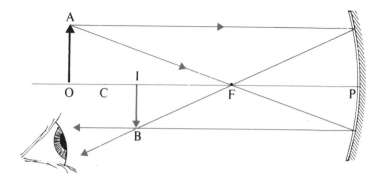

Fig. 2.12   *Formation of an image in a concave mirror when the object is outside C. The image is real, inverted and diminished*

$$i = r \qquad \text{(second law of reflection)}$$
$$i = \theta \qquad \text{(alternate angles)}$$
$$=> \quad r = \theta$$
$$=> \quad |BF| = |FC|$$

But if B is near to P, i.e. if the incident ray is near the axis (this condition is known as the paraxial condition), then

$$|BF| \approx |PF|$$
$$=> \quad |PF| = |FC|$$
$$|PC| = 2|FP|$$

i.e.
$$\boxed{R = 2f}$$

Note: This relationship is an approximation. It is an acceptable approximation if the error introduced by its use is, in practice, less than the experimental error. In most situations this is the case. The making of acceptable approximations is a very important part of the work of the physicist.

Formation of Images

Fig. 2.12 shows an object (*e.g.* a pin) OA placed in front of a concave mirror. Where will the image of the pin be formed? To find the image of any particular point, *e.g.* A, we must follow at least two rays from that point and find where they meet or appear to meet (*cf.* plane mirrors, *Fig. 2.5*). As there is an infinite number of rays from A, we choose two which are easy to follow. We know that light travelling parallel to the principal axis is reflected through the principal focus. Likewise, since light can travel equally well in either direction, light passing through the principal focus will be reflected parallel to the axis. To the observer, these two rays appear to be coming from the point B. Thus B is the image of A. We could repeat this procedure for other points on the object but this is not necessary. A ray of light from the point O on the

object travelling along the principal axis will be reflected back along the axis since the axis is a normal to the mirror. (The angle of incidence is zero so the angle of reflection is also zero.) Therefore, the image of O must lie on the axis. Thus, assuming that there is no distortion in the mirror, since the object is perpendicular to the axis the image will also be perpendicular to the axis. Therefore, the image of OA is IB. In this case, unlike the plane mirror, the image is not imaginary. If you were to place a sheet of paper at I perpendicular to the axis you would see an image of the object on the paper. In this case the image is a **real** image — the light actually passes through it. Note also that the image is **inverted**, *i.e.* upside down and back to front. It is also **diminished**, *i.e.* smaller than the object.

The nature of the image formed by a concave mirror depends on the position of the object. Look at your image in a concave mirror when it is close to your face, then gradually move it away and notice how the nature of the image changes. We have just considered the case of an object placed outside the centre of curvature of a mirror. The other possibilities are shown in *Fig. 2.13*. In each case the position of the image is found by following two rays from a point on the top of the object and finding where they meet or appear to meet. The rays chosen are two of the following: parallel to the axis; through the focus; through the centre of curvature. You should study these diagrams and practise drawing them for yourself. Note that erect (upright) images are laterally inverted, as in a plane mirror.

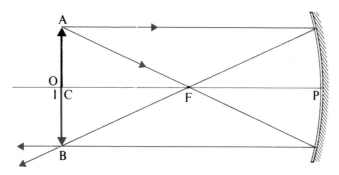

Fig. 2.13(a) Concave mirror, object at C: The image is real, inverted and the same size as the object

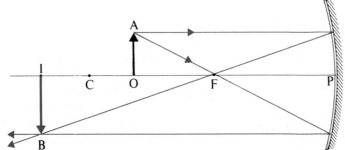

Fig. 2.13(b) Concave mirror, object between C and F: The image is real, inverted and magnified

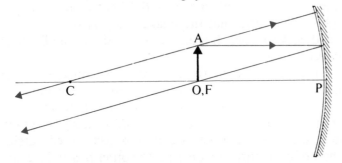

Fig. 2.13(c) Concave mirror, object at F: The image is at infinity

The Mirror Formula

All problems involving mirrors may be solved by drawing ray diagrams to scale and taking measurements from the diagrams. However, it is normally easier to use a formula which relates the positions of the object and image to the focal length of the mirror. Referring to *Fig. 2.14*:

The object distance, u, is $|OP|$.
The image distance, v, is $|IP|$.
The focal length, f, is $|PF|$.

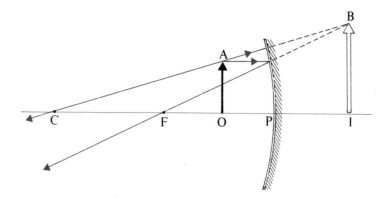

Fig. 2.13(d) Concave mirror, object inside F: The image is virtual, erect and magnified

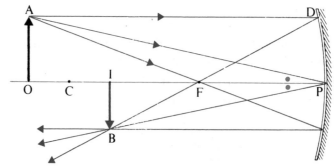

Fig. 2.14

The triangles AOP and BIP are similar.

$$\therefore \qquad \frac{|IB|}{|OA|} = \frac{|IP|}{|OP|} = \frac{v}{u}$$

But,
$$\frac{|IB|}{|OA|} = \frac{\text{height of image}}{\text{height of object}}$$
$$= \text{magnification, } m.$$

Therefore,
$$\boxed{m = \frac{v}{u}}$$

If the paraxial condition is satisfied, DPF may be treated as a right-angled triangle. Then, the triangles DPF and IFB are similar.

$$\therefore \qquad \frac{|IB|}{|DP|} = \frac{|IF|}{|PF|}$$

But,
$$|DP| = |OA|$$

$$\frac{|IB|}{|OA|} = \frac{|IF|}{|PF|}$$

$$= \frac{|IP| - |PF|}{|PF|}$$

i.e. $\quad \dfrac{v}{u} = \dfrac{v - f}{f}$

Multiplying both sides by uf gives

$$vf = uv - uf$$

Dividing both sides by uvf gives

$$\frac{1}{u} = \frac{1}{f} - \frac{1}{v}$$

Rearranging gives

$$\boxed{\frac{1}{f} = \frac{1}{u} + \frac{1}{v}}$$

We have derived this formula for one particular case. It can, however, be applied to all cases involving mirrors (and, as we shall see later, lenses) if we adopt a sign convention. We shall agree to make real distances positive and virtual distances negative. This is called the *'real is positive'* sign convention. The following examples should make clear the use of this convention.

Example
An object is placed 15 cm from a concave mirror of focal length 10 cm. Find the position, magnification and nature of the image.

$$u = +15 \text{ cm}$$

$$f = +10 \text{ cm (a concave mirror has}$$
$$\text{a real focus: } f \text{ is positive)}$$

$$v = ?$$

$$\frac{1}{f} = \frac{1}{u} + \frac{1}{v}$$

$$\frac{1}{10} = \frac{1}{15} + \frac{1}{v}$$

$$\frac{1}{v} = \frac{1}{10} - \frac{1}{15}$$

$$= +0.0333$$

$$=> \quad v = +30 \text{ cm}$$

The image is real (v is positive) and it is 30 cm in front of the mirror.

To calculate the magnification,

$$m = \frac{v}{u}$$

$$= \frac{+30}{+15}$$

$$= +2$$

The image is twice as large as the object and it is inverted.

Note: Under the 'real is positive' convention a **positive** value for m indicates an **inverted** image; a **negative** value indicates an **erect** image.

Example
An object is placed 10 cm from a concave mirror of focal length 20 cm. Find the position, magnification and nature of the image.

$$u = +10 \text{ cm}$$

$$f = +20 \text{ cm}$$

$$v = ?$$

$$\frac{1}{f} = \frac{1}{u} + \frac{1}{v}$$

$$\frac{1}{20} = \frac{1}{10} + \frac{1}{v}$$

$$\frac{1}{v} = \frac{1}{20} - \frac{1}{10}$$

$$= -0.05$$

$$\Rightarrow \quad v = -20 \text{ cm}$$

The image is virtual (v is negative) and it is 20 cm behind the mirror.

To calculate the magnification,

$$m = \frac{v}{u}$$

$$= \frac{-20}{+10}$$

$$= -2$$

The image is twice as large as the object and it is erect (m is negative).

EXPERIMENT 2.3
To Find the Focal Length of a Concave Mirror
Method 1:
Apparatus: Concave mirror, screen (*e.g.* sheet of white cardboard).

Procedure:
1. Focus an image of a distant object, *e.g.* a tree, onto the screen held in front of the mirror, *Fig. 2.15*.
2. Measure the distance from the mirror to the page. This distance is approximately equal to the focal length of the mirror.

RESULTS

Distance from mirror to page . . .   =   cm
Approximate focal length of mirror   =   cm

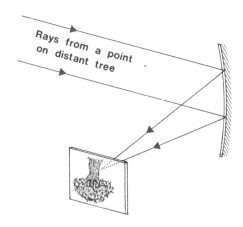

Fig. 2.15 The image of a distant object is formed at the focus

Questions:
1. Why is the measured distance only approximately equal to the focal length?
2. How far away should the object be to ensure that a satisfactory degree of accuracy is obtainable?

Method 2:
Apparatus: Concave mirror, lamp box, screen, metre stick.

Procedure:
1. Arrange the lamp box, mirror and screen so that an image of the slit in the lamp box is formed on the screen, *Fig. 2.16*.

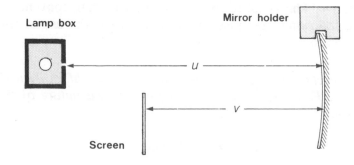

Fig. 2.16

2. Measure the distance from the lamp box to the mirror, u, and the distance from the screen to the mirror, v.
3. Calculate the focal length of the mirror from

$$\frac{1}{f} = \frac{1}{u} + \frac{1}{v}$$

4. Repeat the procedure for different positions of the lamp box. Calculate the value of f in each case and hence find an average value for f. Present your results in the form of a table as shown below.

Note:
If the screen is beside the lamp box $u = v = 2f$.

RESULTS

u/cm	v/cm	f/cm

Focal length of mirror =   cm.

Questions:

1. In what circumstances will it not be possible to obtain an image on the screen?

Method 3:

Apparatus: Concave mirror, two pins, metre stick.

Procedure:
1. Place one pin in front of the mirror. This pin serves as an object.
2. Move the second pin until it coincides with the image of the first. (Use the no-parallax technique.)
3. Measure the object distance and the image distance and calculate the focal length of the mirror from

$$\frac{1}{f} = \frac{1}{u} + \frac{1}{v}$$

4. Repeat the procedure for different positions of the object pin. Calculate the value of f in each case and hence find

an average value for f. Present your results in the form of a table as shown below.

RESULTS

u/cm	v/cm	f/cm

Focal length of mirror =   cm.

Questions:

1. What are the advantages of this method compared with the previous one?

Uses of Concave Mirrors

Concave mirrors are used as shaving and make-up mirrors because they are capable of producing a magnified image. Obviously, when you use such a mirror your face must be inside the focus (*see Fig. 2.13*).

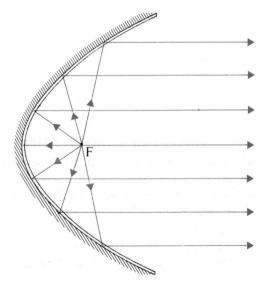

Fig. 2.17 A parabolic reflector

Concave mirrors can be used as reflectors in car headlamps and in spotlights since, when a lamp is placed at the focus, a parallel beam of light is produced. In these cases a parabolic mirror is generally used rather than a spherical one since the parabolic mirror has a sharper focus and hence gives a more accurately parallel beam of light, *Fig. 2.17*.

As we shall see in *Chapter 4*, concave mirrors are also used in reflecting telescopes.

2.4 Convex Mirrors

For a convex mirror, the centre of curvature is behind the mirror, *Fig. 2. 18*. As shown in *Fig. 2.19* rays which come in parallel to the principal axis are reflected in such a way that the reflected rays seem to come from a point on the axis behind the mirror. In other words, the principal focus is behind the mirror. As in the case of the concave mirror, we can show that the focus is half way between the centre of curvature and the pole of the mirror. However, unlike the concave mirror, the reflected rays only seem to pass through the focus, *i.e.* the focus of a convex mirror is virtual.

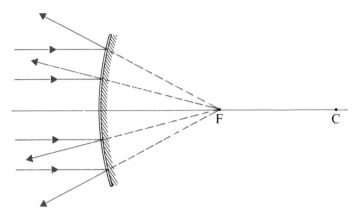

Fig. 2.19 *All rays parallel to the principal axis are reflected as if they came from the focus*

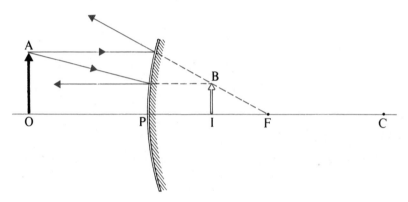

Fig. 2.20 *Formation of an image in a convex mirror. The image is virtual, erect and diminished*

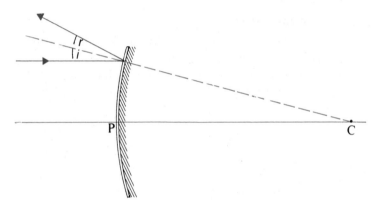

Fig. 2.18 *Reflection at a convex mirror*

Formation of Images in Convex Mirrors

As we did with concave mirrors we can find the position and nature of the image formed in a convex mirror by following two rays from a point on the object and finding where they appear to meet, *Fig. 2.20*. Since the centre of curvature is behind the mirror, the nature of the image is not affected by the position of the object. Thus, the image in a convex mirror is always virtual, erect and diminished. It is also laterally inverted.

EXAMPLE

An object is placed 20 cm from a convex mirror of focal length 5 cm. Find the position and magnification of the image.

$$u = + \, 20 \text{ cm}$$

$$f = - \, 5 \text{ cm (the focus is virtual: } f \text{ is negative)}$$

$$v = ?$$

$$\frac{1}{f} = \frac{1}{u} + \frac{1}{v}$$

$$- \frac{1}{5} = \frac{1}{20} + \frac{1}{v}$$

$$\frac{1}{v} = -\frac{1}{5} - \frac{1}{20}$$

$$= -0.25$$
$$=> \quad v = -4 \text{ cm}$$

The image is virtual (*v* is negative) and it is 4 cm behind the mirror.

To calculate the magnification,

$$m = \frac{v}{u}$$

$$= \frac{-4}{+20}$$

$$= -\frac{1}{5}$$

The image is one fifth the size of the object and it is erect (*m* is negative).

EXPERIMENT 2.4
To Find the Focal Length of a Convex Mirror
Method 1:

Apparatus: Two pins, convex mirror, (cork).

Procedure:
1. Place one pin in front of the mirror. Place the second pin behind the mirror and move it around until in coincides with the image of the first pin (use the no-parallax technique).

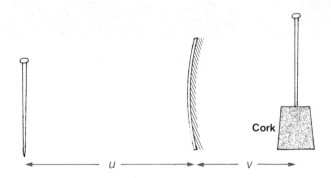

Fig. 2.21

2. Measure the object and image distances and calculate the focal length of the mirror from

$$\frac{1}{f} = \frac{1}{u} + \frac{1}{v}$$

(Remember: Virtual distances are given a negative sign.)
4. Repeat the procedure for different positions of the object pin. Calculate the value of *f* in each case and hence find an average value for *f*.

Note:
The image in a convex mirror is always behind the mirror. The image pin should be placed so that the top of it is visible over the top of the mirror (use the cork if necessary). When it coincides with the image of the object pin it should appear as a continuation of the image in the mirror no matter what angle it is viewed from.

RESULTS

u/cm	v/cm	f/cm

Focal length of mirror =   cm.

Method 2:

Apparatus: Convex mirror, small plane mirror, one pin, metre stick.

Procedure:
1. Place the plane mirror in front of the convex mirror as shown in *Fig. 2.22*.
2. Move the pin until the images of the pin in the two mirrors coincide.
3. Measure the distance from the pin to the plane mirror, *x*, and the distance between the mirrors, *y*. Then $u = x + y$ and $v = x - y$.

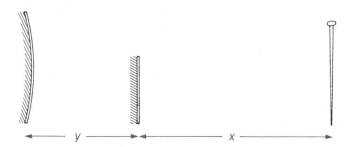

Fig. 2.22

4. Calculate the focal length of the mirror from

$$\frac{1}{f} = \frac{1}{u} + \frac{1}{v}$$

(Remember: Virtual distances are given a negative sign.)

5. Repeat the procedure for different positions of the plane mirror and hence find an average value for f.

RESULTS

x/cm	y/cm	u/cm	v/cm	f/cm

Focal length of mirror =   cm.

Uses of Convex Mirrors

The most common use of convex mirrors is as rear view mirrors in cars. As rear view mirrors they have two advantages over other types of mirror, *viz*.

1. They always give an erect image.
2. They have a wide field of view, *Fig. 2.23*.

The fact that the image is always diminished is something of a disadvantage as it can give a false sense of distance.

Convex mirrors are also used in shops, on the upper deck of buses and at dangerous bends on some roads. In each of these cases the main advantage of using convex mirrors is that they have a wide field of view.

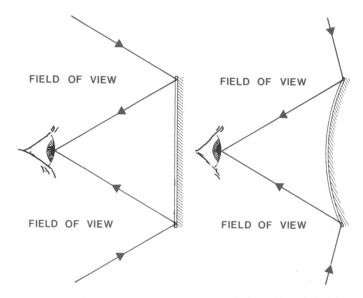

Fig. 2.23 *Convex mirror has a wide field of view*

Fig. 2.24 *A convex mirror at a dangerous bend in the road. The wide field of view allows drivers to see around the bend*

SUMMARY

Light is a form of energy. It travels in straight lines called rays. Light undergoes diffuse reflection at rough surfaces and regular reflection at smooth surfaces. The two Laws of Reflection are:

1. The incident ray, the normal and the reflected ray are in the same plane.
2. The angle of incidence equals the angle of reflection.

The image in a plane mirror is always virtual, the same size as the object, as far behind the mirror as the object is in front of it and laterally inverted.

The position of the image in a spherical mirror may be determined by tracing any two rays from a point on the object and finding where they meet, or *appear* to meet. The incident rays normally chosen are two of:

1. Ray parallel to the principal axis — reflected through the principal focus.
2. Ray through the principal focus — reflected parallel to the principal axis.
3. Ray through the centre of curvature — reflected back along the same path.

If the reflected rays do not actually meet the image is virtual.

The position and nature of an image may be found by using the formula

$$\frac{1}{f} = \frac{1}{u} + \frac{1}{v}$$

with the sign convention that virtual distances are negative.

The magnification of the image is given by:

$$m = \frac{v}{u}$$

For a concave mirror the nature of the image depends on the position of the object *(see Fig. 2.13, p.14)*. Convex mirrors always give virtual, diminished, erect, laterally inverted images of real objects.

Concave mirrors are used as make-up mirrors, shaving mirrors and in reflecting telescopes. Convex mirrors are used as rear view mirrors in cars and also in shops, buses and at dangerous bends on the road.

Questions 2

Section I

1. What is a light ray? ....
   ....

2. What is the difference between a luminous body and a non-luminous body? ....
   ....

3. State the laws of reflection....................................
   ....
   ....

4. An image in a plane mirror appears 'back-to-front'. What is this phenomenon called?.....................................
   ....

5. The focal length of a concave mirror is equal to . . . . . the radius of curvature.

6. For a concave mirror to form a real image the object must be ....
   ....

7. A virtual image in a concave mirror is always............
   ....

8. Give two uses of concave mirrors...........................
   ....

9. An object is placed so that its distance from a concave mirror is twice the focal length of the mirror. Where is the image formed? ....
   ....

10. Why are image and object distances measured from the back of a mirror rather than the front? ....

....

....

11. A white surface, *e.g.* a page, reflects most of the light falling on it. Why does such an object not form an image of objects placed in front of it? ....

....

12. What is meant by saying that an image is formed at infinity? ....

....

13. The image of a distant object formed by a concave mirror is taken to be at the focus of the mirror. Explain why this is so. ....

....

....

14. The image in a convex mirror is always ....

....

15. What is the sign convention used with the mirror formula? What is the purpose of using such a convention?.......

....

....

Section II

1. *Fig. I* shows a ray of light being reflected from two mirrors

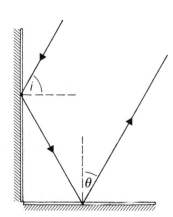

Fig. I

at right angles to each other. If the value of the angle *i* is 60°, what is the value of the angle θ?

2. Calculate the position and magnification of the image formed by a concave mirror of focal length 15 cm when the object is (i) 40 cm; (ii) 20 cm; (iii) 10 cm, from the mirror. Give the nature of the image in each case.

3. Draw a ray diagram showing how an erect, virtual image is formed by a concave mirror. Is it possible for such a mirror to form an erect, real image?

4. A concave mirror of focal length 40 cm forms a virtual image of a real object. If the image is 20 cm behind the mirror where is the object? If the object is a coin, 2.5 cm in diameter, what is the diameter of the image?

5. A concave mirror forms a real image whose magnification is 2. If the object is 30 cm from the mirror, where is the image and what is the focal length of the mirror?

6. An object, 30 cm in front of a convex mirror, produces an image 10 cm behind the mirror. What is the focal length of the mirror?

7. A convex mirror of focal length 10 cm forms an image of a pin which is placed 30 cm from it. Find the position and magnification of the image.

8. Give three uses of convex mirrors and explain why a convex mirror is used in each case rather than a plane or concave mirror.

9. A convex mirror of focal length 15 cm forms an image which is one fifth the size of the object. Find the positions of the object and the image.

10. A convex mirror, a plane mirror and a pin are arranged so that the images of the pin in both mirrors coincide *(see Experiment 2.4, p.19)*. Given that the distance between the two mirrors is 5 cm and the distance between the pin and the plane mirror is 15 cm calculate the focal length of the convex mirror.

11. A mirror forms an image which is twice the size of the object when the object is 10 cm from the mirror and again when the object is 30 cm from the mirror. What type of mirror is it and what is its focal length?

3 Refraction of Light

3.1 Refractive Index

Fig. 3.1 shows a ray of light travelling through a block of glass. Notice that, at the points where the light enters and leaves the glass, it changes direction. This phenomenon is known as **refraction** — light travelling from one medium into another of different density changes direction. Generally, if the second medium is denser than the first, the light is bent towards the normal and *vice versa*. Thus, when the light enters the glass it is bent towards the normal and when it leaves the glass it

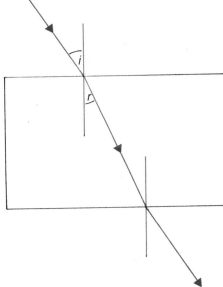

Fig. 3.1 Refraction of light on entering and leaving a block of glass

is bent away from the normal, *Fig. 3.1*. There is one exception to this. If the ray of light strikes the surface of the glass at right angles it is not refracted, it passes straight through. The angle between the incident ray and the normal (*i* in the diagram) is called the **angle of incidence** and the angle between the refracted ray and the normal (*r* in the diagram) is called the **angle of refraction**. Experiments lead to two conclusions regarding the way in which light is refracted. These are known as the **Laws of Refraction:**

> 1. **The incident ray, the normal and the refracted ray are in the same plane.**
> 2. **The sine of the angle of incidence is proportional to the sine of the angle of refraction.**

The second law is usually known as **Snell's Law** in honour of the Dutch mathematician Willebrord Snell (1591 — 1626) who discovered it in 1621. It may be written as

$$\frac{\sin i}{\sin r} = \text{a constant}$$

If the light is passing from vacuum into another medium the constant is called the **refractive index**, *n*, of the other medium. (In practice the refractive index of a medium is usually measured for light going from air into the medium. This is acceptable because very little refraction occurs when light is travelling from vacuum into air.) In general, if light passes from Medium 1 into Medium 2 the constant is called the relative refractive index, $_1n_2$, of the two media.

Example

Fig. 3.2 shows a ray of light travelling from air into glass, then into water and back into air. Using the values given on the diagram calculate (i) the refractive index of the glass, (ii) the relative refractive index from glass to water, (iii) the refractive index of water.

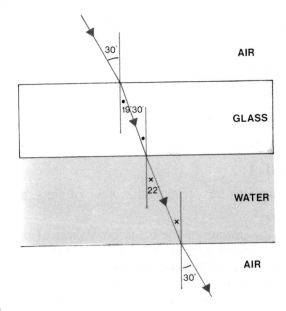

Fig. 3.2

$$n = \frac{\sin i}{\sin r}$$

$$= \frac{\sin 30°}{\sin 19° \ 30'}$$

$$= 1.5$$

$$_{glass}n_{water} = \frac{\sin 19° \ 30'}{\sin 22°}$$

$$= 0.89$$

$$n_{water} = \frac{\sin 30°}{\sin 22°}$$

$$= 1.3$$

Ans. (i) 1.5
(ii) 0.89
(iii) 1.3

Notes:
1. The refractive index of water is given by the sin of the angle in air divided by the sin of the angle in water even though in this case the light is incident in the water.
2. The initial ray in air and the final ray in air are parallel to each other.

EXPERIMENT 3.1

To Verify Snell's Law and hence Determine the Refractive Index of Glass

Method 1:

Apparatus: Ray box, rectangular block of glass, protractor.

Procedure:
1. Draw the outline of the block on a sheet of paper.
2. Direct a ray of light at the block and mark the positions of the rays entering and leaving.
3. Remove the block and join the points where the light entered and left the block. Draw normals at these points and measure the angles of incidence and refraction.
4. Repeat the procedure for different values of the angle of incidence.
5. Prepare a table of results as shown below and plot a graph of sin i against sin r. The slope of this graph is equal to the refractive index of the glass.

Questions:
1. What can you say about the incident and emergent rays? Explain your observations.
2. Why should the ray box give a narrow, parallel beam of light? How is this beam produced?

Method 2:

Apparatus: Four pins, rectangular block of glass, protractor.

Procedure:
1. Draw an outline of the block on a sheet of paper.
2. Draw a line at an angle to the block and place two of the pins on it.
3. Place the other two pins on the other side of the block so that they appear in line with the first two when viewed through the glass. Remove the pins and mark their positions.
4. Remove the block, draw the incident ray and emergent ray and join the points where the light entered and left the block. Draw normals at these points and measure the angles of incidence and refraction.
5. Repeat the procedure for different values of the angle of incidence.
6. Prepare a table of results as shown and plot a graph of sin i against sin r. The slope of this graph is equal to the refractive index of the glass.

RESULTS

$i/°$	$r/°$	sin i	sin r

Slope of graph =   .
Refractive index =   .

Questions
1. What can you say about the incident and emergent rays? Explain your observations.

2. How does the distance between the pins affect the accuracy of the measurements?

Real and Apparent Depth

It is common knowledge that a swimming pool, for example, appears to be less deep than it actually is. This phenomenon is due to refraction at the surface of the pool, as explained by *Fig. 3.3*. To the observer, the point O on the bottom of

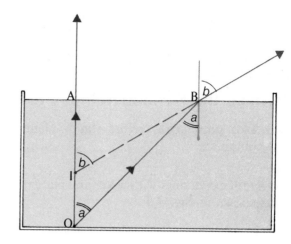

Fig. 3.4

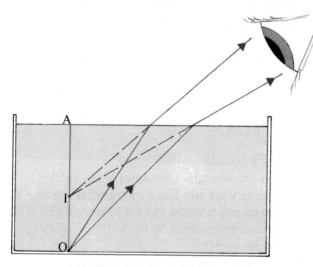

Fig. 3.3

the pool appears to be at I. The same is true for all points on the bottom of the pool. Thus, the **apparent depth** of the pool is AI while the **real depth** is AO. A simple relationship can be deduced between the real and apparent depths and the refractive index of the water. *Fig. 3.4* shows two rays from a point O on the bottom of a tank of water. The ray OA strikes the surface at right angles and so is not refracted. From the diagram,

Refractive index of water, $n = \dfrac{\sin b}{\sin a}$

$$\sin a = \frac{|AB|}{|BO|}$$

and

$$\sin b = \frac{|AB|}{|BI|}$$

$$\frac{\sin b}{\sin a} = \frac{|AB|}{|BI|} \times \frac{|BO|}{|AB|}$$

$$= \frac{|BO|}{|BI|}$$

But,

$$n = \frac{\sin b}{\sin a}$$

$$\therefore \quad n = \frac{|BO|}{|BI|}$$

In practice |AB| is small

$$\Rightarrow |BO| \approx |AO| \quad \text{and} \quad |BI| \approx |AI|$$

$$\therefore n = \frac{|AO|}{|AI|}$$

$$i.e. \quad \boxed{n = \frac{\text{real depth}}{\text{apparent depth}}}$$

EXPERIMENT 3.2
To Determine the Refractive Index of Water

Apparatus: Two pins, cork, retort stand, plane mirror, selection of containers.

Procedure:
1. Completely fill a container with water and place the mirror on top as shown in *Fig. 3.5*.

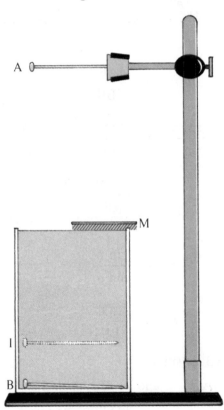

Fig. 3.5

2. Place one pin on the bottom of the container and fix the other in the cork in the clamp of the retort stand.
3. Adjust the height of the pin in the cork until its image in the mirror coincides (by no-parallax) with the image of the pin in the water.

4. Measure the distance from the pin in the cork to the back of the mirror (this is equal to the apparent depth) and the depth of the water in the container (this is the real depth). The refractive index is equal to the real depth divided by the apparent depth.
5. Repeat using the other containers and find an average value for the refractive index.

RESULTS

Real Depth/cm	Apparent Depth/cm	n

Refractive index =   .

Questions

1. Why should the container be completely full?
2. If the container were not full explain what measurements should be taken and how the calculations would be affected.
3. Why are measurements made to the back of the mirror rather than the front?

3.2 Total Internal Reflection

We have seen that light is refracted away from the normal on passing from a denser medium into a less dense medium, *i.e.* the angle of refraction is always greater than the angle of

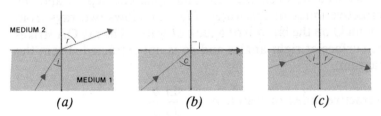

Fig. 3.6

incidence, *Fig. 3.6(a)*. If the angle of incidence is increased, the angle of refraction will also increase until it reaches 90°, *Fig. 3.6(b)*. The angle of incidence in this case is called the **critical angle,** c.

> **The critical angle is the angle of incidence in the denser medium for which the angle of refraction in the less dense medium is 90°.**

If the angle of incidence is further increased, no light crosses the surface; it is all reflected back into the denser medium, *Fig. 3.6(c)*. This phenomenon is known as **total internal reflection.**

> **Total internal reflection occurs when the angle of incidence in the denser medium exceeds the critical angle.**

(Note: For any given angle of incidence some light will be reflected at the surface but for angles greater than c all the light is reflected, hence *total* internal reflection.) In undergoing reflection at the surface the light obeys the laws of reflection *(p.9)*.

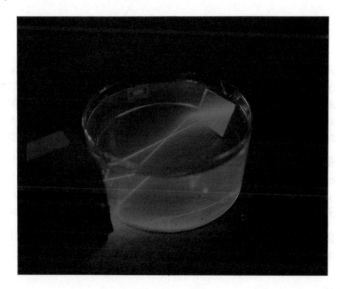

Fig. 3.7 Total internal reflection in water

From *Fig. 3.6(b)*,

$$_1n_2 = \frac{\sin c}{\sin 90°}$$

If Medium 2 is a vacuum (or air) the refractive index of Medium 1 is

$$n = \frac{\sin 90°}{\sin c} = \frac{1}{\sin c}$$

$$\boxed{\sin c = \frac{1}{n}}$$

Example:
Find the critical angle for glass of refractive index 1.5.

$$\sin c = \frac{1}{n}$$

$$= \frac{1}{1.5}$$

$$c = 41° 49'$$

Ans. 41° 49′

This means that if light, travelling in the glass, strikes the surface of the glass at an angle of incidence greater than 41° 49′, it will be reflected back into the glass. In other words, the surface of the glass acts like a mirror. In fact the surface is a better reflector than an ordinary mirror which always absorbs a certain amount of the light falling on it. In addition, ordinary mirrors tend to give multiple images due to reflection from the front surface of the glass as well as from the silver at the back, *Fig. 3.8*.

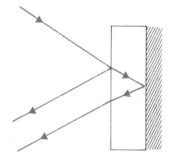

Fig. 3.8

Prisms

An optical prism is a transparent body which has at least two sides which are plane and non-parallel. *Fig. 3.9* shows a common shape of prism. Common values for the refracting angle, θ, are 60°, 90° and 45°

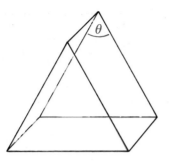

Fig. 3.9 Prism

Fig. 3.10 shows how a prism changes the direction of a ray of light passing through it. The angle between the ray entering the prism and the ray leaving the prism is called the angle of deviation, *d*. Its value depends on the value of the angle of incidence, *i*, as shown in *Fig. 3.11*. Note that *d* has a minimum value, called the angle of minimum deviation, *D*.

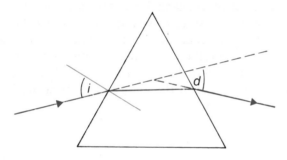

Fig. 3.10 Deviation of a ray of light by a prism

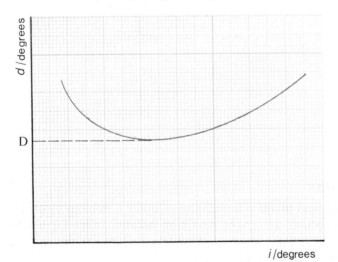

Fig. 3.11   Variation of angle of deviation with angle of incidence

EXPERIMENT 3.3
To Investigate the relationship between the Angle of Incidence and the Angle of Deviation

Apparatus: Triangular glass prism, four pins, protractor.

Procedure:
1. Draw the outline of the prism on a sheet of paper.

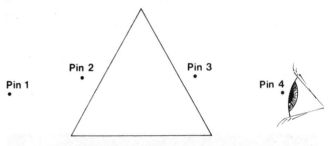

Fig. 3.12

2. Place two of the pins on one side of the prism. Place the other two pins on the other side of the prism so that they are in line with the images of the first two, *Fig. 3.12*. (Make sure the images seen are not the result of total internal reflection in the prism.) Mark the positions of the pins.

3. Remove the prism, draw in the incident ray and the emergent ray and join the points where the light entered and left the prism. Complete the diagram as shown in *Fig. 3.10* and measure the angles of incidence and deviation.

4. Repeat for different values of the angle of incidence between about 35° and 60° and plot a graph of *i* against *d*.

Note:
If the refracting angle, *A*, of the prism is given or is measured (using a protractor) the refractive index of the glass of the prism may be calculated from

$$n = \frac{\sin \dfrac{A + D}{2}}{\sin \dfrac{A}{2}}$$

where *D* is the angle of minimum deviation.

RESULTS

i/°	d/°

Angle of minimum deviation =

Prisms can also be used to deviate light rays through various angles by arranging them so that the light undergoes total internal reflection in the prism. *Fig. 3.13* shows a prism of refracting angle 90° being used to reflect light through 180°. The light is reflected at A and B because the angles of incidence (45°) are greater than the critical angle for the glass. Note that the image is upside down relative to the object.

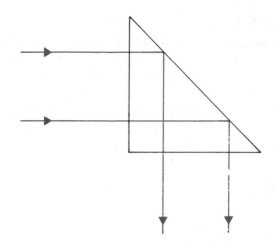

Fig. 3.14 Using a prism to reflect light through 90°

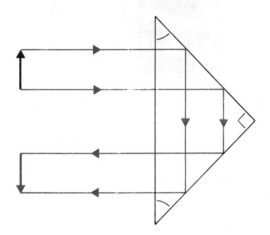

Fig. 3.13 Using a prism to reflect light through 180°

Fig. 3.14 shows how the same shape of prism can be used to reflect light through 90°. Glasses containing prisms in this position allow patients who are unable to move their head to see objects placed in front of them.

Prisms, rather than mirrors, are used in high quality periscopes, *Fig. 3.15*. Prisms give a better quality image than ordinary mirrors and they do not deteriorate with age. Binoculars *(p.45)* also use prisms to reflect light.

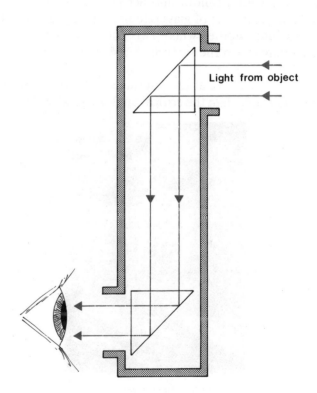

Light from object

Fig. 3.15 Periscope

Optical Fibres

If a narrow beam of light is passed into a glass rod it will travel along it, even if the rod is not straight, provided it always strikes the surface of the glass at an angle of incidence greater than the critical angle, *Fig. 3.16*. Glass rods can thus be used to 'pipe' light around corners and into inaccessible places.

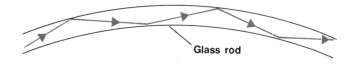

Glass rod

Fig. 3.16 Light passing along an optical fibre

An optical fibre consists of a very fine strand of high quality glass coated with a film of another glass of lower refractive index. (The outer layer of glass is to prevent light crossing from one fibre into another where two come in contact. It also prevents damage to the surface of the core.) Light will then travel along the fibre, no matter how it is bent, provided that the angle of incidence at the interface between the two types of glass is greater than the critical angle (typically about 60°).

Fig. 3.17 Installing an optical fibre telephone link (Courtesy of Telecom Éireann)

Optical fibres can be used for the transmission of information, *e.g.* telephone conversations. Sound is converted into pulses of light which are then transmitted along the fibre. At the other end of the fibre the pulses of light are converted back to sound. The system has a number of advantages over the conventional telephone system. Optical fibres occupy much less space — one fibre, approximately 0.1 mm in diameter, can carry 1920 simultaneous telephone conversations. Since the system is optical, it is not subject to electrical interference and it cannot be 'tapped'. Losses from fibres are low — boosters are required only every 100 km or so compared with every 2 km for copper conductors. Finally, the basic material of the fibres (silica) is cheap and plentiful.

Fig. 3.18 Using a fibre-optic endoscope (Courtesy of the Meath Hospital)

Optical fibre systems are used by doctors to examine, for example, a patient's lungs or stomach. The advantage of this method of examination is that, although it is somewhat uncomfortable, there is no need for surgery.

The Mirage

To a person travelling on a straight road on a hot day, the road some distance ahead often appears wet. This is a mirage. What in fact the person sees is a virtual image of a patch of blue sky. *Fig. 3.19* shows how such an image is formed. On a hot day the surface of the road is hot and so the air immediately above it is also hot. Away from the surface the air becomes gradually cooler. Light from the sky is thus passing through 'layers' of air which are progressively warmer and hence less dense. The light is therefore continuously refracted away from the normal. At a certain stage the angle of incidence exceeds the critical angle and total internal reflection occurs.

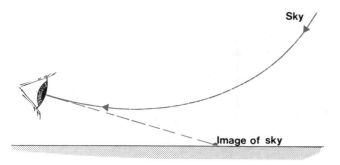

Fig. 3.19 Formation of a mirage

3.3 Converging Lenses

A lens consists of a piece of transparent material — usually glass - the thickness of which varies from the middle to the edges. At least one of the surfaces is usually spherical. If the lens is thicker at the middle than at the edges, it is a **converging lens**.

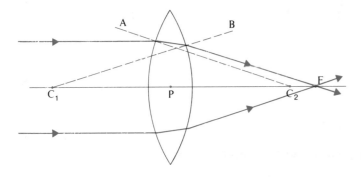

Fig. 3.20 Converging lens

Fig. 3.20 shows a common type of converging lens. Both surfaces of the lens are spherical. C_1 and C_2 are the **centres of curvature**, *i.e.* they are the centres of the spheres of which the surfaces of the lens form a part. (Lenses of this type are sometimes referred to as biconvex, or simply convex, lenses.) The line through C_1 and C_2 is the **principal axis**. A ray of light on entering the lens is refracted towards the normal C_2A (since C_2 is the centre of curvature of the surface any line

through C_2 is a normal to the surface). On leaving the lens the ray is refracted away from the normal C_1B. However since the surfaces are inclined towards each other, the ray is bent towards the principal axis in each case (*cf.* deviation by a prism, *p.28)*. The overall effect is that rays which enter the lens parallel to each other *converge* towards a point. If the incident rays are parallel to the principal axis, as in *Fig. 3.20*, this point is called the **principal focus**. Since light can travel equally well in either direction, there are two foci, one on either side of the lens. Both are equidistant from the lens.

Note that, unlike spherical mirrors, there is not a simple relationship between the focal length and the radii of curvature.

Formation of Images in Converging Lenses

We can determine the position and nature of the image formed by a converging lens using ray diagrams in basically the same way as we did for mirrors. We trace two rays from the 'top' of the object and find where they meet or appear to meet. An incident ray which is parallel to the principal axis is refracted through the principal focus and *vice versa*. A ray passing through the centre of the lens may, if the lens is thin, be taken to pass straight through. The type of image formed by a converging lens depends on the position of the object *(cf. concave mirrors)*. You may verify this by looking at an object through a converging lens and varying the distance between the object and the lens. *Fig. 3.21* shows the formation of an image for different positions of an object. You should study these diagrams carefully and practise drawing them for yourself.

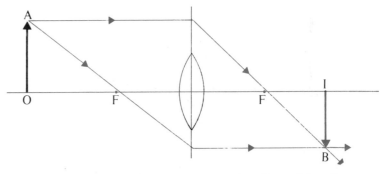

Fig. 3.21(a) Formation of image in converging lens, object distance greater than 2f: The image is real, inverted and diminished

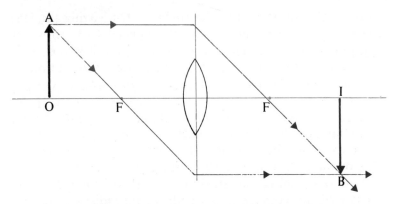

Fig. 3.21(b) *Converging lens, object distance equal to 2 f: The image is real, inverted and the same size as the object*

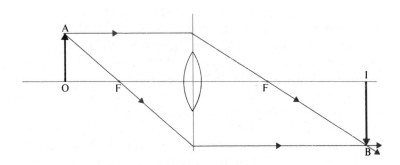

Fig. 3.21(c) *Converging lens, object distance greater than f but less than 2 f: The image is real, inverted and magnified*

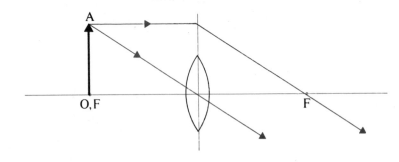

Fig. 3.21(d) *Converging lens, object at F: The image is at infinity*

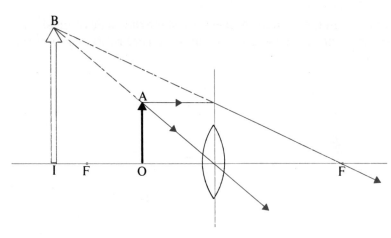

Fig. 3.21(e) *Converging lens, object inside F: The image is virtual, erect and magnified*

The Lens Formula

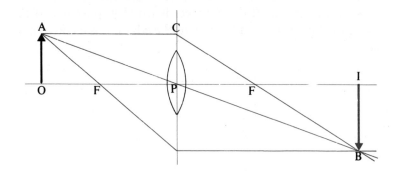

Fig. 3.22

We can show that the formula

$$\frac{1}{f} = \frac{1}{u} + \frac{1}{v}$$

applies to lenses as well as to mirrors. Referring to *Fig. 3.22*, the triangles BIP and PAO are similar,

$$\therefore \quad \frac{|IB|}{|OA|} = \frac{|IP|}{|OP|} = \frac{v}{u}$$

But, $\quad \dfrac{|IB|}{|OA|} =$ magnification, m

i.e.

$$m = \frac{v}{u}$$

The triangles BIF and PCF are similar,

$$\therefore \quad \frac{|\text{IB}|}{|\text{PC}|} = \frac{|\text{IF}|}{|\text{PF}|}$$

But, $\quad |\text{PC}| = |\text{OA}|$

$$\therefore \quad \frac{|\text{IB}|}{|\text{OA}|} = \frac{|\text{IF}|}{|\text{PF}|}$$

$$= \frac{|\text{IP}| - |\text{PF}|}{|\text{PF}|}$$

i.e. $\quad \dfrac{v}{u} = \dfrac{v - f}{f}$

Multiplying both sides by uf gives

$$vf = uv - uf$$

Dividing both sides by uvf gives

$$\frac{1}{u} = \frac{1}{f} - \frac{1}{v}$$

Rearranging gives

$$\frac{1}{f} = \frac{1}{u} + \frac{1}{v}$$

As with mirrors, in order to make the formula apply in all cases, the 'real is positive' sign convention must be adopted.

Example

A pin is placed 5 cm from a converging lens of focal length 10 cm. Find the position, magnification and nature of the image.

$$f = + 10 \text{ cm}$$

$$u = + 5 \text{ cm}$$

$$v = ?$$

$$\frac{1}{f} = \frac{1}{u} + \frac{1}{v}$$

$$\frac{1}{10} = \frac{1}{5} + \frac{1}{v}$$

$$\frac{1}{v} = \frac{1}{10} - \frac{1}{5}$$

$$= - 0.1$$

$$v = - 10 \text{ cm}$$

Ans. The image is virtual (v is negative) and is 10 cm from the lens on the same side as the pin.

$$m = \frac{v}{u}$$

$$= \frac{- 10}{5}$$

$$= - 2$$

Ans. The image is twice the size of the pin and it is erect (m is negative).

EXPERIMENT 3.4

To Determine the Focal Length of a Converging Lens

Method 1:

Apparatus: Converging lens, screen (sheet of cardboard), metre stick.

Procedure:
1. Using the lens focus an image of a distant object, *e.g.* a tree, onto the screen.
2. Measure the distance from the lens to the screen. This is approximately equal to the focal length of the lens.

RESULTS

Distance from lens to screen . . .	=	cm
Approximate focal length of lens . .	=	cm

Questions

1. Why is the distance from the screen to the lens only approximately equal to the focal length?
2. In what circumstances is the approximation valid?

Method 2:

Apparatus: Converging lens, lamp box, screen, metre stick.

Procedure:

1. Set up the lens between the lamp box and the screen, *Fig. 3.23*. Adjust the position of the lens until a sharp image of the slit in the lamp box is seen on the screen.
2. Measure the distance from the lamp box to the lens, *u*, and the distance from the lens to the screen, *v*. Calculate the focal length from the formula

$$\frac{1}{f} = \frac{1}{u} + \frac{1}{v}$$

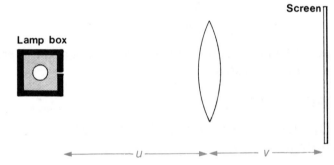

Fig. 3.23

3. Repeat for different positions of the lamp box and find an average value for *f*.

RESULTS

u/cm	v/cm	f/cm

Focal length of lens =   cm.

Questions

1. In what circumstance will it not be possible to obtain an image on the screen?
2. Explain why, for given positions of the lamp box and screen, there are generally two positions of the lens for which an image is formed on the screen? Compare the images produced in a given case and discuss the differences between them.

Method 3:

Apparatus: Converging lens, two pins, metre stick.

Procedure:

1. Place one pin outside the focus (use Method 1 to find the approximate position of the focus).
2. Place the second pin on the other side of the lens so that it coincides with the image of the first pin as seen through the lens (use the no-parallax technique).
3. Measure the object distance and the image distance and calculate the focal length from

$$\frac{1}{f} = \frac{1}{u} + \frac{1}{v}$$

4. Repeat for different positions of the object, including some inside the focus, and find an average value for *f*.

Note that when the object is inside the focus the image is on the same side of the lens, *Fig. 3.21(e)*. In these cases the image is virtual and the value of *v* must be given a negative sign in the formula.

RESULTS

u/cm	v/cm	f/cm

Focal length of lens =   cm.

3.4 Diverging Lenses

A diverging lens is one which is thicker at the edges than in the middle (lenses of this type are sometimes referred to as biconcave, or simply concave, lenses). C_1 and C_2 are the centres of curvature of the two surfaces and so the line passing through C_1 and C_2 is the principal axis.

A ray of light entering the lens, *Fig. 3.24*, is refracted towards the normal C_1A. On leaving the lens the ray is refracted away from the normal C_2A. However, because of the shape of the surfaces, the ray is refracted away from the principal axis in both cases. As can be seen from *Fig. 3.24*, rays which enter this type of lens parallel to each other *diverge* after passing through the lens. Note that the focus is **virtual** and therefore when using the lens formula the focal length must be given a negative value.

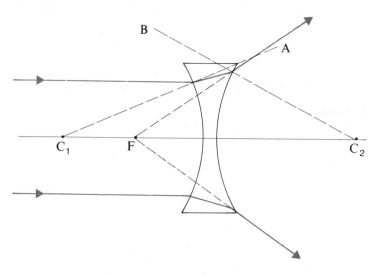

Fig. 3.24 Diverging lens

Formation of Images in Diverging Lenses

Fig. 3.25 shows how an image is formed in a diverging lens. As in the case of the convex mirror the nature of the image does not depend on the position of the object — it is always virtual, erect and diminished. You should verify this for yourself by drawing ray diagrams for different positions of an object and by looking at an object through a diverging lens while varying the distance between the object and the lens.

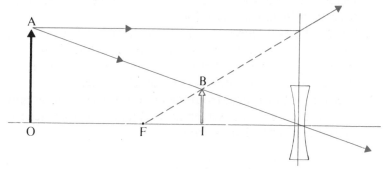

Fig. 3.25 Formation of an image in a diverging lens. The image is virtual, erect and diminished

Example

A diverging lens of focal length 20 cm forms an image 5 cm from the lens. What is the position of the object?

$$f = -20 \text{ cm (Focus is virtual: } f \text{ is negative.)}$$

$$v = -5 \text{ cm (Image is virtual: } v \text{ is negative.)}$$

$$u = ?$$

$$\frac{1}{f} = \frac{1}{u} + \frac{1}{v}$$

$$-\frac{1}{20} = \frac{1}{u} - \frac{1}{5}$$

$$=> \quad \frac{1}{u} = -\frac{1}{20} + \frac{1}{5}$$

$$= 0.15$$

$$=> \quad u = 6.7 \text{ cm}$$

Ans. 6.7 cm from the lens

EXPERIMENT 3.5

To Determine the Focal Length of a Diverging Lens

Method 1:

Apparatus: Diverging lens, converging lens of known focal length (if such is not available see previous experiment), lamp box, screen, metre stick.

Procedure:

1. Determine the focal length of the converging lens as described in *Experiment 3.4*.
2. Set up the two lenses in contact with each other between the lamp box and the screen as shown in *Fig. 3.26*.
3. Adjust the positions of the lenses until a sharp image of the slit in the lamp box is seen on the screen.
4. Measure the distances from the lamp box to the lenses and from the lenses to the screen and calculate the focal length of the combined lenses from

$$\frac{1}{f} = \frac{1}{u} + \frac{1}{v}$$

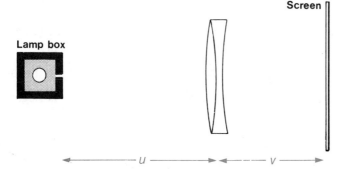

Fig. 3.26

5. Repeat for different positions of the lamp box and find an average value for the focal length of the combination. Calculate the focal length of the diverging lens from

$$\frac{1}{f} = \frac{1}{f_1} + \frac{1}{f_2}$$

where f_1 is the focal length of the converging lens and f_2 is the focal length of the diverging lens*.

Note: The focal length of the converging lens must be shorter than that of the diverging lens.

*The derivation of this formula is outside the scope of this course and is left as an additional exercise for the student.

RESULTS

u/cm	v/cm	f/cm

Focal length of converging lens =   cm
Focal length of combination   =   cm
Focal length of diverging lens   =   cm

Questions

1. Why is it not possible to measure the focal length of a diverging lens directly as was done for the converging lens?
2. Why must the focal length of the converging lens used in this experiment be shorter than that of the diverging lens?

Method 2:

Apparatus: Diverging lens, two pins, metre stick.

Procedure:

1. Place one pin in front of the lens.
2. Place the second pin on the same side of the lens so that it coincides with the image of the first pin as seen through the lens (use the no-parallax technique), see *Fig. 3.25*.
3. Measure the object distance and the image distance and calculate the focal length from

$$\frac{1}{f} = \frac{1}{u} + \frac{1}{v}$$

4. Repeat for different positions of the object and find an average value for *f*.

RESULTS

u/cm	v/cm	f/cm

Focal length of lens =   cm.

Questions

1. Are the images formed in this experiment real or virtual? What effect does this have on the calculation of the focal length?

Uses of Lenses

Converging lens are widely used in a variety of optical instruments, some of which are described in detail in the next chapter. Both types of lens are used as glasses and contact lenses to compensate for defects in the human eye *(see p.39).*

SUMMARY

When light travels from one medium into another of different density it changes direction. This phenomenon is known as refraction. The laws of refraction are:

1. The incident ray, the normal and the refracted ray are in the same plane.
2. The sine of the angle of incidence is proportional to the sine of the angle of refraction (Snell's law).

The refractive index of a medium is given by

$$n = \frac{\sin i}{\sin r}$$

where i is measured in a vacuum (or air).

For light travelling from a denser to a less dense medium, the angle of incidence which makes the angle of refraction 90° is called the critical angle, c. If the angle of incidence is greater than the critical angle total internal reflection occurs, *i.e.* all of the light is reflected back into the denser medium. The critical angle for a particular material in vacuum (air) is related to its refractive index by

$$\sin c = \frac{1}{n}$$

If u is less than f, the image formed by a converging lens is virtual, magnified and erect. If u is greater than f, the image is real and inverted and is magnified or diminished, depending on whether u is less than or greater than $2f$.

A diverging lens has a virtual focus and always forms a virtual, erect, diminished image.

Lenses are used in a variety of optical instruments and as glasses and contact lenses to compensate for defects in the human eye.

Questions 3
Section I

1. State the laws of refraction.....................................

....

....

2. The second law was discovered by the.....................

mathematician, ...

3. What is meant by refractive index?..........................

....

....

4. What is meant by total internal reflection?...............

....

....

5. If the angle of.......................................exceeds the

.........................angle total internal reflection occurs.

6. Give two advantages of using prisms instead of mirrors in optical instruments...

....

7. The real depth of a swimming pool is......................

than the apparent depth because of..........................

at the surface of the water.

8. Give two uses of optical fibres................................

....

9. An object is placed 30 cm from a converging lens of focal length 15 cm. Where is the image formed and what is its magnification? ....

....

10. A lens produces a virtual magnified image of an object placed in front of it. What type of lens is it?............

....

Section II

1. A ray of light enters a block of ice at an angle of incidence of 30°. Calculate the angle of refraction in the ice. (Refractive index of ice = 1.3.)

2. A student measured the angles of incidence for five different rays of light entering a rectangular glass block from air. The values she obtained were 30°, 40°, 50°, 60° and 70°. She measured the corresponding angles of refraction and obtained 20°, 25° 30′, 32°, 35° and 39° 30′, respectively. Calculate the refractive index of the glass.

3. The critical angle for a certain type of glass is 45° 35′. What is the refractive index of the glass? If the real depth of a rectangular block of this type of glass is 4.0 cm what is its apparent depth?

4. A student focuses a travelling microscope on a speck of dust on the bench. When he places a block of glass of thickness 1.0 cm over the dust he finds that he must raise the microscope 3.7 mm in order to bring the dust back into focus. What is the refractive index of the glass?

5. The critical angle for water is 48° 45′. What is the refractive index of water?

6. A point source of light is placed 75 cm below the surface of a pond. Given that the refractive index of water is 1.33 calculate the area of the circle at the surface through which the light can pass.

7. Use a ray diagram to show how a prism may be used to reflect light through 90°. What is the minimum value for the refractive index of the material of the prism which will allow this to occur?

8. Calculate the position and magnification of the image formed by a converging lens of focal length 10 cm when the object is (i) 15 cm; (ii) 5 cm; (iii) 25 cm from the lens. State the nature of the image in each case.

9. A converging lens of focal length 20 cm forms a real image which is 25 cm from the lens. Calculate the position of the object and the magnification of the image.

10. A converging lens of focal length 15 cm forms a real image which is three times the size of the object. Find the position of the object and of the image. If the object were placed at the image position what would be the magnification of the image?

11. A converging lens has a focal length of 40 cm. For what positions of an object will the magnification of the image be 3? In each case draw a ray diagram to show how the image is formed.

12. Calculate the position and magnification of the image formed by a diverging lens of focal length 15 cm when a pin is placed (i) 20 cm; (ii) 40 cm; (iii) 10 cm from the lens. State the nature of the image in each case.

13. *Fig. I* shows an image being formed by a combination of a converging and a diverging lens in an experiment to measure the focal length of the diverging lens. Given that the focal length of the converging lens is 5.0 cm and using the distances given in the diagram, calculate the focal length of the diverging lens.

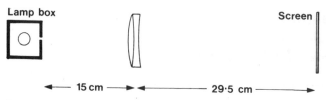

Fig. 1

14. A diverging lens forms an image which is one sixth the size of the object. Given that the focal length of the lens is 20 cm determine the positions of the object and image.

4 Optical Instruments

4.1 The Human Eye

Of all optical instruments the eye is undoubtedly the oldest. It is also the most important since, without it, all other optical instruments are useless!

Fig. 4.1(a) *The diameter of the pupil changes as the brightness of the light changes*

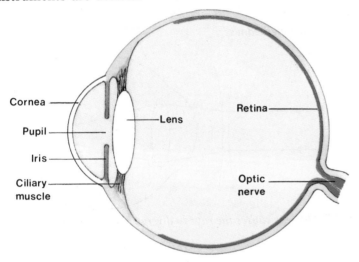

Fig. 4.1 Optical structure of the eye

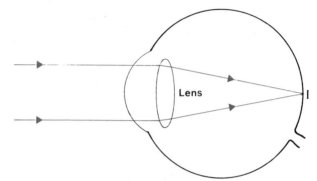

Fig. 4.2 Viewing a distant object

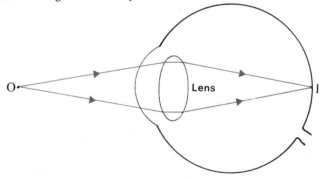

Fig. 4.3 Viewing an object close to the eye

Fig. 4.1 shows the main structures of the eye. Light from an object is refracted at the cornea and again, to a lesser extent, at the lens. An image of the object is thus formed on the retina. In the retina, the light energy is converted into electrical energy which is transmitted to the brain via the optic nerve. The amount of light entering the eye is determined by the size of the pupil (the black part of the eye). The iris (the coloured part) varies the size of the pupil from about 2 mm diameter in very bright light to about 8 mm in darkness. (You can see the pupils of your eyes changing size if you look in a mirror in subdued light while switching a bright light on and off.) The ciliary muscle controls the shape, and hence the focal length, of the lens. When the muscle is relaxed, the lens is thinnest and rays from a distant object are brought to a focus on the retina, *Fig. 4.2*. (Rays from a point on a distant object are effectively parallel to each other, *i.e.* the angle between them is approximately zero.)

When we wish to look at an object near to us the muscle contracts, allowing the lens to become fatter, *Fig. 4.3*. This ability of the eye to change its focal length and so to focus on objects at different distances is known as the **power of accommodation**.

Defects of Vision

People with normal vision can focus clearly on objects between infinity (*i.e.* very far away) and about 10 — 15 cm from the eye. We say that their **far point** is infinity and their **near point** is, for example, 12 cm.

A person who can see distant objects clearly but cannot focus on near objects is said to be **long-sighted**. When such a person tries to focus on a point near the eye the lens remains too thin

39

and the image of the point is formed behind the retina, *Fig. 4.4*. To correct this such a person would have to wear glasses with *converging* lenses, *Fig. 4.5*. The converging lens causes the rays to start to converge before they enter the eye so that the image is now formed on the retina.

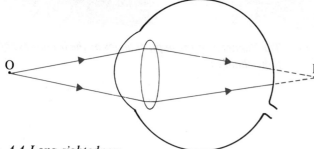

Fig. 4.4 Long-sightedness

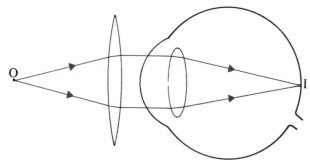

Fig. 4.5 The lens causes the rays to converge before entering the eye

A person who can see near objects clearly but cannot focus on distant objects is said to be **short-sighted**. When such a person tries to focus on a point distant from the eye the lens remains too fat and the image of the point is formed in front of the retina, *Fig. 4.6*. To correct this such a person would have to wear glasses with *diverging* lenses, *Fig. 4.7*. The diverging lens causes the rays to start to diverge before they enter the eye so that the image is now formed on the retina.

The Blind Spot

Close your left eye and look at the cross in *Fig. 4.8*. Slowly move the page towards you. At a certain distance the black dot will disappear. This is because light from the dot is now falling on the part of the retina where the optic nerve leaves your right eye. This area has no light-sensitive cells and is referred to as the blind spot.

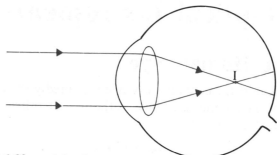

Fig. 4.6 Short-sightedness

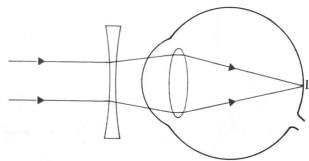

Fig. 4.7 The lens causes the rays to diverge before entering the eye

Fig. 4.8

Persistence of Vision

When an image falls on the retina it remains there for approximately 0.1 s after the object has been removed. This allows us, for instance, to see a film as a continuous motion rather than the series of still photographs which in fact it is. In the cinema, the photographs are projected at a rate of 24 per second and on television, the picture on the screen changes 25 times per second.

The Camera

The camera is very similar to the eye; like the eye the image is formed by a converging lens system, *Fig. 4.9*. The film takes the place of the retina and the aperture corresponds to the pupil. The major difference lies in the method of focusing: in the eye the lens changes shape, in the camera it changes position.

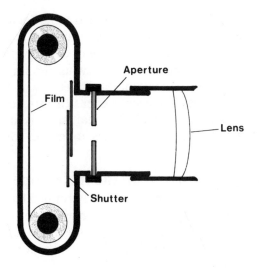

Fig. 4.9 Camera

Fig. 4.10 Using a simple microscope (Courtesy of the Inst. of Physics)

4.2 The Microscope

The simple microscope, commonly known as a magnifying glass, consists of a single converging lens with the object to be viewed being placed inside the focus. This gives a magnified, erect, virtual image, *Fig. 4.10 (c.f. Fig. 3.21(e), p.32)*. The shorter the focal length of the lens the greater the magnification obtained.

The Compound Microscope

The compound microscope, *Fig. 4.11,* consists of two converging lenses — the **objective** and the **eyepiece**. The object, O, to be viewed is placed just outside the focus of the objective lens. The objective lens thus forms a real, inverted, magnified image, I_1 *(cf. Fig. 3.21(c), p.32)*. This image serves as the object for the eyepiece lens which is arranged so that I_1 is just inside its focus. The image, I_2, formed by the eyepiece lens is thus virtual, erect (relative to I_1) and magnified *(see Fig. 3.21(e), p.32)*. The final image is therefore inverted relative to the original object. It can be shown that the shorter the focal lengths of the two lenses the greater the magnification obtained.

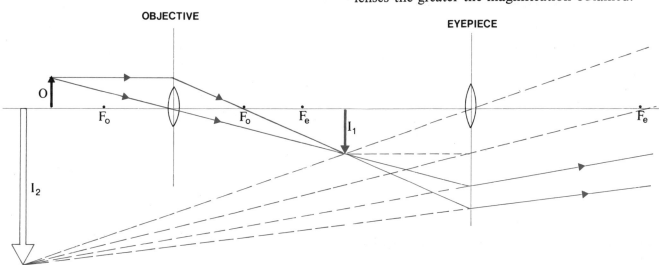

Fig. 4.11 Formation of an image in a compound microscope

Fig. 4.12 A compound microscope

4.3 The Astronomical Telescope

Like the compound microscope the astronomical telescope has two converging lenses, *Fig. 4.13*. In this case the object, say the moon, is very far from the objective lens so that rays from a particular point on the moon may be considered to be parallel to each other. The image, I_1, formed by the objective lens is real, inverted and diminished. (Diminished means that it is smaller than the original object, the moon. It appears to be much larger because it is much nearer the eye. In other words, it is effectively magnified.) In **normal adjustment**, the eyepiece

lens is arranged so that I_1 is formed at its focus. This means that the image formed by the eyepiece lens is at infinity *(cf. Fig. 3.21(d), p.32)*. The advantage of this arrangement is that the eye is most relaxed when viewing the image *(see p.39)*. The longer the focal length of the objective lens and the shorter the focal length of the eyepiece lens the greater the effective magnification *(see below)*. The final image is inverted relative to the original object but this does not matter for astronomical observations.

Magnifying Power of the Telescope

The apparent size of an object is determined by the size of the angle it subtends at the eye. The closer one is to the object the larger it appears and the larger the angle subtended at the eye, *Fig. 4.14*. The telescope effectively increases the angle

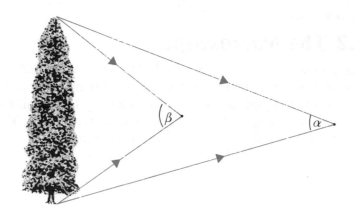

Fig. 4.14 The apparent size of an object depends on the position of the observer

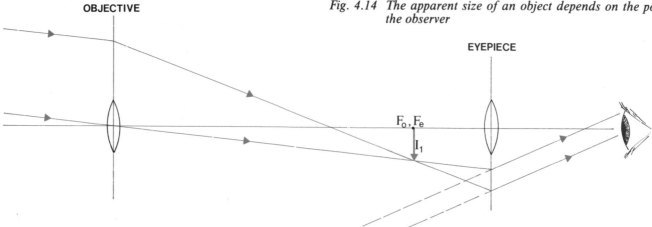

Fig. 4.13 Formation of an image in an astronomical telescope in normal adjustment

subtended at the eye so the image is larger than the apparent size of the object. Referring to *Fig. 4.15* we see that the angle subtended by the image, *b,* is larger than the angle subtended by the object, *a.* The magnifying power of a telescope is defined by:

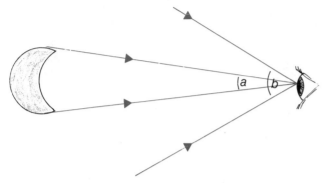

Fig. 4.15 *The angle subtended with the telescope (b), is greater than the angle subtended without the telescope (a)*

$$M = \frac{\text{angle subtended with telescope}}{\text{angle subtended without telescope}}$$

That is,

$$M = \frac{b}{a}$$

Referring to the simplified diagram shown in *Fig. 4.16,*

$$\tan a = \frac{h}{f_o} \quad \text{and} \quad \tan b = \frac{h}{f_e}$$

$$\frac{\tan b}{\tan a} = \frac{h}{f_e} \times \frac{f_o}{h}$$

$$= \frac{f_o}{f_e}$$

But, since *b* and *a* are very small,

$$\tan b \approx b \quad \text{and} \quad \tan a \approx a$$

$$\therefore \quad \frac{b}{a} = \frac{f_o}{f_e}$$

But,

$$M = \frac{b}{a}$$

Therefore,

$$\boxed{M = \frac{f_o}{f_e}}$$

Note: This formula applies only when the telescope is in normal adjustment.

The Terrestrial Telescope

We noted above that the final image in an astronomical telescope is inverted. This makes it unsuitable for use on earth, *i.e.* as a terrestrial telescope. To make a terrestrial telescope a third lens is placed between the objective lens and the eyepiece lens. This third lens, called an erecting lens, is arranged as shown in *Fig. 4.17*. The image, I_1, formed by the objective lens is at a distance of *2f* from the erecting lens, where *f* is the focal length of the erecting lens. I_1 serves as an object for the erecting lens which forms an image, I_2, on the other side of the erecting lens and at a distance of *2f* from it. I_2 is the same size as I_1 but is inverted relative to I_1, (*see Fig. 3.21(b)*, *p.32*). This means that I_2 is erect relative to the original object. When the telescope is in normal adjustment the eyepiece lens is arranged so that I_2 is formed at its focus so that the final image is at infinity. Note that the erecting lens does not

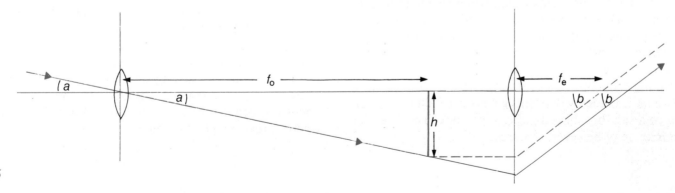

Fig. 4.16

43

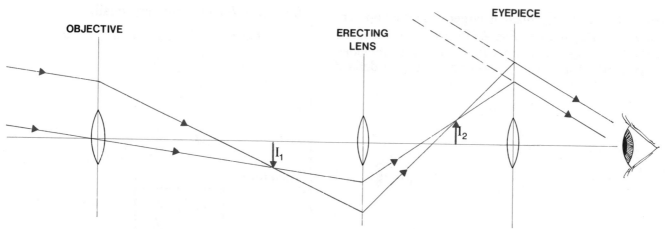

Fig. 4.17 Terrestrial telescope

change the magnifying power of the telescope, it only inverts the image. The disadvantage of this system is that it makes the overall length of the telescope rather large. The overall length is equal to the sum of the focal lengths of the objective and eyepiece lens plus four times the focal length of the erecting lens. In binoculars this problem is solved by using two prisms instead of the erecting lens *(see p.45)*.

The Reflecting Telescope

We have seen that the magnifying power of a refracting telescope is given by

$$M = \frac{f_o}{f_e}$$

From this it would seem that we could make the magnification as large as we liked simply by making f_o very much greater than f_e. However, as the image is made larger, it also becomes dimmer. To counteract this a larger objective lens, which will allow in more light, is needed. Unfortunately, it is difficult to make a large lens which will give a high-quality image. Even if such a lens were made it would become distorted under its own weight and so would produce a distorted image. For this reason, most astronomical work is now done using reflecting telescopes.

A reflecting telescope has as its objective a large concave mirror, which is made by depositing a layer of aluminium on the *front* of a suitably shaped piece of perspex. The diameter of the mirror may be several metres.

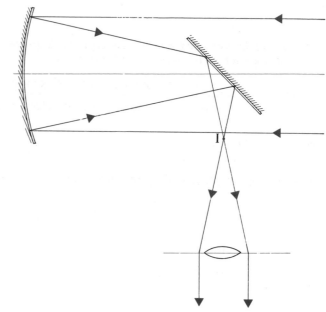

Fig. 4.18 Reflecting telescope

The first reflecting telescope was made by Isaac Newton around 1670. In this type of telescope, *Fig. 4.18*, a small plane mirror, set at an angle of 45° to the principal axis of the concave mirror, is used to bring the light to a convenient focus, I. A converging lens, arranged so that I is at its principal focus, gives a virtual image at infinity. If the image is to be photographed, as is commonly done, the film is placed at I, where a real image is formed.

Fig. 4.19 The La Palma 1 m telescope, partially owned by Ireland (Courtesy of Dunsink Observatory)

4.4 Prism Binoculars

Binoculars are really a pair of telescopes placed side by side. Each has two converging lenses, exactly as in an astronomical telescope. However, in addition to the lenses, each also has two 90° prisms arranged as shown in *Fig. 4.20*. The first prism turns the image the right way around while the second turns it the right way up. (Remember, an astronomical telescope gives an inverted image, *i.e.* upside down *and* back to front.) In addition the prisms reduce the overall length of the instrument, making it more compact.

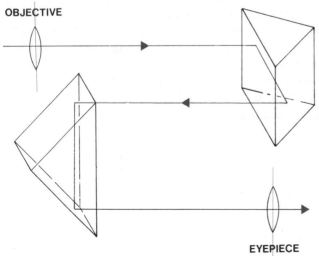

Fig. 4.20 The use of prisms in prism binoculars

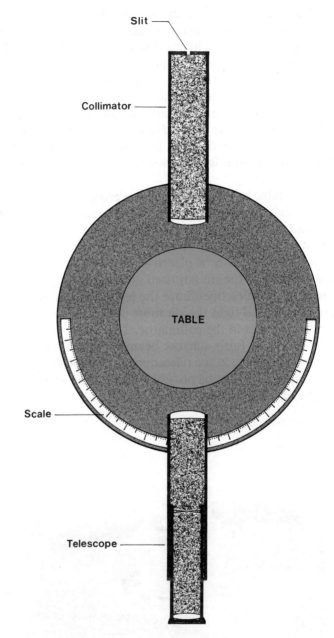

Fig. 4.21 Spectrometer

4.5 The Spectrometer

The spectrometer is an instrument used in laboratories to examine the light emitted from different types of substance *(see Chapter 13)*. It may also be used to measure the angles of a prism and the refractive index of the material of a prism.

45

The main parts of a spectrometer, *Fig. 4.21*, are the collimator, the telescope, the table and the scale, around which the telescope can move. The collimator consists of a circular tube with an adjustable slit at one end and a converging lens at the other. In use, the slit is arranged to be at the focus of the lens so that light entering the collimator through the slit emerges from the lens as a parallel beam. The object to be viewed is placed on the table, *Fig. 4.22*. The table may be rotated and is fitted with three levelling screws. The scale is marked in degrees and is fitted with a vernier scale for more accurate readings.

In use, the table is first levelled. The telescope is then focused for parallel light by focusing it on a distant object; the eyepiece of the telescope is adjusted so that the image of the object falls on the crosswires (this is done using the no — parallax technique). A lamp is now placed in front of the collimator and the position of the slit adjusted until a clear image of the slit is seen in the telescope. Since the telescope was originally focused for parallel light the slit must now be at the focus of the converging lens in the collimator. The width of the slit is adjusted to give a fairly narrow beam of light. If the slit is too narrow the image in the telescope may be too faint; if the slit is too wide it will not be possible to determine the position of the image with reasonable accuracy.

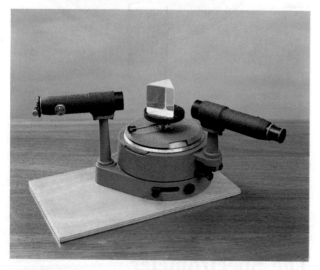

Fig. 4.22 A spectrometer

William Parsons, 3rd Earl of Rosse (1800 — 1867)

William Parsons was born in York, England. He was educated at Oxford and in 1821 was elected to Parliament as Lord Oxmantown. He resigned his seat after 12 years and in 1841 he inherited his father's earldom and took his seat in the House of Lords as one of the Irish peers.

Rosse was obsessed with the idea of building a large reflecting telescope. His first success was a 91 cm diameter mirror, made of 16 plates of a copper/tin alloy soldered to a brass framework. He followed this with a 91 cm solid mirror. In 1842 he started work on a 183 cm mirror and in 1845 the 'Leviathan of Parsonstown' was installed at Birr Castle, the family seat. The mirror had a mass of 4100 kg, the whole telescope was 16.5 m in length and is said to have cost £30,000, a considerable sum at that time. The Leviathan was the largest telescope in the world until it was dismantled in 1908 and it was not until 1917 that a larger telescope, the 2.54 m reflector at Mt. Wilson Observatory in California, was built.

The Irish climate is not really suitable for serious astronomical work. However, Lord Rosse was able to discover the spiral shape of galaxies beyond the Milky Way. He also investigated the Crab Nebula (which he named) and the Great Nebula in Orion.

Rosse was elected to the Royal Society in 1848 and he died at Monkstown, Co. Cork, in 1867.

SUMMARY

In the human eye, light enters through the pupil and is focused on light-sensitive cells in the retina. The amount of light entering the eye is controlled by the iris which varies the size of the pupil. The eye focuses on objects at different distances by changing the shape, and hence the focal length, of the lens — a process known as accommodation. The blind spot is the part of the retina where the optic nerve leaves the eye — there are no light-sensitive cells at this point. An image remains on the retina for a short time after the object has been removed; this is called persistence of vision.

A simple microscope consists of a single converging lens with the object being placed inside the focus.

A compound microscope consists of two lenses, the objective and the eyepiece, each of short focal length. The object is placed just outside the focus of the objective lens and the final image is virtual and inverted.

The astronomical telescope consists of two lenses —

an objective of long focal length and an eyepiece of short focal length. The final image is virtual and inverted. A terrestrial telescope has a third lens placed between the objective and eyepiece lenses to produce a final image which is erect.

The magnifying power of a telescope in normal adjustment is given by

$$M = \frac{\text{focal length of objective}}{\text{focal length of eyepiece}}$$

A spectrometer consists of a collimator, a table and a telescope which moves around a circular scale. The collimator has a narrow slit at one end and a converging lens at the other, the slit being at the focus of the lens so that the light leaves the collimator as a parallel beam. The scale has a vernier scale associated with it to allow more accurate reading of the angles moved through by the telescope.

Questions 4

Section I

1. Name the principal parts of the eye.......................
....
....

2. The ability of the eye to focus on objects at different distances is known as...
....

3. What is the blind spot in the eye?.......................
....

4. ....
makes possible 'moving' films.

5. A . . . . is another term for a magnifying glass.

*6. The final image in a compound microscope is inverted. Is this a disadvantage?.......................................
....

7. What is meant by the term 'normal adjustment'?
....
....

8. Why are reflecting telescopes more commonly used for serious astronomical work than refracting telescopes?
....
....

9. Why are *two* prisms required in binoculars?............
....
....

*10. Name the principal parts of a spectrometer.............
....
....

Section II

1. Draw a ray diagram showing how an image is formed in the human eye. What type of lens is used to correct (i) short sightedness, (ii) long sightedness?

2. A magnifying glass can be used on a sunny day to set fire to dry grass, twigs, *etc.* Explain how this happens.

3. A simple microscope of focal length 10 cm is used to examine a biological specimen. If the image is five times the size of the specimen what is the distance between the specimen and the lens?

*4. A compound microscope and an astronomical telescope both consist, essentially, of two converging lens in a tube. What are the differences between them? What is the difference between an astronomical telescope and a terrestrial telescope?

5. What are the functions of the prisms in prism binoculars? What would be the disadvantages of using plane mirrors instead of the prisms and how many such mirrors would be required?

*6. An astronomical telescope has lenses of focal length 60 cm and 5 cm, respectively. When the telescope is in normal adjustment, calculate (i) the total distance between the lenses; (ii) the magnifying power of the telescope.

*7. The final image in an astronomical telescope is inverted and diminished. Explain why the inversion of the image is unimportant and why there is no contradiction in a telescope giving a diminished image. The total distance between the lenses of an astronomical telescope in normal adjustment is 1.00 m. If the magnifying power of the telescope is 40 what are the focal lengths of the lenses?

*8. A terrestrial telescope has an objective lens of focal length 40 cm and an eyepiece lens of focal length 4 cm. If the distance between the two lenses is 64 cm what is the focal length of the erecting lens?

*9. The distance from the earth to the moon is 3.8×10^8 m and the diameter of the moon is 3.5×10^6 m. What is the angle subtended by the moon at a point on the surface of the earth? If a telescope, having lenses of focal length 10 cm and 2 m respectively, is used (in normal adjustment) to view the moon what is the angle subtended at the observer's eye?

*10. Describe the adjustments which must be made to a spectrometer before it is used. *Fig. I* shows a photograph of a spectrometer scale. What is the reading on the scale?

Fig. I

REVISION EXERCISES A

Section I

1. In an experiment to determine the focal length of a concave mirror the following values for the object and the real image distances were obtained.

u/cm	12.0	16.0	20.0	24.0	28.0	32.0	36.0	40.0
v/cm	60.0	27.2	20.1	17.3	15.6	14.5	13.8	13.3

From the data given calculate the focal length of the mirror. Plot a graph of $1/u$ against $1/v$ and determine the focal length of the mirror from the graph. Compare your two answers for the the focal length.

2. In an experiment to determine the focal length of a convex mirror the following values for the object and image distances were obtained.

u/cm	2.0	4.0	6.0	8.0	10.0	12.0	14.0	16.0
v/cm	1.8	3.2	4.3	5.2	6.1	6.7	7.2	7.7

From the data given calculate the focal length of the mirror. Plot a graph of $1/u$ against $1/v$ and determine the focal length of the mirror from the graph. Compare your two answers for the the focal length.

3. In an experiment to determine the refractive index of glass a ray of light was passed through a rectangular block of glass and the following values for the angles of incidence and refraction were recorded.

i/°	6	12	17	27	37	49	61	80
r/°	4	8	11	17	22	30	36	39

Plot a suitable graph and hence determine the refractive index of the glass.
Explain how the angle of refraction might have been determined in this experiment.

4. In an experiment to determine the refractive index of water by the real and apparent depth method the following values were obtained for the depth of the water, X, and the distance from the search pin to the back of the mirror, Y.

X/cm	4.5	9.1	14.6	18.3	24.0	28.8
Y/cm	3.3	6.8	10.9	13.8	18.1	21.5

Plot a graph of X against Y and hence determine the refractive index of the water.
If, in each case, the water had been 2 cm below the back of the mirror how would the final result have been affected? Sketch a graph to confirm your answer.

5. A lamp box and a screen were used to find the focal length of a converging lens and the following values for the object and image distances were obtained.

u/cm	6.0	8.0	10.0	12.0	14.0	16.0	18.0	20.0
v/cm	33.0	10.9	8.9	7.8	7.1	6.5	6.4	6.7

From the data, what is the maximum value which the focal length of the lens could have had?
Use the data to calculate a value for the focal length of the lens. Plot a suitable graph and use it to find the focal length.

6. Two pins were used to find the focal length of a diverging lens and the following values for the object and image distances were obtained.

u/cm	2.0	6.0	10.0	14.0	18.0	22.0	26.0	30.0
v/cm	1.8	4.6	6.7	8.3	9.5	10.5	11.3	12.0

Use the given data to find the focal length of the lens. Determine the focal length of the lens from a suitably drawn graph. Compare the values you obtain by the two methods.

Section II

1. State the laws of reflection of light. How would you show experimentally that, for a plane mirror, the object and image are at the same distance from the mirror?

2. Draw a diagram to show how the final image is formed in a periscope using two plane mirrors. Is the final image laterally inverted? Why are prisms often used in preference to mirrors?

3. Draw ray diagrams to show how an image is formed by a concave mirror when the object is (i) between the centre of curvature and the focus; (ii) inside the focus.

4. Describe an experiment to determine the focal length of a concave mirror. Mention some possible sources of error and say what steps you would take to obtain a more accurate result.

5. A pin of height 2.0 cm is placed 16 cm from a concave mirror of focal length 10 cm. Determine (i) the position of the image; (ii) the height of the image; (iii) the nature of the image.

6. A concave mirror has a focal length of 20 cm. Find the position, magnification and nature of the image of an object 7.5 cm from the mirror.

7. A concave mirror produces a real image which is 2.5 times the size of the object. If the object is 21 cm from the mirror, what is the focal length of the mirror?

8. Draw a ray diagram to show how an image is formed in a convex mirror. Does the nature of the image change for different positions of the object? Explain your answer.

9. Describe an experiment to measure the focal length of a convex mirror. What are the most likely sources of error in the experiment and what steps would you take to eliminate these errors?

10. A pin is placed 12 cm from a convex mirror of focal length 10 cm. Calculate the position and magnification of the image.

11. A convex mirror produces an image which is 0.25 times the size of an object placed 24 cm from the mirror. What is the focal length of the mirror?

12. Give two uses of concave mirrors. What are the advantages of using the particular mirror in each case?

13. State the laws of refraction. Describe an experiment to verify *Snell's Law* using a rectangular block of glass.

14. Describe an experiment to determine the refractive index of water. State possible sources of error and precautions necessary to obtain an accurate result.

15. Explain how total internal reflection occurs. The refractive index of a certain type of glass is 1.45. What is the critical angle for this glass?

16. Explain, with the aid of a diagram, why a dry road sometimes appears wet on a hot day. What is the name given to this type of optical illusion?

17. Draw a ray diagram to show how an image is formed by a converging lens when the object is (i) outside the focus; (ii) inside the focus.

18. Describe an experimental method of determining the focal length of a converging lens. State possible sources of error in the experiment and explain how the effect of these errors might be minimised.

19. A converging lens produces a virtual image which is 5.2 times the size of the object, placed 4.0 cm from the lens. What is the focal length of the lens? Give one possible use of a converging lens used in this way.

20. How would you determine experimentally the focal length of a diverging lens? Why is it not possible to measure the focal length of a diverging lens directly, as it is for a converging lens?

21. A lamp box is placed 14 cm from a diverging lens of focal length 20 cm. Calculate the position and magnification of the image.

22. Draw a ray diagram showing how the final image is formed in an astronomical telescope in normal adjustment. How may an astronomical telescope be modified to give an erect image?

*23. Draw a ray diagram showing how the final image is formed in a compound microscope. What is the nature of this image?

*24. The total distance between the lenses in an astronomical telescope in normal adjustment is 88 cm. If the magnifying power of the telescope is 10, what is the focal length of the objective lens?

*25. Name the main parts of a spectrometer and give the function of each.

5 Linear Motion

5.1 Scalar and Vector Quantities

Physical quantities may be divided into two types: those which have a direction associated with them and those which do not.

Scalar Quantities

Scalar quantities are those which have no direction associated with them. Examples of scalar quantities are mass, time, energy, *etc*. Scalar quantities may be added and subtracted using the normal rules of algebra.

Vector Quantities

Vector quantities are those which have a direction associated with them. Examples are velocity, acceleration, force, *etc*. Vectors may only be added and subtracted algebraically if they are in same direction. Two vector quantities which have opposite directions may be added or subtracted algebraically if one of them is first given a negative sign. Vector quantities which make an angle of other than 180° with each other must be added using the triangle law or the parallelogram law *(see Chapter 9)*.

5.2 Displacement, Velocity and Acceleration

Everyone is at least vaguely familiar with these terms. No doubt, you have some idea of what each one means. However, in physics, "some idea" is not good enough! It is important to know exactly the meaning of the terms we use. This is particularly true of terms which are in common use. In everyday language a word can have several meanings, depending on the context. In physics each term is very carefully defined and can have no other meaning. As well as learning the definitions of the various quantities you should note in each case the unit of the quantity, whether it is a vector or scalar quantity and the symbols for the quantity and its unit. This information is given after the definition in each case.

Displacement

If a person walks from A to B, *Fig. 5.1*, and then from B to C the total distance which he has walked is |AB| + |BC|. However, when he is at C his distance from his starting point is only |AC|. We say that his displacement from A is |AC| in the direction shown.

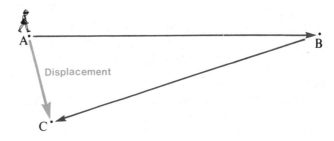

Fig. 5.1

> **The displacement of a particle from a fixed point is its distance from that point in a particular direction.**

Displacement, *s*, is a **vector** quantity and its unit is the **metre**, m.

If a person leaves her home in Ballinasloe and drives to Dublin, her resulting displacement is 129 km due E from her home. If she now drives home again, the total *distance* she has travelled is 258 km but her displacement from her home is 0 km. (In fact the distance travelled would be considerably greater (302 km) since the road is not a straight line between the two points.)

Velocity

If a car undergoes a displacement of 60 km due N in 2 hours then its average velocity is 30 km per hour due N, *i.e.* the average velocity is the change in displacement divided by the time taken. The average velocity does not tell us the velocity of the car at any given instant; its velocity may have changed several times during the journey. To find the velocity at a given instant we calculate the average velocity between two points which are very near together, *i.e* we find the change in displacement over a very short time. If the time is very small the velocity cannot change much during it. So the average velocity over a very small period of time must be the same as the actual velocity at any instant during that time. The change in displacement divided by the time, when the time is very small, is called the rate of change of displacement with respect to time, or the **velocity.**

Velocity is the rate of change of displacement with respect to time.

Velocity, v, is a **vector** quantity and its unit is the **metre per second**, m s^{-1}. The magnitude of the velocity is called the **speed**. Speed is thus a **scalar** quantity.

Example
The displacement of Sligo from Cork is 265 km N. The distance by road is 330 km. A motorist travels from Cork to Sligo in 5 h. Find (i) her average speed, (ii) her average velocity.

$$\text{Average speed} = \frac{\text{distance}}{\text{time}}$$

$$= \frac{330}{5}$$

$$= 66 \text{ km h}^{-1}$$

$$\text{Average velocity} = \frac{\text{displacement}}{\text{time}}$$

$$= \frac{265}{5}$$

$$= 53 \text{ km h}^{-1} \text{ N}$$

Ans. 66 km h^{-1}; 53 km h^{-1} N

Acceleration
When the velocity of a body is changing it is said to be accelerating. If the velocity of a car increases from 5 m s^{-1} S to 13 m s^{-1} S in 4 s its average acceleration is 2 m s^{-1} per second S, *i.e.* the average acceleration is the *change* in velocity divided by the time taken for the change to occur. As in the case of velocity, the acceleration at a particular instant is determined by finding the change in velocity over a very small time.

Acceleration is the rate of change of velocity with respect to time.

Acceleration, a, is a **vector** quantity and its unit is the **metre per second per second**, m s^{-2}. Note that acceleration is a change in velocity. This change may be an increase or a decrease or a change of direction, *(see below)*.

Example
The velocity of a car changes from 10 m s^{-1} E to 25 m s^{-1} E in 10 s. Find its average acceleration.

$$a = \frac{\text{change in velocity}}{\text{time}}$$

$$= \frac{25 - 10}{10}$$

$$= 1.5 \text{ m s}^{-2} \text{ E}$$

Ans. 1.5 m s^{-2} E

In this example the magnitude of the velocity, *i.e.* the speed, changes while the direction remains the same. In some cases the direction changes while the speed remains the same, *e.g.* a car going around a corner at a constant speed has an acceleration *(see Chapter 7)*. It is also possible for both speed and direction to be changing at the same time. (The definitions of velocity and acceleration are given in calculus notation in *Appendix A1*.)

5.3 Equations of Motion

The behaviour of moving bodies can be summarised very neatly by equations. Using such equations we can easily work out speeds, distances covered, times taken, *etc*. For the moment we shall confine ourselves to equations which describe motion in which the acceleration is constant.

If a body starts with a velocity u and has a velocity v after a time t, then its acceleration is

$$a = \frac{v - u}{t}$$

Multiplying both sides by t gives

$$at = v - u$$

Rearranging gives

$$\boxed{v = u + at} \qquad \text{(i)}$$

If we know the values of three of the quantities in this equation we can find the value of the fourth. The velocity at the start, u, is usually called the **initial velocity**.

Example

A car starts with a velocity of 5 m s^{-1} N. What is its velocity after 10 s if its acceleration is 2 m s^{-2} N?

$$u = 5 \text{ m s}^{-1}$$
$$t = 10 \text{ s}$$
$$a = 2 \text{ m s}^{-2}$$
$$v = ?$$

$$v = u + at$$
$$= 5 + (2 \times 10)$$
$$= 25 \text{ m s}^{-1} \text{ N}$$

Ans. 25 m s^{-1} N

In solving problems involving displacement, velocity and acceleration we must always remember that we are dealing with vector quantities. If they are all in the same direction this does not present a problem. If two quantities are in opposite directions we must choose one direction as positive and the other is then negative. For example, if you walk 100 m N and then 80 m S your resulting displacement is 20 m from your starting point. The sum of your two displacements is 100 + (−80), *i.e.* 20 m.

Example

A ball is thrown with an initial velocity of 12 m s^{-1} vertically upwards. Given that the acceleration (caused by gravity) is 9.8 m s^{-2} vertically downwards, find the velocity of the ball after (i) 1.0 s, (ii) 2.0 s.

Taking upwards to be positive,

$$u = + 12 \text{ m s}^{-1}$$
$$a = - 9.8 \text{ m s}^{-2}$$
$$t = 1.0 \text{ s and } 2.0 \text{ s}$$
$$v = ? \text{ and } ?$$

(i)
$$v = u + at$$
$$= 12 + (-9.8 \times 1.0)$$
$$= 12 - 9.8$$
$$= 2.2 \text{ m s}^{-1}$$

(ii)
$$v = u + at$$
$$= 12 + (-9.8 \times 2.0)$$
$$= 12 - 19.6$$
$$= - 7.6 \text{ m s}^{-1}$$

Ans. (i) 2.2 m s^{-1} upwards, (ii) 7.6 m s^{-1} downwards

Eqn. (i) relates the velocity to the time, We shall now derive an equation which relates the displacement, s, to the time.

$$s = (\text{average velocity}) \times t$$

Since acceleration is constant

$$\text{average velocity} = \frac{v + u}{2}$$

$$\therefore \quad s = \frac{v + u}{2} \times t$$

$$= \tfrac{1}{2} (v + u)t$$

But,
$$v = u + at$$

$$=> \quad s = \tfrac{1}{2} (u + at + u)t$$

$$= \tfrac{1}{2} (2ut + at^2)$$

$$= ut + \tfrac{1}{2} at^2$$

$$\boxed{s = ut + \tfrac{1}{2} at^2} \qquad \text{(ii)}$$

(For an alternative derivation of *Eqns. (i)* and *(ii)* see *Appendix A2*.)

Example

A ball is thrown with a velocity of 10.0 m s^{-1} vertically upwards. Given that the acceleration is 9.8 m s^{-2} vertically downwards, find the times at which the ball is 2.5 m above the ground. Find also the velocity of the ball at these times.

$$u = + 10.0 \text{ m s}^{-1}$$
$$a = - 9.8 \text{ m s}^{-2}$$
$$s = + 2.5 \text{ m}$$
$$t = ?$$
$$v = ?$$

$$s = ut + \tfrac{1}{2} at^2$$

$$2.5 = 10t + \tfrac{1}{2}(-9.8)t^2$$

$$= 10t - 4.9t^2$$

$$4.9t^2 - 10t + 2.5 = 0$$

$$\Rightarrow \quad t = \frac{10 \pm \sqrt{(10^2 - 4 \times 4.9 \times 2.5)}}{2 \times 4.9}$$

$$= \frac{10 \pm \sqrt{(100 - 49)}}{9.8}$$

$$= 0.292 \text{ s or } 1.749 \text{ s}$$

To find velocity

(i)
$$v = u + at$$
$$= 10 + (-9.8 \times 0.292)$$
$$= 10 - 2.862$$
$$= 7.138 \text{ m s}^{-1}$$

(ii)
$$v = u + at$$
$$= 10 + (-9.8 \times 1.749)$$
$$= 10 - 17.14$$
$$= -7.14 \text{ m s}^{-1}$$

Ans. 0.29 s; 1.7 s;
7.1 m s^{-1} upwards
7.1 m s^{-1} downwards

While it is possible to solve all problems involving motion in a straight line with constant acceleration using only *Eqns. (i)* and *(ii)*, it is sometimes convenient to combine them to form a third equation which does not contain *t*.

$$v = u + at \quad \textit{(Eqn. (i))}$$
$$\Rightarrow \quad v^2 = (u + at)^2$$
$$= u^2 + 2uat + a^2t^2$$
$$= u^2 + 2a(ut + \tfrac{1}{2} at^2)$$

But, $ut + \tfrac{1}{2} at^2 = s$ *(Eqn. (ii))*

$$\therefore \quad \boxed{v^2 = u^2 + 2as} \qquad \text{(iii)}$$

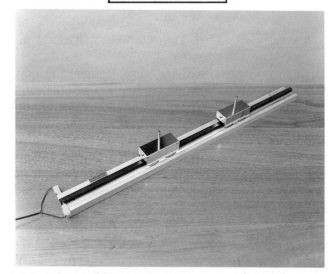

Fig. 5.2 Powder track timer

Fig. 5.3(a)   Pattern on track when trolley travels at constant speed

Fig. 5.3(b)   Pattern when trolley is accelerating

EXPERIMENT 5.1
To Measure Velocity and Acceleration

Method 1:
Apparatus: Powder track timer, *(Fig. 5.2)* trolley.

Procedure:
1. Connect one of the leads from the adaptor to the base of the track and the other to the rail. Sprinkle a fine layer of sulphur powder along the rail.
2. Adjust the slope of the track until the trolley will run at a constant speed.
3. Place the trolley at the top of the track and give it a gentle push.
4. Find the average distance from the start of one patch of powder to the start of the next. This gives the distance travelled in 1/50 s. Calculate the average speed of the trolley from $v = s/t$. Repeat for different speeds.

RESULTS

s/cm	t/s	v/cm s^{-1}

5. Increase the slope of the track so that the trolley will accelerate. Give the trolley a gentle push from the top of the track.
6. Find the average speed of the trolley over a short distance near the top of the track and again near the bottom of the track. Measure the time taken from the middle of the first distance to the middle of the second (count the number of patches of powder — each corresponds to 0.02 s).
7. Calculate the average acceleration of the trolley from $a = (v - u)/t$. Repeat for different accelerations.

RESULTS

u/cm s^{-1}	v/cm s^{-1}	t/s	a/cm s^{-2}

Questions
1. Why does adjusting the slope of the track allow the trolley to run at constant speed? How do you know that the trolley is running at constant speed?
2. How may the average distance between patches of powder be determined?
3. Suggest another method for calculating the acceleration. What are the advantages or disadvantages of this method?

Other Methods
There are a number of other methods for measuring speed and acceleration. The method of measuring time varies from one to the other but the procedure is essentially the same in each.

Ticker Tape Timer:
This device produces a series of dots at equal time intervals on a paper tape attached to the trolley. The time interval between dots may or may not be 1/50 s depending on the model. If the time interval is not known it is taken as unit time and speeds and accelerations are expressed in terms of this time rather than the second.

Fletcher's Trolley
In this method a pen is attached by a spring to the trolley. The pen is set vibrating at right angles to the trolley and as the trolley moves the pen draws a sine wave on a strip of paper fixed to the bench. The time taken to travel from the crest of one wave to the crest of the next is taken as unit time.

5.3 Momentum

A very useful concept is that of momentum. Before defining momentum, however, we must consider what is meant by the word "mass"

Mass

It is more difficult to push a truck than it is to push a car — we say that the **mass** of the truck is greater than the mass of the car. Likewise, if the car and truck are both moving at the same speed, it is more difficult to stop the truck than it is to stop the car.

> **The mass of a body is a measure of its ability to resist changes in its velocity.**

Mass, m, is a **scalar** quantity and its unit is the **kilogram**, kg. The ability to resist changes in velocity is a property of all matter and is called **inertia**. ·

The mass of a body does not depend on where it is . A body which has a mass of 10 kg on earth will also have a mass of 10 kg on the moon or out in space. In the absence of friction, it is just as difficult to change the velocity of a body on the moon or in space as it is on earth.

Momentum

> **The momentum of a body is the product of its mass and its velocity.**

Momentum, p, is a **vector** quantity and its unit is the kg m s^{-1}, *i.e.* the unit of mass multiplied by the unit of velocity. The definition may be stated in symbols as

$$p = mv$$

Conservation of Momentum

The principle of conservation of momentum is one of the most fundamental laws of nature. There are no known exceptions to it. Before stating this principle we must first consider the idea of a closed system. The word system means the body or bodies with which we are concerned at a given time. For example, if we are interested in what happens when a tennis ball is struck with a racket the system consists of the ball and the racket. All other bodies are outside the system. A closed system is one in which bodies inside the system are not affected

in any way by bodies outside the system, *i.e.* there is no interaction between bodies in the system and bodies outside the system. Thus, two billiard balls colliding on a table may be regarded as a closed system since there is no (or, at least, a negligible) interaction between the balls (the system) and the table which is outside the system. On the other hand, the tennis ball and racket referred to above is not a closed system since there is an interaction between the racket (part of the system) and the person holding it (not part of the system).

We may now state the principle of conservation of momentum as

> **In any interaction within a closed system the total momentum before the interaction is equal to the total momentum after.**

Thus, in the case of the billiard balls colliding on the table, the total momentum of the balls before they collide is equal to their total momentum after the collision. However, the total momentum of the tennis ball and racket will not be the same after they collide as before.

Example

A bullet of mass 0.02 kg and velocity of 400 m s^{-1} E strikes a block of wood of mass 2.0 kg which is at rest. After the collision the block and bullet move together, Fig. 5.4. Assuming that the interaction between the block and its support is negligible, calculate the velocity of the block and bullet after the impact.

Taking East to be positive,

$$\text{Momentum before} = (0.02 \times 400) + (2.0 \times 0)$$
$$= 8$$
$$\text{Momentum after} = 2.02 \times v$$
$$\text{Momentum after} = \text{Momentum before}$$
$$\Rightarrow \quad 2.02v = 8$$
$$\Rightarrow \quad v = \frac{8}{2.02}$$
$$= 3.96 \text{ m s}^{-1}$$

Ans. 4.0 m s^{-1} E

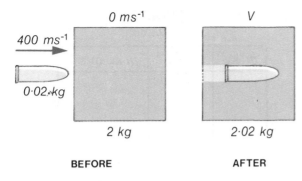

BEFORE   AFTER

Fig. 5.4

Example

A rifle of mass 5 kg fires a bullet of mass 10 g with an initial speed of 700 m s^{-1}. Ignoring the interaction between the rifle and its support calculate the speed of recoil of the rifle.

$$\text{Momentum before} = 5.01 \times 0$$
$$= 0$$
$$\text{Momentum after} = (5 \times v) + (0.01 \times 700)$$
$$= 5v + 7$$
$$\text{Momentum after} = \text{Momentum before}$$
$$=> \quad 5v + 7 = 0$$
$$=> \quad 5v = -7$$
$$=> \quad v = \frac{-7}{5}$$
$$= -1.4 \text{ m s}^{-1}$$

The minus sign indicates that the direction of recoil of the rifle is opposite to the direction in which the bullet is moving.

Ans.   1.4 m s^{-1}

In solving the previous problem we ignored the mass of gas expelled from the explosion. In a recoilless rifle this gas is expelled backwards, *i.e.* in the opposite direction to that of the bullet. The momentum of the gas is equal in magnitude but opposite in direction to the momentum of the bullet, so the rifle gains no momentum, *i.e.* it remains at rest.

The principle of conservation of momentum has many important applications. As we have just seen, it must be taken into account in the design of firearms. The principle is also applied in rocket propulsion. The rocket engines expel a mass of gas with momentum in the "backward" direction; the rocket then gains an equal amount of momentum in the "forward" direction. Although calculations of rocket velocities are complicated by the fact that the total mass of the rocket changes continuously as the fuel is used up the basic principle is still the same. The same principle is used in "steering" a spacecraft in space. In order to turn a craft to the right a rocket motor expels a small mass of gas to the left, and *vice versa*. Similarly, to slow down a spacecraft a mass of gas is expelled in the same direction as the craft is travelling. The craft thus loses the momentum gained by the gas and so it slows down.

Fig. 5.5

Fig. 5.6

The propulsion of jet aircraft is also based on the principle of conservation of momentum. In this case, the gas expelled by the engines is air. The air is drawn in at the front of the engine at a low speed. It is compressed in the engine and expelled backwards with a much higher speed. The air thus gains momentum in the backward direction so the aircraft gains an equal amount of momentum in the forward direction.

EXPERIMENT 5.2

To Verify the Principle of Conservation of Momentum

Method 1:
Apparatus: Powder track timer, two trolleys with magnets*, metre stick.

Procedure:
1. Connect one terminal of the adaptor to the track and the other to the rail. Sprinkle some powder on the track. Adjust the slope of the track so that a trolley, given a gentle push, will roll at a constant speed.
2. Attach a magnet to each of the trolleys and place one of the trolleys near the top of the track and the other half-way down, *Fig. 5.7*.
3. Give the trolley at the top a gentle push. Measure the speed (see previous experiment) of this trolley before it collides with the second trolley and the speed of both trolleys after the collision.
4. Measure the mass of each trolley. Calculate the momentum before the collision and the momentum after the collision. Repeat with different masses on each trolley.

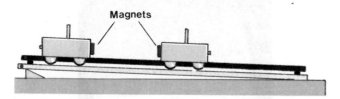

Magnets

Fig. 5.7

*There are various other ways of ensuring that the trolleys stay together after they collide, *e.g.* attaching a cork to one and a pin to the other, sticking a piece of "Blu Tack" or similar substance on each, *etc.*

RESULTS
TROLLEY 1

m/kg	v/m s^{-1}	p/kg m s^{-1}

TROLLEY 1 + TROLLEY 2

m/kg	v/m s^{-1}	p/kg m s^{-1}

Other Methods:
The speeds of the trolleys may be measured by any of the methods referred to in the previous experiment, *e.g.* ticker tape timer, Fletcher's trolley, *etc*. The procedure is essentially the same in each case. With some types of apparatus, *e.g.* the powder track timer, the experiment may also be carried out by placing the two trolleys together near the middle of the track with a compressed spring between them. When the spring is released the momentum of each trolley is determined. In this case the total momentum before the interaction is zero so the total momentum after the interaction should also be zero, *i.e.* the momenta of the two trolleys should be equal in magnitude but opposite in direction. Note that for this method the track must be level. (Why?)

Questions
1. Does the fact that the magnets (Method 1) exert a force on each other result in a change in the momentum of the system? Explain.
2. Should the slope of the track be changed each time extra masses are added to the trolleys? Explain your answer and check it experimentally.

Galileo Galilei (1564 — 1642)

1564   1642

Galileo Galilei

POSTE ITALIANE L. 70

Galileo was born in Pisa, Italy. His father, a mathematician, wanted the boy to study medicine, but Galileo chose instead to study mathematics and science.

Unlike the ancient Greeks, with the notable exception of Archimedes, Galileo was not content to make observations and then devise abstract theories. Instead, he insisted on designing experiments to test and refine his theories as he developed them.

In 1581 he discovered that the time of swing (the period) of a pendulum is independent of the amplitude of the swing. He did this by using his pulse to measure the time of swing of a chandelier in Pisa Cathedral. He later showed that all bodies fall at the same speed, regardless of their weight, provided that the effects of air resistance are ignored. He also demonstrated that a body rolling down an inclined plane accelerates continuously. This latter discovery was very important since it contradicted the universally held belief that a force was required simply to keep a body moving with constant speed.

In 1609 Galileo constructed a telescope which he used to discover the mountains on the moon, four of the moons of Jupiter and many stars invisible to the naked eye. He also established, by observing the motion of sun spots, that the sun rotates with a period of 27 days. (Observing the sun through a telescope is, of course, extremely dangerous. Galileo seriously damaged his eyes and in later life went completely blind.)

In Galileo's time it was generally believed that the earth was the centre of the universe and that the sun, and all the other stars, moved around the earth. The view that it was the earth which moved around the sun was first proposed by the Greek astronomer Aristarchus almost 2000 years previously and was developed into a detailed description of the structure and behaviour of the solar system by the Polish astronomer Copernicus around 1512. Galileo's observations of the heavens led him to the belief that, as Copernicus had suggested, the earth moved around the sun rather than the sun around the earth. However, in 1616 the Copernican system was declared a heresy by Pope Pius V and when Galileo persisted with his view that the earth moved he was eventually brought before the Inquisition. A frail man of nearly 70 years of age, he recanted under the threat of torture.

Galileo laid the foundations of modern mechanics and his discoveries were used by Newton as the basis for his laws of motion. He died in the year that Newton was born, 1642.

SUMMARY

Displacement, *s*, is the distance from a particular point in a particular direction; its unit is the metre, m. Velocity, *v*, is the rate of change of displacement; its unit is the m s^{-1}. The magnitude of the velocity is called the speed. Acceleration, *a*, is the rate of change of velocity; its unit is the m s^{-2}. All three are vector quantities.

The three equations of motion relate velocity, time, displacement and acceleration to each other. These equations apply only to linear motion in which acceleration is constant. The equations are:

$$v = u + at$$
$$s = ut + \tfrac{1}{2}at^2$$
$$v^2 = u^2 + 2as$$

Mass, *m*, is a measure of a body's inertia, *i.e.* of its ability to resist changes in its velocity. The unit of mass is the kilogram, kg. Momentum, *p*, is equal to the product of mass and velocity; its unit is the kg m s^{-1}. Mass is a scalar quantity; momentum is a vector quantity.

The principle of conservation of momentum states that in a closed system the total momentum is constant. The propulsion of rockets and jet aircraft is based on this principle.

Questions 5

DATA: Acceleration due to gravity, $g = 9.8$ m s^{-2}.

Section I

1. Give three examples of vector quantities and three examples of scalar quantities.......................................

2. The distance travelled in a particular direction is called the
....

3. Velocity is the rate of change of.........................with respect to time.

4. Acceleration is the rate of change of........................ with respect to time.

5. Is it possible for a body travelling at a constant speed to have an acceleration? Explain...............................

6. In the equation $v = u + at$, what does each of the symbols represent? ....

7. When a body, thrown vertically upwards, is at its greatest height its velocity is....................and its acceleration is..

8. What is meant by the inertia of a body?...................

9. Define momentum and give its unit..........................

10. State the principle of conservation of momentum.......

11. Why does a gun recoil when a bullet is fired from it?.

12. Is momentum conserved when a returning spacecraft lands on the earth? Explain...

Section II

1. A car starts from rest and reaches a speed of 20 m s^{-1} after 15 s. Calculate the magnitude of its acceleration.

2. A car starts from rest with a uniform acceleration of magnitude 1.5 m s^{-2}. Find (i) the speed, (ii) the distance travelled after 15 s.

3. A car is travelling due South at a speed of 30 m s^{-1} when the brakes are applied. If it travels 100 m in coming to rest, calculate the time taken and the acceleration (assumed uniform).

4. A man swims upstream from A to B at an average speed of 2 m s^{-1}. He then swims back to A at an average speed of 6 m s^{-1}. What is his average speed for the whole journey?

5. A stone is dropped from the top of a tower 50 m high. Find the height of the stone above the ground and its speed after 3.0 s. (Neglect air resistance.)

6. A ball is thrown vertically upwards with an initial speed of 12 m s^{-1}. Find (i) the greatest height reached, (ii) the times at which the height is 5.0 m, (iii) the velocity after 0.8 s.

7. A stone is dropped from the top of a cliff 20 m high. At the same instant another stone is thrown vertically upwards with an initial speed of 30 m s^{-1}. At what height will the two stones meet?

8. A ball is dropped from the top of a building 10 m high. 0.3 s later another ball is thrown vertically downwards with an initial speed of 15 m s^{-1}. Which ball will strike the ground first? With what initial speed would the second ball have to be thrown in order that both balls would strike the ground at the same time?

9. A tennis player strikes a ball of mass 40 g travelling towards her with a horizontal velocity of 20 m s^{-1} and returns it with a horizontal velocity of 30 m s^{-1}. Calculate the change in momentum of the ball. This

momentum may be considered as a gain for the ball. Which body, or bodies, lose momentum?

10. A bullet of mass 0.05 kg and travelling at a speed of 500 m s^{-1} strikes a block of wood of mass 10 kg which is at rest. If the bullet becomes embedded in the block, calculate the initial speed of the block after the collision.

11. A bullet of mass 40 g and travelling at a speed of 600 m s^{-1} enters a block of wood of mass 4.0 kg which is at rest. The bullet passes through the block and emerges with a speed of 100 m s^{-1}. What is the initial speed of the block?

12. A car of mass 800 kg is travelling due North at 30 m s^{-1} when it is involved in a head-on collision with a truck of mass 30 t travelling due South at 25 m s^{-1}. Calculate (i) the initial velocity of the wreckage, (ii) the magnitude of the change in velocity of (a) the truck, (b) the car.

13. A tennis ball of mass 40 g, dropped from a height of 2.0 m, bounces to a height of 60 cm. Calculate the amount of momentum lost by the ball when it strikes the ground.

14. A spacecraft of mass M slows down by expelling a quantity of gas of mass m with a speed v. Calculate the change in the speed of the craft. (Assume that m is negligible compared with M.)

15. A cyclist starts from rest and reaches a speed of 8.0 m s^{-1} in 15 s. She continues at this speed for 5 minutes. She then brings the bicycle to rest in a distance of 20 m. Calculate (i) the magnitude of the acceleration in the initial and final stages of the journey, (ii) the total distance travelled.

6 Force I

6.1 Newton's Laws

In 1687 there appeared what many regard as the greatest scientific work ever published. It has come to be known simply as the *Principia* (Principles) and its author was Sir Isaac Newton. In this book Newton put down formally, for the first time, the fundamental principles of mechanics. The ideas put forward in the book are, with only some slight modifications, still valid to-day. (The modifications are unimportant in everyday life. They are only significant if we are dealing with very small masses and/or very high speeds and are accounted for in Einstein's *Theory of Relativity*.)

Force

The concept of force is basic to the study of mechanics and, indeed to much of physics. Force is loosely understood as being a "push" or a "pull". It is more precisely defined as follows.

> **A force is that which causes acceleration.**

Force, F, is a **vector** quantity and its unit is the **newton**, N.

Resultant Force

The resultant force on a body is the sum of all the forces acting on it. Since force is a vector quantity only forces which are in the same direction may be added arithmetically. Forces which are in opposite directions are added algebraically, *i.e.* forces in one direction are given a positive sign and those in the opposite direction are give a negative sign. For example, the resultant of an upward force of 20 N and a downward force of 5 N is an upward force of 15 N, *i.e.* $+ 20 - 5 = 15$. In all other cases involving the addition of forces the *Triangle Law* or the *Parallelogram Law* must be used *(see Chapter 9)*.

Newton's Laws of Motion

> **Law I**
> The velocity of a body does not change unless a resultant external force acts on it.
>
> **Law II**
> When a resultant external force acts on a body the rate of change of the body's momentum is proportional to the force and takes place in the direction of the force.
>
> **Law III**
> In any interaction between two bodies A and B, the force exerted by A on B is equal in magnitude but opposite in direction to the force exerted by B on A.

Law I

This law, while simple, is difficult to accept because it seems to contradict our normal experience. If you throw a ball with a certain velocity it does not carry on indefinitely with that velocity. Indeed, its velocity becomes zero in a short time, even though there *seems* to be no force acting on it. In fact, as the ball moves through the air or along the ground the air or ground exerts a force on it. This force, called friction *(see p. 72)*, is always in the opposite direction to the ball's velocity, so the ball slows down and eventually stops.

In the laboratory, various techniques can be used to reduce friction to negligible proportions. One method used is the air track, *Fig. 6.1*. The top of the track has a large number of

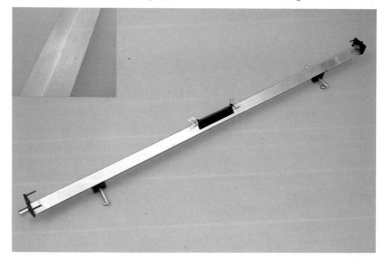

Fig. 6.1 Air track (inset: Close-up of section of track showing holes)

holes (see inset) through which air is forced under pressure. The "riders" are thus sitting on a cushion of air and friction is negligible at low speeds. The first law may be verified by photographing a moving "rider" at equal time intervals. The images on the resulting photograph are equidistant from each other, showing that the "rider" travelled equal distances in equal times, *i.e.* that it had a constant speed. (Note: This experiment does not entirely verify the law since it only verifies that the *magnitude* of the velocity does not change.)

In space there is no friction. An astronaut wishing to move outside a spacecraft must be attached to it by a line. On stepping out of the craft the astronaut acquires a velocity away from it. Since there is no other body which can exert a force on the astronaut he/she would, without the line, carry on with this velocity, perhaps for ever.

Law II

If the force acting on the body is constant, *i.e.* does not change with time, this law may be written in the form

$$F \propto \frac{\text{change in } (mv)}{t}$$

or,

$$F \propto \frac{\Delta mv}{t}$$

If the mass is also constant, as it is at speeds normally encountered in mechanics, then we can write

$$F \propto m \frac{\Delta v}{t}$$

$$=> \quad F \propto ma$$

or,

$$F = kma \qquad \text{(i)}$$

where k is a constant.

The question now arises: "What is the value of k in *Eqn. (i)*?" The answer is that it depends on the definition of the unit of force, the newton. For simplicity, it has been decided to define the newton so that k has the value 1.

> **One newton is that force which gives a mass of 1 kg an acceleration of 1 m s^{-2}.**

Substituting for $k = 1$ in *Eqn. (i)* gives

$$\boxed{F = ma}$$

Example

Find the acceleration produced by a force of 10 N south west, acting on a body of mass 20 kg.

Taking S-W to be positive

$$F = ma$$
$$=> \quad 10 = 20\,a$$
$$=> \quad a = 0.5 \text{ m s}^{-2}$$

Ans.   0.5 m s^{-2} S-W

Example

A car of mass 1 t is travelling with a velocity of 40 m s^{-1} N-W. What force must be applied to the car in order to bring it to rest in a distance of 100 m?

Firstly, we must find the acceleration.

Taking N-W to be positive

$$v^2 = u^2 + 2as$$
$$=> \quad 0^2 = 40^2 + 2 \times a \times 100$$
$$=> \quad 0 = 1600 + 200a$$
$$=> \quad 200a = -1600$$
$$=> \quad a = -8 \text{ m s}^{-2}$$

Knowing the acceleration we can now find the force

$$F = ma$$
$$= 1 \times 10^3 \times -8$$
$$= -8 \times 10^3 \text{ N}$$
$$= -8 \text{ kN}$$

Since we took N-W to be positive the negative sign means that the force is in the S-E direction.

Ans.   8 kN S-E

EXPERIMENT 6.1

To Verify that $a \propto F/m$

Method 1:

Apparatus: Powder track timer, trolley, metre stick.

Procedure:

1. Connect one of the leads from the adaptor to the base of the track and the other to the rail. Sprinkle a fine layer of sulphur powder along the rail.
2. Adjust the slope of the track until the trolley will run at a constant speed.
3. Attach a length of string to the trolley and attach one of the 10 g masses to the string using the hook. Place the remaining 10 g masses on the trolley.
4. Hold the trolley at the top of the track with the string passing over the pulley and the mass hanging freely, *Fig. 6.2*. Release the trolley.

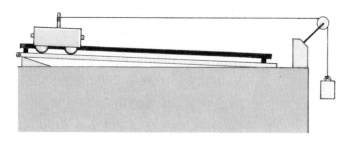

Fig. 6.2

5. Determine the acceleration of the trolley *(see Experiment 5.1)*. The force accelerating the trolley is equal to the weight on the end of the string. The weight, in newtons, is equal to the mass, in kilograms, multiplied by 9.8 *(see p.67)*.

6. Repeat the procedure with different masses (transferred from the trolley) on the end of the string. Plot a graph of acceleration, *a*, against force, *F*.

RESULTS

a/m s^{-2}	F/N

7. Repeat the procedure, keeping the accelerating force constant and using different masses on the trolley.
8. Plot a graph of acceleration, *a*, against the reciprocal of the mass, m^{-1}.

RESULTS

a/m s^{-2}	m/kg	m^{-1}/kg^{-1}

Other Methods:

This experiment may be carried out using any of the methods referred to in Experiment 5.1, *e.g.* ticker tape timer, Fletcher's trolley, *etc.* The procedure is essentially the same in each case.

Questions

1. Why is the track sloped? How would the graphs be affected if this were not done?
2. In the first part of the experiment why are the masses stored on the trolley before being transferred to the end of the string? If this were not done would the effect on the results be significant? Explain.

Law III

This law says, in effect, that forces always occur in pairs. There are many examples of this. A typical example is shown in *Fig.*

6.3(a) — a book resting on a table. *Fig. 6.3(b)* shows the forces acting on the book, while *Fig. 6.3(c)* shows the force on the table due to the book. The book exerts a downward force, of magnitude R, on the table; the table exerts an equal upward force on the book.

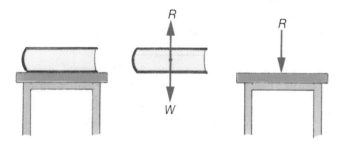

Fig. 6.3

When you walk, your foot exerts a backward force on the ground while the ground exerts an equal forward force on your foot, *Fig. 6.4(a)*. Similarly, when a car or bicycle is driven forward the tyre exerts a backward force on the road while the road exerts an equal forward force on the tyre. An aircraft stays airborne because, as the wings exert a downward force on the air beneath them, the air exerts an equal upward force on the wings, *Fig. 6.4(b)*.

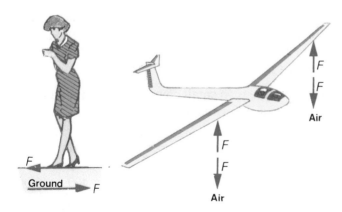

Fig. 6.4

The principle of conservation of momentum and Newton's third law are, in fact, different statements of the same physical principle. In *Chapter 5* we saw how the recoil of a gun could be explained in terms of conservation of momentum. It should now be clear that it could be explained equally well in terms of Newton's third law: the gun exerts a force on the bullet — the bullet exerts an equal opposite force on the gun. Rocket and jet propulsion may also be explained in a similar way. As the rocket engine exerts a force on the gas which it is expelling the gas exerts an equal opposite force on the engine and hence on the rocket.

Isaac Newton (1642 — 1727)

Isaac Newton was born in Lincolnshire on Christmas Day, 1642. He was not particularly bright at school, although he showed a flair for designing and constructing mechanical devices. After a spell on his mother's farm in the late 1650's he went to Cambridge, graduating in 1665.

When the plague hit London in 1665, Newton moved back to his mother's farm. By this time he had already worked out the binomial theorem and discovered the basics of calculus. It was during this period (1665 — 66) that he developed his theory of gravitation and established that white light could be separated into different colours by passing it through a prism.

In 1667, Newton returned to Cambridge and two years later was appointed a professor of mathematics. In 1672, he was elected to the Royal Society where he reported on his work on light and colours. From this work he developed his particle theory of light which held sway for over a century. In 1668 he invented the reflecting telescope.

In 1687, Newton published what is generally accepted to be the greatest scientific work ever written, *Philosophiae Naturalis Principia Mathematica* (Mathematical Principles of Natural Philosophy). In this book, written in Latin, he set down what are now known as Newton's three laws of motion. The book also contained his theory of gravitation, developed over twenty years earlier on his mother's farm.

Although Newton is remembered principally for his work in physics, he also wrote at great length, but with little impact, on chemistry and theology. In 1689 he was elected to Parliament and kept his seat for several years, although he never made a speech. In 1699 he became master of the mint, at a salary in the region of £2000 per year. He devoted himself entirely to his work at the mint, taking a particular interest in the problem of counterfeiting.

Newton was a very sensitive man and was childish in his response to criticism. He was given to violent rages and was utterly ungenerous in his treatment of those he believed to be his enemies. He was an intense thinker and inevitably his health suffered. He had at least two nervous breakdowns.

Newton was undoubtedly one of the greatest scientists, if not the greatest, of all time. The extent of his contribution to the development of scientific knowledge is perhaps best summed up in Alexander Pope's famous couplet:

"Nature and Nature's laws lay hid in night:
God said, Let Newton be! and all was light."

6.2 Gravity

During his enforced holiday on his mother's farm during 1665-66, Newton developed his theory of gravitation. This theory provides an explanation for both the motion of heavenly bodies and of falling bodies on earth.

The theory of gravitation holds that there is a force of attraction between all bodies in the Universe and that the magnitude of this force is related to the masses of the bodies concerned and to the distance between them. This relationship is now known as **Newton's Universal Law of Gravitation** and it may be stated as follows.

> **The force between any two point masses is proportional to the product of the masses and inversely proportional to the square of the distance between them.**

In symbols, Newton's law may be expressed as

$$F \propto \frac{m_1 m_2}{d^2}$$

where F is the magnitude of the force of attraction between two point masses, m_1 and m_2, which are a distance d apart. This expression may also be written as

$$F = G \frac{m_1 m_2}{d^2}$$

where G is a constant, called the universal constant of gravitation. The value of G is found by experiment to be 6.7 x 10^{-11} N m^2 kg^{-2}. Because the value of G is so small we do not notice the force of attraction between, say, two tennis balls lying side by side. It is only when we are dealing with very large masses, like those of stars and planets, that the force becomes appreciable.

Note that this law refers to point masses, *i.e.* to masses which have no dimensions. However, it can be shown that, with regard to the gravitational force on another body, a uniform spherical body behaves as if its mass were concentrated at its centre, *i.e.* as if it were a point mass. Thus, Newton's law may be applied to planets and stars, since they are approximately spherical. When applying Newton's law to such bodies the effective distance between them is the distance between their centres.

Example
Calculate the gravitational force between the earth and a satellite of mass 400 kg at a height of 2.0 x 10^6 m above the surface of the earth, given that the radius of the earth is 6.4 x 10^6 m and its mass is 6.0 x 10^{24} kg.

Since the satellite is so small compared with the earth we may consider it as a point mass. Then

$$F = G \frac{m_1 m_2}{d^2}$$

$$= \frac{6.7 \times 10^{-11} \times 400 \times 6.0 \times 10^{24}}{(2.0 \times 10^6 + 6.4 \times 10^6)^2}$$

$$= 2.3 \times 10^3 \text{ N}$$

$$= 2.3 \text{ kN}$$

Ans.   2.3 kN

Weight

For a body on, or near, the earth, the gravitational force between it and the earth is known as the **weight** of the body.

> **The weight of a body is the gravitational force exerted on it by the earth.**

Thus, the weight of a body of mass m at a point on the earth's surface where the radius of the earth is r is, *Fig. 6.5*,

$$\cdot W = \frac{GMm}{r^2}$$

where M is the mass of the earth.

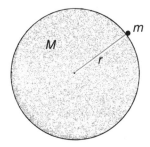

Fig. 6.5

From this equation we see that the weight of a body of given mass depends on its distance from the centre of the earth. The further it is from the centre of the earth the smaller is its weight. (Note: This relationship applies only to bodies on, or above, the surface of the earth.) Since the earth is not quite spherical but is "flatter" at the Poles this means that a body of a given mass would have a greater weight at one of the Poles than at the Equator.

Note that weight is a force. It is therefore a **vector** quantity and its unit is the **newton**, N.

Example

Given that the mass of the earth is 6.0 x 10²⁴ kg and its radius is 6.4 x 10⁶ m, find the weight of a boy whose mass is 50 kg.

$$W = \frac{GMm}{r^2}$$

$$= \frac{6.7 \times 10^{-11} \times 6.0 \times 10^{24} \times 50}{(6.4 \times 10^6)^2}$$

$$= 491 \text{ N} \qquad\qquad \text{Ans. 491 N}$$

Acceleration Due to Gravity, g

We have just seen that the weight of a body of mass m is given by

$$W = \frac{GMm}{r^2} \qquad (i)$$

where M is the mass of the earth and r is the radius of the earth. At a particular point on the earth's surface G, M and r are constants. Thus, the weight of a body is equal to its mass multiplied by a constant, *i.e.*

$$W = m \times \text{(a constant)}$$

or,

$$\boxed{W = m g} \qquad (ii)$$

where g is a constant.

The question now arises, "What is the value of g?". Comparing *Eqns. (i)* and *(ii)* we see that

$$\boxed{g = \frac{GM}{r^2}} \qquad (iii)$$

Using the values given in the previous example:

$$g = \frac{6.7 \times 10^{-11} \times 6.0 \times 10^{24}}{(6.4 \times 10^6)^2}$$

$$= 9.81 \text{ (unit)}$$

The weight of a body is the force of gravity on it and we know, from Newton's second law of motion, that

$$F = ma$$

Therefore,

$$W = m \times \text{(acceleration due to gravity)}$$

Comparing this with *Eqn. (ii)* shows us that the constant g is in fact the acceleration due to gravity, *i.e.* the acceleration of a body falling freely under gravity. The unit of g is therefore m s⁻². You should check that this is so by substituting the appropriate units into *Eqn. (iii)*. In the previous example, knowing the value of g, the weight of the boy could more easily have been calculated from *Eqn. (ii)*.

$$W = mg$$

$$= 50 \times 9.81$$

$$= 491 \text{ N}$$

It should be clear from *Eqn. (iii)* that the value of g at a particular point on the earth's surface depends on the radius of the earth at that point. It follows that g is greater at the Poles than at the Equator and also that g is greater at sea-level than at the top of a mountain. *Eqn. (iii)* may also be used to find the acceleration due to gravity at points above the earth's surface and at points on, or near, other bodies, *e.g.* the moon.

Example
What is the acceleration due to gravity at a point 30 000 km above the surface of the earth?

$$g = \frac{GM}{r^2}$$

In this case,

$$r = \text{(radius of earth + 30 000 km)}$$

$$= 6.4 \times 10^6 + 30 \times 10^6$$

$$= 36.4 \times 10^6 \text{ m}$$

Thus,

$$g = \frac{6.7 \times 10^{-11} \times 6.0 \times 10^{24}}{(36.4 \times 10^6)^2}$$

$$= 0.30 \text{ m s}^{-2}$$

Ans. 0.30 m s^{-2}

EXPERIMENT 6.2

To Measure the Acceleration due to Gravity, *g* (Free fall method).

Apparatus:
Millisecond (or centisecond) timer, metal ball, trapdoor, magnet, "gate", retort stand, thread, leads, metre stick.

Procedure:
1. Using the leads attach the terminals on the "gate" to the start terminals on the timer and attach the stop terminals on the timer to the trapdoor and the magnet, respectively.
2. Arrange the metal ball above the trapdoor as shown in *Fig. 6.6*. Measure the distance, s, from the bottom of the ball to the trapdoor.

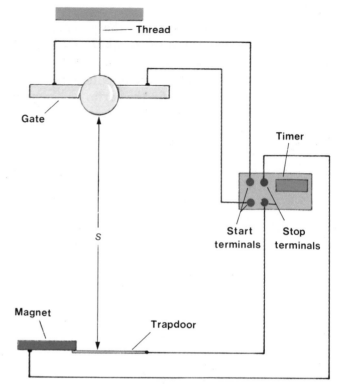

Fig. 6.6

3. Burn the thread holding the ball and note the time recorded on the timer. Repeat a number of times and note the shortest time recorded. Repeat the procedure for different values of s.
4. The value of g is calculated from the equation s = ½gt². Plot a graph of s against t² and calculate the value of g from the graph.

RESULTS

s/m	Values of t/s	Smallest value of t/s	t²/s²

Slope of graph =   m s^{-2}
Value of g . . . =   m s^{-2}

Note: *See also Experiment 11.1.*

Questions

1. Why is the smallest value of t, rather than the average value, used?
2. Why should the thread holding the ball be burned rather than cut or broken?
3. Rather than the ball being suspended from a thread it may be held by an electromagnet. How might this be arranged?

SUMMARY

Force is that which causes acceleration. It is a vector quantity and its unit is the newton, N.

Newton's laws of motion are:

1. The velocity of a body does not change unless a resultant external force acts on it.
2. When a resultant external force acts on a body the rate of change of the body's momentum is proportional to the force and takes place in the direction of the force.
3. In any interaction between two bodies A and B, the force exerted by A on B is equal in magnitude but opposite in direction to the force exerted by B on A.

Newton's Universal Law of Gravitation states that the gravitational force between two point masses is proportional to the product of the masses and inversely proportional to the square of the distance between them, i.e.

$$F = G \frac{m_1 m_2}{d^2}$$

where G is the universal constant of gravitation.

The weight, W, of a body of mass m is the gravitational force exerted on it by the earth.

$$W = mg$$

g is called the acceleration due to gravity and on the surface of the earth it is given by

$$g = \frac{GM}{r^2}$$

where r is the radius of the earth. Since the radius of the earth varies from one place to another the value of g also varies but is always approximately equal to 9.8 m s^{-2}. The value of g may be determined by measuring the time taken for a metal ball to fall a measured distance (*Experiment 6.2*).

Questions 6

DATA: $g = 9.8$ m s^{-2}
$G = 6.7 \times 10^{-11}$ N m^2 kg^{-2}.

Section I

1. Define force and give its unit...................................
....
....

2. What it the resultant of an upward force of 80 N and a downward force of 55 N?...............................
....

3. The velocity of a body does not change unless...........
....
....

4. Give two cases in which a force acting on a body would *not* produce a change in its momentum....................
....
....

5. State Newton's second law of motion.......................
....
....

6. If body A exerts a force on body B then B exerts.......
....
....

7. What acceleration is produced by a force of 40 N acting on a body of mass 25 kg?...
....

8. Define the unit of force..
....
....

9. What two conditions must be satisfied before the equation $F = ma$ can be applied to the motion of a body?......
....
....

10. How may Newton's third law be used to explain the acceleration of a spacecraft?.................................
....
....

11. Give three achievements for which Isaac Newton is remembered..
....
....

12. State Newton's Universal Law of Gravitation............
....
....

13. The gravitational force between a body and the earth is known as...
....

14. What is the value of g at a height above the earth equal to the radius of the earth?...
....

15. A girl has a weight of 400 N on the surface of the earth. At a height above the earth equal to three times the radius of the earth her weight would be...........................
and her mass would be...

Section II

1. A car experiences a forward horizontal force of 3400 N due to the engine and a backward horizontal force of 1200 N due to friction. Calculate the resultant force on the car. If the mass of the car is 800 kg what is its acceleration?

2. A car of mass 500 kg starts from rest and reaches a speed of 15 m s^{-1} after 10 s. Calculate the magnitude of the acceleration (assumed uniform) and the magnitude of the resultant force on the car.

3. A car of mass 800 kg, travelling at 25 m s^{-1}, is braked to rest over a distance of 100 m. Find the average magnitude of the resultant force on the car.

4. A body of mass 2.8 kg starts from rest and travels with a constant acceleration of magnitude 4.2 m s^{-2}. What is the magnitude of the force acting on the body and what is its speed after it has travelled 1.6 m? An additional force is now applied to the body and as a result it is brought to rest after 2.0 s. What is the magnitude of the second force?

5. A car starts from rest and travels 80 m northwards in 20 s with uniform acceleration. Find its velocity at the end of the 20 s. If the mass of the car is 1400 kg what constant force would be required to bring it to rest in 4.0 s and how far would it travel in that time?

6. What is the minimum distance in which a car of mass 700 kg, travelling at 25 m s^{-1}, can be brought to rest if the maximum force exerted on the car due to the brakes is 5.5 kN?

7. A body passes a point A with a constant velocity of 24 m s^{-1} in a certain direction. 15 s later a second body of mass 2.2 kg at rest at A is acted on by a constant force of 5.5 N in the same direction as the velocity of the first body. How long will it take the second body to overtake the first? At what distance from A will this event occur?

8. A body of mass 12.6 kg, initially at rest, is acted on by a constant force of magnitude 8.2 N for a period of 22 s. What is the momentum at the end of the period? Is it necessary in this case to know the mass of the body in order to calculate its momentum?

9. A woman of mass 56 kg stands on the floor of a lift. Calculate the force exerted on the woman by the floor of the lift when the lift is (a) stationary, (b) moving upwards with an acceleration of 1.4 m s^{-2}, (c) moving upwards with a constant speed of 4.0 m s^{-1}, (d) moving downwards with an acceleration of 1.4 m s^{-2}.

10. Calculate the gravitational force between (i) the earth and the moon, (ii) the sun and the earth, given that the masses of the sun, earth and moon are, respectively, 2.0×10^{30} kg, 6.0×10^{24} kg and 7.3×10^{22} kg and that the distances between the earth and the sun and the earth and the moon are 1.5×10^{11} m and 3.8×10^{8} m respectively.

11. Calculate the value of the acceleration due to gravity at a height of 2.4×10^{4} m above the surface of the earth.

At what height would the acceleration due to gravity be half its value on the surface? (Radius of earth = 6.4×10^{6} m.)

12. Given that the mass of the moon is 7.3×10^{22} kg and the radius of the moon is 1.7×10^{6} m, calculate the magnitude of the acceleration due to gravity on the surface of the moon.

13. A man has a weight of 650 N on the surface of the earth. What would be his weight on the surface of a planet whose mass is 318 times that of the earth and whose radius is 11 times that of the earth?

14. A spacecraft is travelling from the earth to the moon. Assuming that it is travelling on a straight line between the earth and the moon and that its engines are switched off determine its distance from the earth when its acceleration is zero. (The distance from the earth to the moon is 3.8×10^{8} m and the mass of the earth is 81 times the mass of the moon.)

7 Force II

7.1 Friction

When a body, *e.g.* a desk, is at rest, it is possible to apply a force to it without moving it. Since the table is not accelerating under the action of the force, there must be another force, of equal magnitude but opposite direction, acting on it. This force is called a **frictional** force and it comes into effect when two bodies which are in contact try to move relative to each other. As we have seen, its direction is always such as to oppose the relative motion of the bodies.

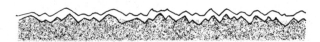

Fig. 7.1 Two surfaces in contact; no surface is perfectly smooth

The frictional force is due to the fact that no surface is perfectly smooth. On a microscopic scale even a highly polished surface is a series of peaks and valleys. When two surfaces attempt to move relative to each other the peaks come into contact, exerting forces on each other, *Fig. 7.1*.

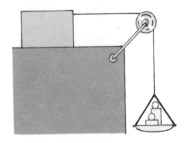

Fig. 7.2

The magnitude of the frictional force may be investigated using an arrangement like that shown in *Fig. 7.2*. It is found that the block remains at rest as weights are added to the pan, until a certain critical weight is reached, *i.e.* the frictional force increases with the applied force up to a certain maximum value. This maximum value of the frictional force is known as **limiting friction**. If further weights are added to the pan the block will begin to move. *Fig. 7.3* shows the forces acting on the block when it is on the point of moving. R is the normal force exerted

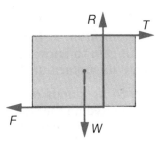

Fig. 7.3

on the block by the table and is equal to W, the weight of the block (assuming that the table is horizontal). While the block is at rest $F = T$, where T is the force applied to the block due to the weights in the pan. By placing additional weights on top of the block, it can be shown that the maximum frictional force, F, is proportional to the normal force, R, *i.e.*

$$F \propto R$$

or

$$\boxed{F = \mu R}$$

where μ is a constant called the **coefficient of (static) friction**, *i.e.*

> **The coefficient of static friction is the ratio of the maximum frictional force to the normal force.**

The value of μ depends on the nature of the surfaces in contact, being greater the rougher the surfaces. It is independent of the area and shape of the surfaces. Note that μ, being a ratio between two forces, has no unit. It is simply the number of times that one force is as big as the other.

If the value of the applied force is increased so that the block starts to move the frictional force usually decreases slightly. The **coefficient of dynamic friction** is defined as the ratio of the new (lower) frictional force to the normal force.

Example

A wooden block, of mass 3.0 kg, rests on a rough horizontal surface. If the coefficient of static friction between the block and the surface is 0.4 what is the minimum horizontal force required to move the block?

The minimum force required is equal to (in fact, negligibly greater than) the maximum frictional force. Then,

$$F = \mu R$$

Since the block is not accelerating vertically,

$$R = W$$
$$= mg$$
$$=> \quad F = \mu mg$$
$$= 0.4 \times 3.0 \times 9.8$$
$$= 11.8 \text{ N}$$

Ans. 11.8 N

EXPERIMENT 7.1
To Measure the Coefficient of Dynamic Friction.

Apparatus: Block of wood, pulley, weights (or sand), string, balance.

Procedure:
1. Arrange the block as shown in *Fig. 7.2.* Add weights (or sand) to the pan until the block will move at a constant speed when given a small push. Record the weight of the block and the weight in the pan, including the weight of the pan itself.
2. Place weights on the block and repeat the procedure for each increase in the weight of the block.
3. The coefficient of dynamic friction is calculated from $F = \mu R$, where F is equal to the weight in the pan and R is equal to the weight of the block. Plot a graph of F against R and calculate the value of μ from the graph.

RESULTS

F/N	R/N

Coefficient of dynamic friction =

Questions

1. In what circumstances would R, the normal reaction on the block, not be equal to the weight of the block?
2. Suggest ways in which the coefficient of dynamic friction between the block and the bench might be reduced. Check your conclusions experimentally.

Effects of Friction

The effects of frictional forces are vitally important in everyday life. While it is true that frictional forces oppose motion, it is also true that without them, no motion would be possible. When you walk, it is the frictional force which keeps your foot from sliding backwards when you try to move forwards (*cf. Fig. 6.4 (a)*) — try walking quickly on ice or marbles where the coefficient of friction is low! Likewise, bicycles, cars and trains could not move if there were no frictional forces between their wheels and the ground or rails.

In many cases, frictional forces are a nuisance. Even the simplest piece of machinery has parts which must slide over each other, *e.g.* the wheel of a bicycle slides on the axle. In cases like this, the frictional forces are reduced by placing bearings — small smooth spheres or cylinders, usually made of steel — between the moving parts. Friction is further reduced by coating the moving parts with a lubricant, usually oil or grease.

Fig. 7.4 Hovercraft

When a ship is moving through water its speed is relatively low because of the large frictional force between the hull and the water. A hovercraft, *Fig. 7.4*, can travel at much higher speeds because it travels on a cushion of air produced by a large fan and kept in position by a rubber skirt. (This is yet another example of Newton's third law. The fan exerts a downward force on the air, the air exerts an equal upward force on the fan and hence on the hovercraft.)

7.2 Moments

In *Chapter 6* we noted that force is a vector quantity, *i.e.* that it has direction as well as magnitude. However, as well as magnitude and direction a force has a third characteristic. This is the position of its line of action, *i.e.* the line along which it acts. *Fig. 7.5(a)* and *(b)* shows a force of 5 N being applied

Fig. 7.6 Turning effect of a force

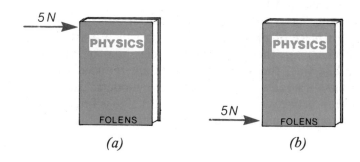

(a)   *(b)*

Fig. 7.5

to a book. In both cases the direction of the force is the same, so in each case the book will experience an acceleration from left to right. However, in the first case the book will also start to rotate in a clockwise direction while in the second case it will rotate in an anticlockwise direction. Thus, not only does the force produce an acceleration from left to right (a linear acceleration), it also produces an acceleration in a circular sense (an angular acceleration).

The turning effect of a force is called the **moment** of the force. The magnitude of the moment depends on the distance between the line of action of the force and the axis around which the book is turning. (Try opening a door by pushing first at the edge and then at a point a few centimetres from the hinges!) Thus we define the moment of a force as follows (*cf. Fig. 7.6*).

> The moment, *T*, of a force about any axis is the product of the force and the perpendicular distance between the axis and the line of action of the force.

i.e. $$T = Fr$$

The moment of a force is also called the **torque** and its unit is the **newton metre**, N m.

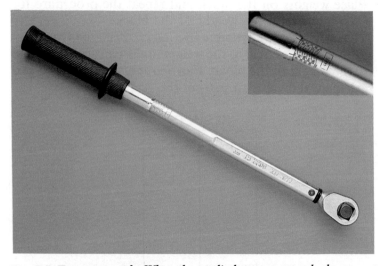

Fig. 7.7 Torque wrench. When the applied torque exceeds the pre-set value the handle slips, preventing over-tightening of the nut

Example
Fig. 7.8 shows a body acted upon by a number of forces. Calculate the resultant torque on the body about an axis through O.

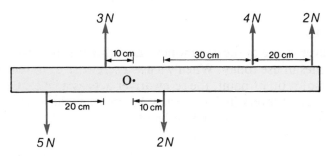

Fig. 7.8

Some of the forces tend to turn the body clockwise while the others tend to turn it anticlockwise. In finding the resultant torque anticlockwise moments are given a negative sign while clockwise moments are given a positive sign. The resultant torque is then

$$T = -(5 \times 0.3) + (3 \times 0.1) + (2 \times 0.1) - (4 \times 0.4) - (2 \times 0.6)$$

$$= -1.5 + 0.3 + 0.2 - 1.6 - 1.2$$

$$= -3.8 \text{ N m}$$

Ans. 3.8 N m anticlockwise

Principle of Moments

Before considering this principle we must first examine what is meant by the term **equilibrium.**

A body is in equilibrium when its acceleration is zero.

From this definition of equilibrium it follows immediately that, when a body is in equilibrium the resultant force on it must be zero.

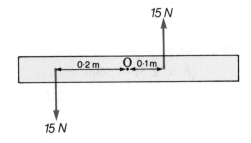

Fig. 7.9

Now consider the situation shown in *Fig. 7.9*. The sum of the forces acting on the body is zero, yet there is a resultant torque of 4.5 N m in an anticlockwise direction about an axis through O. The body therefore has an angular acceleration. Thus, we see that, for a body to be in equilibrium, not only must the sum of the forces be zero but the sum of the moments must be zero also. This fact is known as the **principle of moments.**

When a body is in equilibrium the sum of the moments, about any axis, of the external forces acting on the body is zero.

Example
A uniform metre stick is suspended at its mid-point, O, and is in equilibrium with forces of 5 N, 2 N, 4 N and X applied vertically downwards at the 10 cm, 40 cm, 60 cm and 80 cm marks, respectively, Fig. 7.10. By taking moments about O, assumed to be at the 50 cm mark, calculate the value of X.

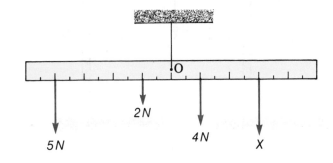

Fig. 7.10

By the principle of moments

$$-(5 \times 0.4) - (2 \times 0.1) + (4 \times 0.1) + (X \times 0.3) = 0$$

$$=> \qquad -2 - 0.2 + 0.4 + 0.3X = 0$$

$$=> \qquad -1.8 + 0.3X = 0$$

$$0.3X = 1.8$$

$$X = 6 \text{ N}$$

Ans.   6 N

EXPERIMENT 7.2
To Verify the Principle of Moments

Apparatus: Metre stick, two spring balances, two retort stands, weights, string.

Procedure:
1. Find the weight of the metre stick and its centre of gravity. Fix the spring balances in the stands and arrange the metre stick and weights as shown in *Fig. 7.11*.

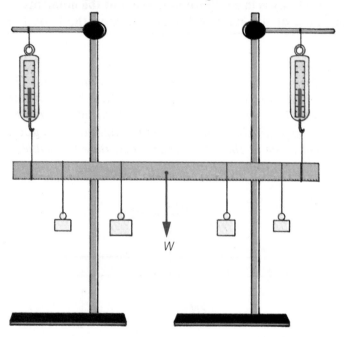

Fig. 7.11

2. Adjust the positions of the weights and/or balances until the metre stick is horizontal and in equilibrium.
3. Calculate the moment, *T*, of each of the forces acting on the metre stick, including its own weight, about any point on the metre stick.

RESULTS

F/N	d/m	T/N m

Sum of moments =   N m.

Questions

1. Why should the metre stick be horizontal when measuring the moments? Explain how you would measure the moments if the metre stick were not horizontal.

Levers

The lever is one of the commonest, and also one of the oldest, machines in use today. When we use a screwdriver to remove the lid of a tin of paint, the screwdriver acts as a lever. Other examples of levers include a claw hammer, a pliers, an oar, the brake pedal of a car, a door handle, and so on. Considering what these have in common leads us to a definition of a lever.

> **A lever is any rigid body which is free to rotate about a fixed axis.**

The axis about which a lever rotates is called a **fulcrum**.

Example
A crowbar is to be used to lift a boulder, Fig. 7.12. The distance from the boulder to the fulcrum is 10 cm and from the person's hands to the fulcrum is 2.5 m. What force must the person exert if the weight of the boulder is 600 N?

Fig. 7.12

From the principle of moments

$$(F \times 2.5) - (600 \times 0.1) = 0$$
$$2.5F = 60$$
$$F = 24 \text{ N}$$

Ans. 24 N

The weight of the boulder is called the **load** and the force exerted by the person is called the **effort**. From this example we see that a lever can magnify the force that we can apply to something. It does not, however, magnify the amount of work which we can do. It allows us to apply a *greater* force but over a *shorter* distance *(see Chapter 8)*.

Couples

The system of forces shown in *Fig. 7.13* is known as a **couple.**

> A couple is a system of forces which has a turning effect only, *i.e.* the resultant of the forces is zero.

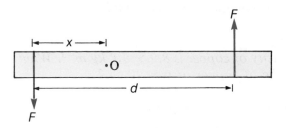

Fig. 7.13 Couple

Very often a couple consists of a pair of equal but opposite forces like those shown in *Fig. 7.13* but this is not always the case. A couple may consist of any number of forces as long as the resultant of the forces is zero.

To establish a general expression for the moment, or torque, of a couple, consider the forces in *Fig. 7.13*. The total torque, in the anticlockwise direction, is

$$T = Fx + F(d - x)$$
$$= Fx + Fd - Fx$$
$$= Fd$$

That is, the moment of a couple consisting of two forces is given by

$$T = Fd$$

where F is the magnitude of each force and d is the perpendicular distance between them. Since x does not appear in the final equation its value does not matter, *i.e* the torque is the same for all axes perpendicular to the plane containing the forces.

Example

In turning the steering wheel of a car, a driver exerts a force of 5 N at opposite ends of a diameter of the wheel. If the diameter is 50 cm what is the torque on the wheel?

$$T = Fd$$
$$= 5 \times 0.5$$
$$= 2.5 \text{ N m}$$

Ans. 2.5 N m

7.3 Pressure

When the surfaces of two bodies come into contact they exert a force on each other. The effect which these forces will have on the surfaces in contact depends, not only on the nature of the surfaces, but also on the area of the surfaces in contact. Thus, it is easy to break an egg by squeezing it between the point of your finger and thumb but almost impossible to break it in the palm of your hand using all of your fingers (not just the points). Similarly, it is easy to push the point of a pencil into a cork but very difficult to push the unsharpened end of the pencil into the cork. The physical quantity which depends on force and the area over which the force is applied is called **pressure.**

> The pressure, p, at a point is the force per unit area at that point.

i.e.
$$p = \frac{F}{A}$$

Pressure is a **scalar** quantity and its unit is the **pascal**, Pa. $1 \text{ Pa} = 1 \text{ N m}^{-2}$.

Example

A cube of wood of side 40 cm and weight 32 N rests on a horizontal bench. What is the pressure under the block?

$$p = \frac{F}{A}$$
$$= \frac{32}{0.4 \times 0.4}$$
$$= 200 \text{ Pa}$$

Ans. 200 Pa

Measuring Pressure

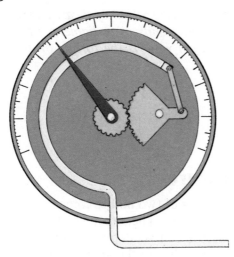

Fig. 7.14 Bourdon Gauge

The most common device for measuring pressure is the Bourdon gauge, *Fig. 7.14*. When the pressure in the tube increases the tube tends to straighten out. This movement causes a needle to move around a scale. When the pressure is released the tube returns to its original shape and the needle goes back to zero.

Fig. 7.15 Mechanism of Bourdon gauge

Fluid Pressure

Before proceeding with this section it is convenient to introduce the concept of **density**. It is defined as follows.

> **The density of a substance is the mass of unit volume of it.**

Density, ρ, is a **scalar** quantity and its unit is the **kilogram per metre cubed**, kg m^{-3}.
In symbols, the definition may be written as

$$\rho = \frac{m}{V}$$

Knowing the density of a substance allows us to calculate the mass, and hence the weight, of any given volume of it. The higher the density of a substance the greater the mass of any given volume, and *vice versa*.

Example
The density of copper is 8.9×10^3 kg m^{-3}. What is the mass of a cube of copper of side 10 cm?

The volume of the copper is

$$\begin{aligned}
V &= 10^3 \\
&= 1000 \text{ cm}^3 \\
&= 1 \times 10^{-3} \text{ m}^3 \\
\rho &= \frac{m}{V} \\
\Rightarrow \quad m &= \rho V \\
&= 8.9 \times 10^3 \times 1 \times 10^{-3} \\
&= 8.9 \text{ kg}
\end{aligned}$$

Ans. 8.9 kg

Now, consider a rectangular tank full of a fluid, *e.g.* water. There is a force on the bottom of the tank due to the weight of the water. (There are also forces on the sides of the tank and, indeed, between adjacent layers of water in the tank.) What then is the pressure of the water at the bottom of the tank?

$p = \dfrac{F}{A}$ where A is the area of the bottom of the tank

$ = \dfrac{W}{A}$ where W is the weight of the water

$ = \dfrac{mg}{A}$ where m is the mass of the water

But, density, $\rho = \dfrac{\text{mass}}{\text{volume}} = \dfrac{m}{V}$

Therefore, $m =$ Therefore, $m = \rho V$

Therefore,

$p = \dfrac{\rho Vg}{A}$ where V is the volume of the water and ϱ is its density

$ = \dfrac{\rho hAg}{A}$ where h is depth of the water

$ = \rho gh$

$$\boxed{p = \rho gh} \quad \text{(i)}$$

From *Eqn. (i)* we see that the pressure in a fluid depends on the **depth** and the **density**. This can be verified using the apparatus shown in *Fig. 7.17*. A thin piece of rubber is stretched over the mouth of a thistle funnel and the funnel is connected to a sensitive Bourdon gauge. By placing the funnel at different depths in a vessel of water it can be shown that the pressure increases with the depth. Then, by placing the funnel at the same depth in liquids of different density, it can be shown that the pressure is greater the greater the density.

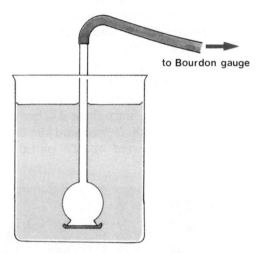

to Bourdon gauge

Fig. 7.17 Measuring pressure at different depths

It follows from *Eqn.(i)* that the pressure at all points at a given depth in a given fluid is the same. This means that the pressure at, say, the bottom of each of the vessels shown in *Fig. 7.18* is the same, since the depth of the liquid is the same in all of them. This may also be verified experimentally as outlined above.

Fig. 7.16   Divers require protective suits since pressure increases with depth

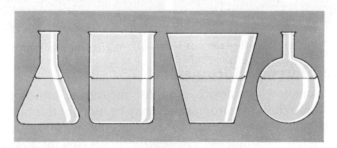

Fig. 7.18 Pressure in a given liquid depends only on the depth

Atmospheric Pressure

The earth is surrounded by a layer of gas several kilometres deep. The pressure resulting from the force of gravity on this gas is known as atmospheric pressure. The value of this pressure varies with altitude and weather conditions but its normal value is taken as 1.0×10^5 Pa.

We can demonstrate the effect of atmospheric pressure in several ways. One way is to fill a beaker or gas jar with water, place a light piece of cardboard over the top of it and turn it upside down. The water does not come out because the force exerted on the outside of the cardboard by the atmosphere is greater than the weight of the water. Another method involves removing the air from a large (*e.g.* 5 litre) can which can be sealed. One way of doing this is to boil a little water in the can until it is full of steam. Then stop heating and screw on the cap tightly. When the can cools the steam condenses inside, leaving a partial vacuum. Since there is (almost) no air inside the can the pressure inside is zero and the can is unable to withstand the forces exerted on the outside of it by the atmosphere, *Fig. 7.19*.

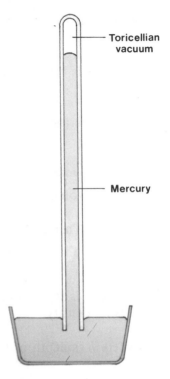

Fig. 7.20 Simple barometer

Fig. 7.19 Can (a) before air was removed, (b) after air was removed

The Mercury Barometer

A barometer is a device used for measuring atmospheric pressure. While a Bourdon gauge may be used for this purpose the mercury barometer is more commonly used.

A mercury barometer, *Fig. 7.20*, consists of a thick-walled glass tube, about 1 m long, inverted in a basin of mercury. The tube is first filled with mercury before being inverted in the basin. When it is inverted the mercury in the tube falls until the pressure due to the column of mercury is equal to atmospheric pressure. (When the mercury falls it leaves a

Fig. 7.21 Fortin barometer

vacuum, known as the Torricellian vacuum, at the top of the tube. The pressure at the top of the mercury in the tube is therefore zero.) Normal atmospheric pressure is taken to be equivalent to a column of mercury 76 cm high *(see Q.11 on p.86)*. *Fig. 7.21* shows the type of mercury barometer usually found in school laboratories. It is known as a Fortin barometer.

The Aneroid Barometer

While the mercury barometer is very accurate it is also very expensive and fragile. The aneroid barometer is much cheaper and more robust, although it is less accurate. It consists essentially of a flat metal box with corrugated sides, *Fig. 7.22*. Most of the air is removed from the box before it is sealed. An increase in the atmospheric pressure causes the sides of the box to move inwards, a drop in the pressure allows them to spring out again. This movement is very small and it has to be magnified by a system of levers attached to one side of the box. The levers are attached to a pointer and arranged so that movements by the sides of the box cause the pointer to move around a scale.

As mentioned above, atmospheric pressure decreases with height above the earth. In fact, the atmospheric pressure decreases by about 1% for every 100 m increase in height. A barometer can thus be used to measure height above the earth. The aneroid barometer is particularly suited to this purpose since it is much more portable than the mercury barometer. An instrument used for measuring height is called an altimeter. It is simply an aneroid barometer with the scale converted to read height rather than pressure.

Archimedes' Principle

Fig. 7.23 shows a block of metal being weighed in air *(a)*, in methylated spirits *(b)*, and in water *(c)*. It seems to weigh less in the methylated spirits than in the air and less again in the water. Since the weight of a body at a particular point on the earth's surface depends only on its mass *(see p.67)* and since the mass of the block does not change, the liquid must exert an upward force on the block in each case. More than 2000 years ago, Archimedes (c. 287 BC — 212 BC), a Greek scientist, discovered that this upward force, called the **upthrust**, was equal in magnitude to the weight of liquid displaced. This fact is now known as **Archimedes' Principle.**

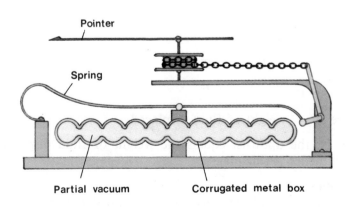

Fig. 7.22 Aneroid barometer

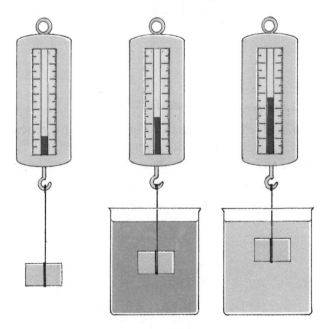

Fig. 7.23 Effect of upthrust in different fluids

When a body is partly or wholly immersed in a fluid, the upthrust is equal in magnitude to the weight of the fluid displaced.

Archimedes' Principle applies equally to gases and liquids (*i.e.* fluids). It can be verified experimentally by first finding the upthrust (the difference between the actual weight and the apparent weight) as illustrated in *Fig. 7.23* and then finding the weight of liquid displaced by placing the block in an overflow can, *Fig. 7.24*. In our example above, the upthrust in the methylated spirits is less than the upthrust in the water since the density of methylated spirits is less than the density of water and the volume displaced is the same in each case.

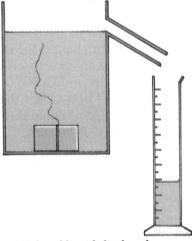

Fig. 7.24 Finding weight of liquid displaced

Archimedes' Principle can also be derived theoretically as follows. Consider a block of height l immersed in a fluid of density ρ, *Fig. 7.25*. The pressure at the top face of the block is

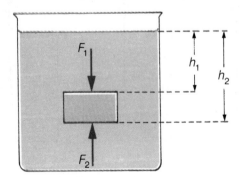

Fig. 7.25

$$p_1 = \rho g h_1$$

and at the bottom face is

$$p_2 = \rho g h_2$$

Since $F = pA$, the corresponding forces are

$$F_1 = \rho g h_1 A$$

and

$$F_2 = \rho g h_2 A$$

where A is the area of each face of the block.

The resultant upward force, the upthrust, is

$$
\begin{aligned}
F &= F_2 - F_1 \\
&= \rho g A (h_2 - h_1) \\
&= \rho g A l \\
&= \rho g V
\end{aligned}
$$

where V = volume of the block
= volume of fluid displaced

$$\therefore \qquad F = \rho g \frac{m}{\rho}$$

where m = mass of fluid displaced

i.e.

$$
\begin{aligned}
F &= mg \\
&= W
\end{aligned}
$$

where W = weight of fluid displaced.

That is, the upthrust equals the weight of fluid displaced (Archimedes' Principle).

Example
A cube of metal of side 20 cm has a weight of 500 N. What is its apparent weight in a liquid whose density is 800 kg m^{-3}?

$$
\begin{aligned}
\text{Volume of cube} &= (0.2)^3 \\
&= 8 \times 10^{-3} \text{ m}^3 \\
\text{Volume of liquid displaced} &= 8 \times 10^{-3} \text{ m}^3 \\
\text{Mass of liquid displaced} &= 8 \times 10^{-3} \times 800
\end{aligned}
$$

$$= 6.4 \text{ kg}$$

$$\text{Weight of liquid displaced} = 6.4 \times 9.8$$

$$= 62.7 \text{ N}$$

$$\text{Upthrust} = 62.7 \text{ N}$$

$$\text{Apparent weight} = (500 - 62.7)$$

$$= 437.3 \text{ N}$$

Ans. 437 N

Law of Flotation

When a body floats in a fluid the resultant force on it is zero (its acceleration is zero). Since the only forces acting on it are its weight and the upthrust of the fluid, these two forces must be equal in magnitude and opposite in direction. From Archimedes' Principle we know that the upthrust is equal to the weight of fluid displaced. It follows, therefore, that the weight of a floating body is equal to the weight of the fluid displaced by it. This is known as the **Law of Flotation** and is a special case of Archimedes' Principle.

> **The weight of a floating body is equal to the weight of fluid displaced.**

Example

A piece of wood has a weight of 2.0 N. It floats in a liquid (i) of density 800 kg m⁻³, (ii) of density 600 kg m⁻³. Calculate the volume of liquid displaced in each case.

(i)

$$\text{Weight of liquid displaced} = 2.0 \text{ N}$$

$$\text{Mass of liquid displaced} = \frac{2.0}{9.8}$$

$$= 0.204 \text{ kg}$$

$$\text{Volume of liquid displaced} = \frac{0.204}{800}$$

$$= 2.55 \times 10^{-4} \text{ m}^3$$

$$= 255 \text{ cm}^3$$

(ii)

$$\text{Mass of liquid displaced} = 0.204 \text{ kg}$$

$$\text{Volume of liquid displaced} = \frac{0.204}{600}$$

$$= 3.40 \times 10^{-4} \text{ m}^3$$

$$= 340 \text{ cm}^3$$

Ans. (i) 255 cm³; (ii) 340 cm³

The Hydrometer

From the previous example it is clear that the volume of liquid displaced by a particular floating object depends on the density of the liquid in which it is floating — the more dense the liquid the smaller the volume displaced. In other words, the more dense the liquid, the more of a floating body will be above the surface.

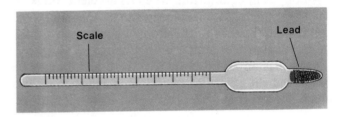

Fig. 7.26 Hydrometer

A hydrometer, *Fig. 7.26*, is a device, based on the Law of Flotation, for measuring the densities of liquids. It consists of a glass bulb with a long glass stem carrying a scale which reads from top to bottom. A smaller bulb at the bottom end, containing lead, helps to make the hydrometer float upright.

Hydrometers are used to measure the density of the acid in a car battery as this gives an indication of the state of charge of the battery. The acid in a fully charged battery has a density of 1280 kg m⁻³, and in a completely "flat" battery the density is just over 1000 kg m⁻³. (Allowing the density to fall below about 1150 kg m⁻³ may permanently damage the battery.) Hydrometers are also used in dairies to measure the density of milk and in breweries to find the density of beer and hence the percentage of alcohol in it.

SUMMARY

Frictional forces occur when two bodies in contact try to move relative to each other. The direction of these forces always opposes the relative motion. They have a maximum value (limiting friction, F) which depends on the nature of the surfaces in contact and the normal force, R, between them. The coefficient of friction, μ, is defined by

$$\mu = \frac{F}{R}$$

The turning effect of a force is called its moment or torque, T. The moment of a force about any axis is the product of the force and the perpendicular distance from the axis to the line of action of the force.

A body is in equilibrium when its acceleration is zero. The Principle of Moments states that when a body is in equilibrium the sum of the moments, about any axis, of the external forces acting on it is zero.

A lever is any rigid body free to rotate about a fixed axis called the fulcrum. A couple is a system of forces which has a turning effect only, *i.e.* its resultant is zero.

The torque of a couple consisting of two forces, each of magnitude F, is given by

$$T = Fd$$

where d is the perpendicular distance between the forces. Pressure is force per unit area.

$$p = \frac{F}{A}$$

The pressure at a depth h in a fluid of density ρ is given by

$$p = g\rho h$$

Archimedes' Principle states that the upthrust on a body immersed in a fluid is equal to the weight of fluid displaced.

The Law of Flotation states that the weight of a floating body is equal to the weight of fluid displaced.

The hydrometer is a device, based on the Law of Flotation, for measuring the densities of liquids.

Questions 7

DATA: $g = 9.8$ m s^{-2}
Density of water $= 1.0 \times 10^3$ kg m^{-3}.

Section I

1. Frictional forces occur when.................................

....

2. The ratio of the maximum frictional force to the normal force is called...

....

3. What is the difference between the coefficient of static friction and the coefficient of dynamic friction? Which

is the greater?..

....

....

4. What determines the magnitude of the frictional force between two surfaces?..

....

5. What is meant by the moment of a force?...............

....

....

6. The unit of torque is..

7. State the principle of moments...............................

....

....

8. What condition must be fulfilled if a body is in

   equilibrium?..

   ....

   ....

9. Define pressure and give its unit......................

   ....

   ....

10. What is the mass of 1 m³ of water?.........................

    ....

11. The pressure in a fluid depends on the...................

    and the...

12. State Archimedes' Principle..................................

    ....

    ....

13. Why is it possible for a ship to carry a greater load in salt

    water than in fresh water?...................................

14. Why does a submarine float on the surface when its ballast
    tanks are empty and sink when they are filled with water?

    ....

    ....

15. A hydrometer is used for determining.....................

    It is based on the...

    ....

Section II

1. A block of wood of mass 20 kg rests on a horizontal surface. Calculate the minimum horizontal force required to set the block in motion if the coefficient of static friction between the block and the surface is 0.4.

2. Define coefficient of dynamic friction. What determines the value of the coefficient for a particular pair of surfaces? Why is it better not to "lock" the wheels of a car when braking in an emergency?
   Why do racing cars have wide tyres?

3. A car travelling at 20 m s^{-1} is braked to rest. If the coefficient of friction between the wheels and the road is 0.7 what is the minimum distance travelled by the car in coming to rest?

4. A car of mass 750 kg is travelling along a level straight road. The total frictional force, assumed constant, is 3.4 kN. Calculate the force on the car due to the action of the engine when (i) the car is travelling at a constant speed of 10 m s^{-1}, (ii) when the car has a constant acceleration of magnitude 1.2 m s^{-2}.

5. A block of mass 16 kg is being pulled on a horizontal surface by a rope. Given that the rope is horizontal and that the tension in it when the block starts to move is 54 N calculate the coefficient of static friction between the block and the surface. If the coefficient of dynamic friction is equal to 90% of the coefficient of static friction calculate the acceleration of the block, assuming that the tension in the rope is maintained at 54 N.

6. A see-saw consists of a plank, 3.0 m long, balanced at its centre point. If a girl of mass 30 kg sits 20 cm from one end, how far from the other end must a boy of mass 35 kg sit in order to balance the see-saw?

7. A simple balance consists of a uniform rigid beam 40 cm long and supported at its centre. A slider of mass 2.0 kg moves along one side of the beam, while the bodies to be weighed are attached to a point, P, 1.0 cm from the other end, *Fig. I*. Calculate the distances of the slider from the centre for which the beam will be balanced when bodies of mass 0.4 kg, 0.8 kg and 1.2 kg, respectively, are attached at P.

Fig. I

8. A crowbar consists of an iron bar 2.0 m long. A man wishes to use it to lift a load of mass 900 kg. If the maximum force which the man can exert is 1000 N, calculate the maximum distance between the point of support of the crowbar and the point of application of the load. (Assume both load and effort to be applied at the ends of the bar.)

9. The pressure on the surface of a table under a box of mass 2.4 kg is 2.4 kPa. What is the area of the base of the box?

10. Define density.
    Calculate the mean density of the earth given that the radius of the earth is 6.4×10^6 m and that the value of G is 6.7×10^{-11} N m^2 kg^{-2}.

11. Calculate the value of normal atmospheric pressure in kPa given that it is equivalent to the pressure due to a column of mercury 76 cm high.
    (Density of mercury $= 1.36 \times 10^4$ kg m^{-3}.)

12. Why does an object appear to weigh less when suspended in a liquid than it does in air? Is its weight actually less? A cubic block of side 15 cm and mass 22 kg is attached to a spring balance and hangs in a tank of water. What is the reading on the balance?

13. The weight of a block of wood is 4.0 N. Its apparent weight when totally immersed in water is 1.0 N. Calculate (i) the weight of water displaced, (ii) the volume of the block, (iii) the density of the wood.

14. A ship displaces 2000 m^3 of sea-water of density 1020 kg m^{-3}. What is the mass of the ship?

15. A submarine has a mass of 1000 t and floats on the surface of the sea with one sixth of its volume above the surface. What mass of water must the submarine pump into its ballast tanks in order to sink beneath the surface? Why will the submarine not continue sinking until it strikes the sea-bed?

16. A cube of wood, of side 20 cm, floats in a liquid of density 800 kg m^{-3} in such a way that 2 cm of the block appears above the surface. What is the density of the wood?

8 Energy

8.1 Work

Everyone has a general idea of what is meant by the term "work" in everyday life. However, as in the case of other words you have already met, physicists need a more precise definition.

> **Work is done when a force moves a body. The amount of work done is equal to the product of the force and the displacement.**

(This definition assumes that the force and the displacement are in the same direction.)
In symbols, the work, W, done by a force, F, in moving a body through a displacement, s, in the direction of the force, is given by

$$W = Fs$$

Work is a **scalar** quantity and its unit is the **joule**, J.

> **1 J is the work done when a force of 1 N causes a displacement of 1 m in the same direction as the force. 1 J = 1 N m.**

Example
A girl pulls a sledge a distance of 100 m. If the force exerted by the girl is 70 N in the direction in which the sledge is moving calculate the work done.

$$
\begin{aligned}
W &= Fs \\
&= 70 \times 100 \\
&= 7000 \text{ J} \\
&= 7 \text{ kJ} \qquad \text{Ans. 7 kJ}
\end{aligned}
$$

8.2 Energy

If you "wind up" a toy car, the spring in the car becomes capable of doing work, *i.e.* it can apply a force which causes a displacement. Instead of saying that the spring can do work we simply say that the spring has **energy**.

> **Energy is the ability to do work.**

Thus, if a body is capable of doing a certain amount of work we say that it possesses that amount of energy. Since energy and work are equivalent both have the same unit, *i.e.* the joule, J.

Forms of Energy

There are many different situations, in addition to the example given above, in which a body is capable of doing work, *e.g.* if it is moving, if it is stretched, if it is held above the ground, *etc.* For convenience, a different name is given to the energy in each case. Thus, the moving body has **kinetic** energy; the stretched body and the body held above the ground have **potential** energy; *etc.* While these are said to be different forms of energy, they are all essentially the same, *viz.* the ability to do work.

Fig. 8.1

Along with kinetic energy and potential energy, other forms of energy are: **internal** ("heat") energy, **sound** energy, **chemical** energy, **electrical** energy, **radiant** (light, *etc.*) energy and **nuclear** energy. As we noted in *Chapter 1*, much of Physics is devoted to the study of energy in its different forms. We shall study kinetic and potential energy later in this chapter and return to the other forms in subsequent chapters.

Conservation of Energy

Perhaps the most important principle in Physics is the **Principle of Conservation of Energy**. This principle may be stated as follows.

> **In any closed system the total amount of energy is constant.**

Compare this statement with the statement of the principle of conservation of momentum, *p.56*.

While the total amount of energy in any closed system does not change, the energy may be changed from one form to another. Not all forms of energy are equally useful to mankind. When we talk of "energy shortages" we mean shortages of energy in readily usable forms — the total amount of energy in the Universe never changes. But, while it is relatively easy to convert chemical energy (*e.g.* in coal or oil) to other forms (*e.g.* internal or electrical), it is very difficult to convert internal energy to other forms. Thus, when coal is burned much of its available chemical energy is converted to internal energy in the atmosphere and is effectively lost.

Work as Energy Conversion

Work may also be considered as the process of converting energy from one form to another. When you push your bicycle, for example, you are doing work (force × displacement); you are also converting chemical energy (of your body) to kinetic energy (of the bicycle and your body). When a stone falls from a height work is done — force (weight) × displacement (distance fallen); while the stone is falling energy is being converted from potential energy to kinetic energy.

Kinetic Energy

The energy which a moving body has is called kinetic energy. We can calculate the amount of kinetic energy a body has in a particular case by calculating the amount of work it would do in coming to rest. When we do this we find that the kinetic energy, E_k, of a body depends on its mass, m, and its speed, v, and is given by

$$E_k = \tfrac{1}{2} mv^2$$

Consider a car of mass m, travelling with an initial speed u. If the brakes are applied and it comes to rest in a distance s, the work done is

$$W = Fs$$

where F is the force stopping the car.

But,
$$F = ma$$
$$\Rightarrow W = mas \qquad (i)$$

But,
$$v^2 = u^2 + 2as$$
$$\Rightarrow 0 = u^2 - 2as$$
$$\Rightarrow 2as = u^2$$

or,
$$as = \frac{u^2}{2}$$

Substituting for as in *Eqn. (i)* gives

$$W = \frac{mu^2}{2}$$

Since this is the work which the car does in coming to rest it is also the energy which it had when travelling with a speed u. That is

$$E_k = \tfrac{1}{2} mu^2$$

This is the initial kinetic energy of the car (u = initial speed). In general, when a body is travelling with speed v its kinetic energy is given by

$$E_k = \tfrac{1}{2} mv^2$$

Example

A block of mass 1.2 kg is acted on by a constant resultant force of magnitude 6.3 N over a distance of 14 m. Calculate (i) the work done, (ii) the kinetic energy of the body at the end of the 14 m.

(i)
$$W = F s$$
$$= 6.3 \times 14$$
$$= 88.2 \text{ J}$$

(ii)
$$E_k = \tfrac{1}{2} mv^2$$

In order to find v we must first find a.

$$a = \frac{F}{m}$$

$$= \frac{6.3}{1.2}$$

$$= 5.25 \text{ m s}^{-2}$$

Now,

$$v^2 = u^2 + 2as$$

$$= 0 + 2 \times 5.25 \times 14$$

$$= 147 \text{ m}^2 \text{ s}^{-2}$$

Substituting for v^2 in

$$E_k = \tfrac{1}{2} mv^2$$

gives

$$E_k = \tfrac{1}{2} \times 1.2 \times 147$$

$$= 88.2 \text{ J}$$

Ans. (i) 88.2 J; (ii) 88.2 J

This example illustrates the fact that the work done in moving a body over a certain distance is equal to the kinetic energy gained by the body provided no other force acts on the body. This result follows immediately from the definition of kinetic energy and the derivation of the formula for kinetic energy given above.

Potential Energy

Potential energy is the energy a body has because of its **condition** or **position**, *e.g.* a compressed spring or a rock at the top of a cliff. In the latter case, the rock has energy due to the gravitational force between it and the earth. This, then, is an example of gravitational potential energy. As in the case of the car in the last section, we can calculate how much energy the rock has by calculating how much work it would do in falling from the cliff to the earth. We shall see that the potential energy, E_p, of a body of mass m, at a height h, is given by

$$\boxed{E_p = mgh}$$

where g is the acceleration due to gravity.

Consider a rock of mass m, at a height h above the ground, *Fig. 8.2*. When the rock falls to the ground the work done is

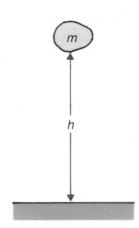

Fig. 8.2

$$W = Fh$$

The force on the rock is its weight, which is equal to mg, where m is its mass. Therefore, the work done is

$$W = mgh$$

Since this is the work done it is also the energy which the rock had before it fell, *i.e.*

$$E_p = mgh$$

Fig. 8.3 Hydroelectric power station: potential energy of the water is converted to electrical energy in the generators

Conservation of Energy and Falling Bodies

As a body falls it loses potential energy (h gets smaller) and it gains kinetic energy (v gets larger). If we ignore the frictional forces between the body and the air, the Principle of Conservation of Energy tells us that the loss of potential energy should be equal to the gain in kinetic energy.

Consider a body of mass m at rest at point A, a height h above the ground, *Fig. 8.4*. At A the sum of its potential and kinetic energies is

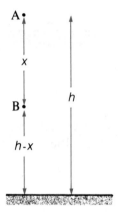

Fig. 8.4

$$E_p + E_k = mgh + 0$$

If the body now falls a distance x its speed is given by

$$v^2 = u^2 + 2as$$
$$= 0 + 2gx$$
$$= 2gx$$

Therefore, the kinetic energy of the body at B is

$$E_k = \frac{1}{2} mv^2$$
$$= \frac{1}{2} m(2gx)$$
$$= mgx$$

The potential energy of the body at B is

$$E = mg(h - x)$$

The total energy of the body at B is

$$E_p + E_k = mg(h - x) + mgx$$
$$= mgh - mgx + mgx$$
$$= mgh$$
$$= \text{total energy at A.}$$

Thus, the sum of the body's potential and kinetic energies is constant, *i.e.* the loss in potential energy is equal to the gain in kinetic energy. We can use this result to calculate the velocity, *etc.*, of a body moving under gravity. This method is often more convenient than a direct application of the equations of motion.

Example

A ball is thrown vertically upwards from ground level with an initial speed of 12 m s^{-1}. Calculate (i) the greatest height reached, (ii) the speed of the ball when it is at a height of 5.0 m above the ground.

(i)
At ground level the potential energy is zero while at the greatest height the kinetic energy is zero (the ball is stopped). Therefore, from conservation of energy we have

$$\frac{1}{2} mu^2 = mgh$$
$$=> \quad h = \frac{u^2}{2g}$$
$$= \frac{144}{2 \times 9.8}$$
$$= 7.3 \text{ m}$$

(ii)
The kinetic energy at ground level is equal to the sum of the potential and kinetic energies at a height of 5 m. Thus, we have,

$$\frac{1}{2} mu^2 = mgh + \frac{1}{2} mv^2$$

Multiplying both sides by 2 and dividing by m gives

$$u^2 = 2gh + v^2$$
$$=> \quad 144 = 2 \times 9.8 \times 5 + v^2$$
$$= 98 + v^2$$
$$=> \quad v^2 = 46$$
$$=> \quad v = \pm 6.8 \text{ m s}^{-1}$$

Ans. (i) 7.3 m; (ii) 6.8 m s^{-1}

Energy and Mass

In the early years of this century Albert Einstein put forward his revolutionary Theory of Relativity. One consequence of this theory is the proposal that energy and mass are equivalent, *i.e.* are different aspects of the same thing. So, for example, when we speak of a car gaining kinetic energy, we could equally say it is gaining mass.

We measure energy in joules and mass in kilograms, so, if we are measuring the same thing in each case, these units must be related. In other words, it must be possible to convert kilograms into joules just as it is possible to convert miles into kilometres. Einstein deduced that the conversion factor was c^2, where c is the speed of light in a vacuum (3.0×10^8 m s^{-1}). Then

$$E(\text{J}) = m(\text{kg}) \times c^2$$
$$\text{or}$$

$$\boxed{E = mc^2}$$

Since c^2 is very large (9.0×10^{16}), a small amount of mass(kg) is equivalent to a large amount of energy (J). Conversely, even quite a large amount of energy is equivalent to a very small amount of mass. This is why we do not notice the increase in the mass of a car, for example, when it gains speed. The following example illustrates just how small this increase in mass is.

Example

A car of mass 1 t accelerates from rest to a speed of 20 m s^{-1}. What is the increase in its mass?

The increase in kinetic energy of the car is

$$
\begin{aligned}
E_k &= \tfrac{1}{2} mv^2 \\
&= \tfrac{1}{2} \times 10^3 \times 20^2 \\
&= 500 \times 400 \\
&= 2 \times 10^5 \text{ J}
\end{aligned}
$$

To convert this to kilograms we use

$$
\begin{aligned}
E &= mc^2 \\
2 \times 10^5 &= m \times 9 \times 10^{16} \\
\Rightarrow \quad m &= \frac{2 \times 10^5}{9 \times 10^{16}} \\
&= 2.2 \times 10^{-12} \text{ kg}
\end{aligned}
$$

That is, the increase in the mass of the car is 2.2×10^{-12} kg. In percentage terms this amounts to $2.2 \times 10^{-13}\%$ (0.22 billionths of one percent) of the original mass of the car.

Ans. 2.2×10^{-12} kg

From this example it is clear that the change in mass in everyday situations is far too small to be measurable. It is only when we come to deal with nuclear energy or with very small particles travelling at high speeds that the change in mass is an appreciable percentage of the original mass.

Albert Einstein (1879 — 1955)

ALBERT EINSTEIN 1879-1955

ITALIA 120
I.P.Z.S.-ROMA-1979   F. TULLI

Einstein was born in Germany and, like Newton, allegedly showed little promise at school, which he left at the age of 15 without a diploma. With some difficulty he resumed his education in Switzerland and studied physics for four years in Zurich. He graduated in 1900 and accepted a job in the patent office in Bern.

In 1905 Einstein published no fewer than five papers, three of which revolutionised our understanding of the Universe. These were his explanation of the photoelectric effect (p.290), his special theory of relativity and his statement of the equivalence of mass and energy, $E = mc^2$. In 1916 he published his general theory of relativity. He was awarded the 1921 Nobel Prize in Physics for his work on the photoelectric effect.

Einstein quickly became an internationally known and respected figure, even among non-scientists. He used his standing to work for various social causes including liberalism, Zionism and, in particular, pacifism. The latter is ironic since his discoveries were to contribute to the development of nuclear weapons. Indeed, he was instrumental in the establishment of the "Manhattan Project" which produced the first atomic bomb, an involvement he bitterly regretted.

In 1933 Einstein moved to Princeton, U.S.A., where he remained for the rest of his life. He devoted himself to a fruitless search for a single theory of gravitation and electromagnetism — the unified-field theory. His failure in this regard does not take away from his standing as one of the greatest scientists, if not the greatest, of all time. He died in his sleep at Princeton Hospital on April 18, 1955.

8.3 Power

In the example on work in the first section of this chapter we calculated the work done by the girl (7000 J) without considering how long it took her to do the work. Suppose it took 100 s. Then the amount of work done in each second is $7000/100 = 70$ J. This quantity — the amount of work done in one second — is called the **power**, P.

Power is the rate at which work is done.

If the work is being done at a constant rate this definition may be expressed in symbols as

$$P = \frac{W}{t}$$

where W is the amount of work done in time t.
Power is a **scalar** quantity and its unit is the **watt**, W.

The power is 1 W if work is being done at the rate of 1 J per second. $$1\ W = 1\ J\ s^{-1}.$$

Since work may be considered as the process of converting energy from one form to another, power may also be defined as follows.

Power is the rate at which energy is converted from one form to another.

This second statement of the definition is more useful in situations where it is not immediately obvious what the force is which is causing the displacement. For example, an electric light bulb may be rated as "100 W". This means that, in such a bulb, electrical energy is being converted to radiant energy at a rate of 100 W, *i.e.* at a rate of 100 J per second.

The power of an engine (car, train, *etc.*) is often quoted as being so many horsepower. This unit was adopted by James Watt after carrying out experiments with strong dray horses around 1790. One horsepower is equivalent to 746 W. The horsepower is not, of course, part of the SI system of units.

James Watt (1736 — 1819)

Watt was born in Greenock, Scotland. He was unable to attend school due to illness and was taught to read and write by his mother. As a teenager he travelled to London where he learned the craft of instrument making.

Watt is usually remembered as the inventor of the steam engine. In fact, the first steam engine was invented by Savery in 1700 and improved by Newcomen in 1712. By the middle of the 18th century Newcomen's engine was widely used for pumping water. However, it was extremely inefficient. James Watt's first achievement in 1769 was to make the steam engine more efficient. Such

a vast improvement was Watt's engine that Newcomen's was all but forgotten in a few years. Even more importantly, in 1781, Watt invented a device for using the to-and-fro motion of the piston to turn a wheel, thus changing the steam engine from being merely a pump into a machine capable of a wide variety of tasks.

In 1800, Watt retired, a wealthy and respected man. He was awarded an honorary doctorate from Glasgow University and was elected to the Royal Society. He died near Birmingham in 1819.

SUMMARY

Work is done when a force causes a displacement. The amount of work, W, is given by

$$W = Fs$$

where the force, F, and the displacement, s, are in the same direction. The unit of work is the joule, J.

Energy is the ability to do work. There are different forms of energy — kinetic, potential, internal, sound, chemical, radiant, electrical, nuclear. The Principle of Conservation of Energy states that the total amount of energy in a closed system is constant — energy cannot be created or destroyed.

The energy of a moving body is called kinetic energy, E_k.

$$E_k = \tfrac{1}{2} mv^2$$

where m and v are the mass and speed, respectively, of the body. The energy which a body has due to its position above the ground is called gravitational potential energy, E_p.

$$E_p = mgh$$

where h is the height of the body above the ground and g is the acceleration due to gravity. Energy and mass are equivalent. Energy in joules is related to mass in kilograms by

$$E = mc^2$$

Power is the rate at which work is done or the rate at which energy is converted from one form to another.

$$P = \frac{W}{t}$$

The unit of power is the watt, W.

Questions 8

DATA: $g = 9.8$ m s^{-2}.

Section I

1. Define the unit of work. Express this unit in terms of kg, m and s..
....
....

2. List five devices in which energy is converted from one form to another, naming the forms of energy in each case.
....
....

3. A body is acted on by a constant force of magnitude 22 N over a distance of 8.0 m. If it starts from rest what is its final kinetic energy?.....................................
....

4. When a body is thrown vertically upwards the loss in .... is equal to the gain in ....

5. Write down an equation which relates mass to energy.
....

6. Albert Einstein is remembered for, among other things, his theories of . . . . . and his explanation of . . . . . In what century did Einstein do most of his work?

7. Power is defined as...
....

8. James Watt is best remembered for.........................
....

9. How much electrical energy is converted in a 100 W light bulb in five minutes? Into what form is this energy converted initially?...

10. Express the watt in terms of the basic units of mass, length and time...

Section II

1. A body of mass 2.0 kg, initially at rest, is acted upon by a constant resultant force of magnitude 10 N over a distance of 20 m. Given that the force and displacement are in the same direction, find (i) the kinetic energy gained by the body, (ii) the speed of the body at the end of the 20 m.

2. A body of mass 5.0 kg rests on a smooth horizontal surface. A horizontal force of magnitude 20 N is applied to it. Calculate (i) the work done in moving it 25 m, (ii) its speed at the end of 25 m, (iii) the magnitude of its acceleration, (iv) the time taken to travel the 25 m.

3. A body of mass 10 kg, initially at rest, is acted upon by a constant force. After covering a distance of 25 m its velocity is 20 m s^{-1} in the direction of the force. Find the magnitude of the applied force.

4. Calculate the potential energy of a stone of mass 2.5 kg at a height of 10 m above the ground. If the stone is dropped from this height, what is its speed just before it strikes the ground?

5. An object of mass 0.5 kg is dropped from a window 10 m above the ground. Calculate (a) its potential energy at the window, (b) its kinetic energy and its potential energy when it is (i) 8 m, (ii) 4 m, above the ground.

6. A ball is thrown upwards from ground level with an initial speed of 10 m s^{-1}. What is its speed at a height of 3.0 m above the ground? What is the greatest height reached?

7. A ball of mass 200 g strikes a wall with a speed of 20 m s^{-1} and rebounds in the opposite direction at the same speed. Calculate the change in momentum of the ball. Is momentum conserved in this collision? Is energy conserved? Explain.

8. A body of mass 2.6 kg is travelling with a velocity of 4.5 m s^{-1} east when it collides with another body of mass 1.4 kg and velocity 6.8 m s^{-1} west. After the collision both bodies move together. Calculate the kinetic energy before the collision and after the collision and account for the difference.

9. A girl sits on a swing of length 2.2 m. At its highest point, the swing makes an angle of 60° with the vertical. Calculate the speed of the girl at the lowest point.

10. A car of mass 1 t is free-wheeling along a straight road which is inclined at 30° to the horizontal. Assuming that the car starts from rest and neglecting friction, calculate (i) the kinetic energy of the car, (ii) the speed of the car, when it has travelled 40 m. Repeat the calculations for the case where there is a constant frictional force of magnitude 1.2 kN acting on the car. Account for the difference in kinetic energy in the two cases.

11. A man carries a case of mass 20 kg up a flight of stairs in 10 s. If the landing is 3.5 m above the bottom of the stairs, what is the power exerted by the man?

12. An engine pulls a train along a level track at a constant speed of 30 m s^{-1}. If the total resistive force due to friction is 12 kN, what is the power exerted by the engine?

13. A particle of mass 4.0 kg is acted upon by a constant resultant force which gives it an acceleration of magnitude 5.5 m s^{-2} for 40 s. What is the work done and the power developed by the force?

14. A weightlifter lifts a set of weights of total mass 80 kg above his head from ground level, a total distance of 2.4 m. Calculate the energy gained by the weights. Where does this energy come from? If the weightlifter holds the weights above his head his muscles quickly become tired. Explain why this is so.

15. A train is travelling at a constant speed of 40 m s^{-1} on a level track. If the total resistive force acting on the train is 35 kN what is the power developed by the engine?

9 Vectors

9.1 Addition of Vectors

As we saw in *Chapter 5*, a vector quantity is one which has a direction associated with it. (A quantity which has no direction is called a scalar quantity.) In the earlier chapters we met four examples of vector quantities, *viz.* displacement, velocity, acceleration and momentum. Such quantities may be represented by a **vector**, *i.e.* a line the length of which is proportional to the magnitude of the vector quantity and which has an arrow indicating the direction of the vector quantity.

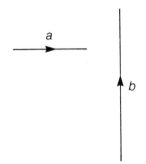

Fig. 9.1 Vectors

For example, the vectors a and b in *Fig. 9.1* might represent displacements of 5 m due East and 10 m due North, respectively. Note that a vector does not have a position. Any line, equal in length to b and parallel to it, could be used to represent the same displacement.

If two, or more, vectors lie on the same straight line they may be added algebraically. In such a case, one direction is chosen as positive; the opposite direction is then negative. The sum of two or more vectors is usually called their **resultant**.

Example

What is the resultant of the following displacements: $s_1 = 10$ m East; $s_2 = 15$ m West; $s_3 = 7$ m East?

Taking East to be positive,

$$s = +10 - 15 + 7$$
$$= +2$$

Ans. 2 m East

If two vectors are not in the same straight line they cannot be added algebraically. Suppose you leave a point A, *Fig. 9.2*, and walk to a point B, a distance of 3 m. You then turn through 90° and walk to C, a distance of 4 m from B. The total distance you have walked is 7 m but your displacement from A is 5 m in the direction AC. This method of adding two displacements can be used for any vector quantity and is known as the **Triangle Law**.

When two vectors form two sides of a triangle their resultant is the third side *(Fig. 9.2).*

Note that the two vectors to be added must be drawn ''nose-to-tail'' as in *Fig. 9.2*.

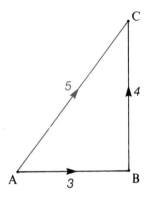

Fig. 9.2

Example

Find the resultant of the following displacements. $s_1 = 9$ m East; $s_2 = 12$ m North.

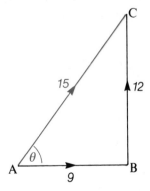

Fig. 9.3 Adding vectors which are at right angles to each other

s_1 and s_2 are represented by two vectors drawn as in *Fig. 9.3*. Their resultant is then represented by the third side of the triangle, AC. From Pythagoras' theorem the magnitude of the resultant displacement is 15 m. The direction of the resultant is found as follows.

$$\sin \theta = \frac{12}{15}$$

$$= 0.8$$

$$\theta = 53° 8$$

Ans. 15 m E 53° 8′ N

Example

Find the resultant of the following velocities. $v_1 = 4 \ m \ s^{-1}$ E; $v_2 = 5 \ m \ s^{-1}$ E 60° N.

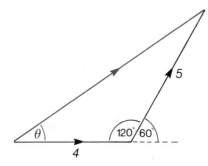

Fig. 9.4 Adding vectors not at right angles to each other

Again, the velocities are represented by vectors, *Fig. 9.4*. In this case, the magnitude of their resultant is found by the Cosine Formula.

$$v^2 = 4^2 + 5^2 - (2 \times 4 \times 5 \cos 120°)$$

$$= 16 + 25 + 20$$

$$= 61$$

$$=> \quad v = 7.8 \ m \ s^{-1}$$

The direction of the resultant may be found using the Sine Rule.

$$\frac{5}{\sin \theta} = \frac{7.8}{\sin 120°}$$

$$=> \quad \frac{\sin \theta}{5} = \frac{\sin 120°}{7.8}$$

$$=> \quad \sin \theta = \frac{5 \sin 120°}{7.8}$$

$$= 0.555$$

$$\theta = 33° 43′$$

Ans. 7.8 m s⁻¹ E 33° 43′ N

An alternative method to the triangle law for the addition of vectors is known as the **Parallelogram Law**.

> **When two vectors form adjacent sides of a parallelogram their resultant is the diagonal of the parallelogram,** *(Fig. 9.5)*

In this method the two vectors to be added must be drawn "tail-to-tail" as shown in *Fig. 9.5*. A parallelogram is then formed by drawing two lines parallel to the two vectors. The resultant of the vectors is then the diagonal of the parallelogram. *Fig. 9.5* shows this method applied to the previous example. The calculations and, of course, the answer are exactly the same as before. Note that addition of vectors is sometimes referred to as **composition** of vectors.

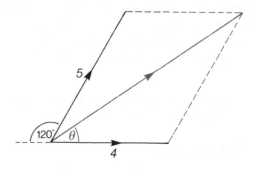

Fig. 9.5 Addition of vectors using the parallelogram law

9.2 Resolution of Vectors

It is sometimes necessary to break up a vector into two smaller vectors, usually at right angles to each other. The two vectors are called **resolved parts** or **components** and the process is called **resolution.** Resolution is exactly the reverse of addition, so to resolve a vector we simple apply the triangle law in reverse. In other words, to resolve a vector into two components at right angles to each other, we construct a right angled triangle using the given vector as hypotenuse. The magnitudes of the components are then found using trigonometry. The following example should make the method clear.

Example

A tennis ball is travelling with a velocity of 10 m s⁻¹ at an angle of 60° to the horizontal. Calculate the horizontal and vertical components of its velocity.

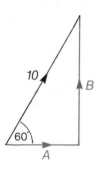

Fig. 9.6 *Resolving a vector into two components*

The required components are A and B in *Fig. 9.6*. Their magnitude is found as follows.

$$\frac{A}{10} = \cos 60°$$

$$A = 10 \cos 60°$$

$$= 5 \text{ m s}^{-1}$$

$$\frac{B}{10} = \sin 60°$$

$$B = 10 \sin 60°$$

$$= 8.7 \text{ m s}^{-1}$$

Ans. 5 m s⁻¹; 8.7 m s⁻¹

Since a vector does not have a position B could equally well have been drawn as shown in *Fig. 9.7*. A and B are now the sides of a rectangle of which the original vector is a diagonal. It should be clear that this is just a case of applying the parallelogram law in reverse.

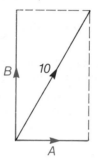

Fig. 9.7

EXPERIMENT 9.1

To Verify the Parallelogram Law for the Addition of Forces

Apparatus: Large board, two retort stands, two pulleys, weights, thread.

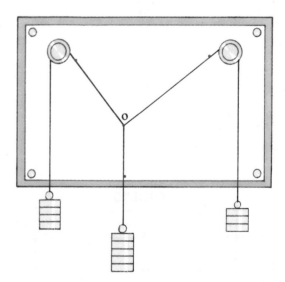

Fig. 9.8(a)

Procedure:
1. Attach a sheet of paper to the board. Fix the pulleys to the board and attach weights to thread passing over the pulleys as shown in *Fig. 9.8(a)*. Make sure that the weights can move freely.
2. When the system has come to rest mark the point o where the three strings meet. Mark also the positions of the three threads by placing a dot behind each thread as far from o as possible.
3. Remove the sheet of paper from the board and draw lines to represent the positions of the threads. Mark off a length on each line proportional to the weight on the corresponding thread. Complete the parallelogram as shown in *Fig. 9.8(b)*.
4. Draw the diagonal od and measure its length. The diagonal should be colinear with oc and equal in length to |oc|. Repeat with different weights.

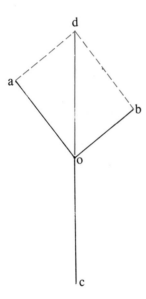

Fig. 9.8(b)

RESULTS

w_1/N	w_2/N	w_3/N	|oa|/cm	|ob|/cm	|oc|/cm	|od|/cm

Questions

1. Discuss the effect of friction at the pulleys on the result of this experiment.
2. This experiment may also be carried out using three spring balances instead of the weights and pulleys. Suggest how this might be done and compare the two methods.

SUMMARY

A vector quantity is one which has a direction associated with it. A vector is a line used to represent a vector quantity. A quantity which does not have a direction associated with it is called a scalar quantity.

Scalar quantities are always added algebraically. Vectors which lie on the same straight line are added algebraically; vectors not on the same straight line are added using the Triangle Law or the Parallelogram Law.

Resolution of a vector is the breaking up of a vector into two components, usually at right angles to each other. Resolution is achieved by applying the Triangle Law or the Parallelogram Law in reverse.

Questions 9

Section I

*1. What is the difference between vector quantities and scalar quantities? Give two examples of each...........

....

....

*2. What is the resultant of a displacement of 40 km N and a displacement of 15 km S?................................

....

*3. What is the resultant of a displacement of 5 m S and a displacement of 12 m W? ....

....

*4. What is the resultant of a horizontal force of 10 N and a vertical force of 20 N? ....

....

*5. Two vectors, A and B, make an angle θ with each other. Write down an expression for the magnitude of their resultant. ....

....

*6. What is the horizontal component of a force of 24 N which makes an angle of 60° with the horizontal? ...

....

*7. A vector A makes an angle θ with the horizontal. Give expressions for the horizontal and vertical components of A. ....

....

*8. A block is being pulled along a smooth horizontal surface by a rope which makes an angle of 60° with the horizontal. If the magnitude of the force is F what horizontal force would be required to do the same amount of work in a given time? ....

....

Section II

*1. An aeroplane leaves Dublin and flies 150 km due South and then flies 200 km due East. Find the total distance travelled by the aeroplane and its final displacement from Dublin.

*2. Find the resultant of a horizontal force of 200 N and a vertical force of 120 N.

*3. A man sets out, walking due East, to walk around the perimeter of a square of side 100 m. Given that he walks Northwards along the second side of the square calculate his displacement from his starting point when he has walked 250 m.

*4. Smoke leaves a vertical chimney with a speed of 0.8 m s^{-1}. If the velocity of the wind is 1.5 m s^{-1} S, calculate the resultant velocity of the smoke.

*5. A girl sets out to swim straight across a river with a velocity of 2.5 m s^{-1} W. If the velocity of the river is 10 m s^{-1} S find the resultant velocity of the girl. If the river is 30 m wide, how long does it take her to reach the opposite bank?

*6. A car is travelling north at 20 m s^{-1}. One minute later it is travelling south at 45 m s^{-1}. What is the average acceleration of the car? If the mass of the car is 1 t what average force would be required to produce this acceleration?

*7. A car is travelling with a velocity of 15 m s^{-1} E. After 5 s its velocity is 20 m s^{-1} S. Calculate its average acceleration in (i) the eastward, (ii) the southward, direction. What is its average resultant acceleration?

*8. A force of 200 N acts at 40° to the horizontal. Calculate the horizontal and vertical components of the force.

*9. Dublin is 276 km N-E of Cork. What is (i) the eastward, (ii) the northward, displacement of Dublin from Cork?

*10. A billiard ball of mass 80 g and travelling with a speed of 2.0 m s^{-1} strikes a cushion at an angle of 30°. What is the momentum of the ball (i) perpendicular to the cushion, (ii) parallel to the cushion?

*11. A railway porter pulls a trolley by a handle which makes an angle of 25° with the horizontal. If the force exerted by the porter is 150 N calculate, neglecting friction, the horizontal force on the trolley. If the porter pulls the trolley a distance of 26 m in 30 s calculate (i) the work done, (ii) the average power.

*12. A tennis ball of mass 40 g is moving downwards at an angle of 20° to the horizontal with a speed of 15 m s^{-1} when it is struck by a racket. Given that the ball and racket are in contact for 0.1 s, what force does the racket exert on the ball if its initial return velocity is 40 m s^{-1} horizontally?

10 Circular Motion

10.1 Angular Speed

Fig. 10.1 shows a particle P, travelling in a circular path of radius *r*, with a constant speed, *v*. Suppose the particle moves from A to B in a time *t* and that θ is the angle subtended at the centre by the arc AB. Then θ is the angle swept out by the radius in time *t, i.e.* θ is the angular displacement of P from OA in time *t*. The average angular speed of P is therefore θ / t. Since the particle is travelling at a constant speed, the speed at any instant is the same as the average speed. Thus, we can define **angular speed**, ω, as

$$\omega \ = \ \frac{\theta}{t}$$

or

> **Angular speed is the rate of change of angular displacement with respect to time.**

Angular displacement is measured in **radians** and angular speed is measured in **radians per second**, rad s^{-1}. Both are **vector** quantities.

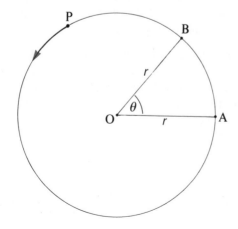

Fig. 10.1

Angular Speed and Linear Speed

The linear speed of the particle in *Fig. 10.1* is given by

$$v \ = \ \frac{|AB|}{t}$$

The angular speed is given by

$$\omega \ = \ \frac{\theta}{t}$$

But,

$$\theta \ = \ \frac{|AB|}{r}$$

Therefore,

$$\omega \ = \ \frac{|AB|}{rt}$$

But,

$$\frac{|AB|}{t} \ = \ v$$

Therefore,

$$\omega \ = \ \frac{v}{r}$$

or

$$v \ = \ r\omega$$

Example

A particle is travelling at a constant speed in a circular path of radius 50 cm. If the particle makes 2 complete revolutions in each second what is (i) its angular speed in rad s^{-1}, (ii) its linear speed?

There are 2π radians in a full circle, so the angular speed is

$$2\pi \times 2 = 4\pi$$

$$= 12.6 \text{ rad s}^{-1}$$

$$v = r\omega$$

$$= 0.5 \times 12.6$$

$$= 6.3 \text{ m s}^{-1}$$

Ans. 12.6 rad s^{-1}; 6.3 m s^{-1}

10.2 Centripetal Acceleration

Although the speed, both linear and angular, of the particle in *Fig. 10.1* is constant the linear velocity is not. The direction of the linear velocity at any point is along the tangent to the circle at that point. Therefore, although the magnitude of the linear velocity remains constant, its direction changes continuously, *i.e.* the particle has a linear acceleration.

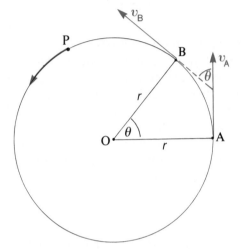

Fig. 10.2

To find the acceleration we need to find the change in velocity in a given time. *Fig. 10.2* again shows a particle, P, travelling at a constant speed in a circular path of radius r. When the particle is at A its velocity is v_A and when it is at B its velocity is v_B. If the particle travels from A to B in a time t, the change in its velocity in that time is

$$\Delta v = v_B - v_A.$$

Its average acceleration is then

$$a_{ave} = \frac{\Delta v}{t}$$

$$= \frac{(v_B - v_A)}{t}$$

$v_B - v_A$ is the difference between two vector quantities. To calculate the difference we can rewrite it as a sum and then use the triangle law. Thus,

$$\Delta v = v_B - v_A$$

$$= v_B + (-v_A)$$

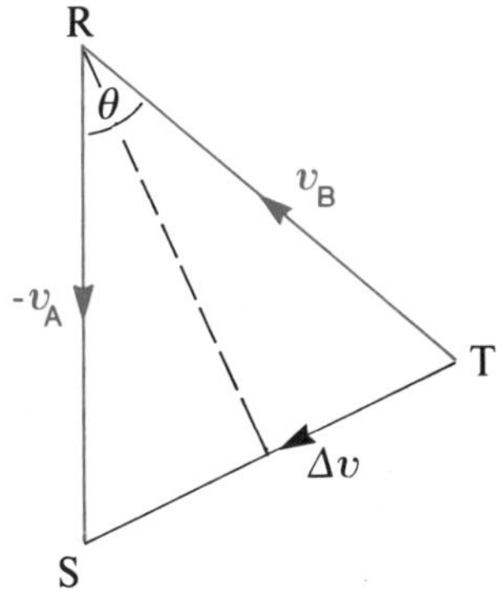

From *Fig. 10.3* the change in velocity, Δv, is represented by the vector TS . The magnitudes of v_A and v_B are equal since the speed of the particle is constant. Therefore RST is an isosceles triangle. Then

$$\frac{\frac{1}{2}\Delta v}{v} = \sin \frac{\theta}{2}$$

$$= \sin \frac{\omega t}{2}$$

$$=> \qquad \frac{1}{2}\Delta v = v \sin \frac{\omega t}{2}$$

$$=> \qquad \Delta v = 2v \sin \frac{\omega t}{2}$$

The average acceleration is

$$a_{ave} = \frac{\Delta v}{t}$$

$$= \frac{2v}{t} \sin \frac{\omega t}{2}$$

Fig. 10.3

101

The acceleration, a, at any instant is found by making t very small, *i.e.*

$$a = \frac{\Delta v}{t} \quad \text{as} \quad t \to 0$$

Therefore, $\quad a = \frac{2v}{t} \sin \frac{\omega t}{2} \quad \text{as} \quad t \to 0$

When $t \to 0 \quad \sin \frac{\omega t}{2} \to \frac{\omega t}{2}$

Therefore, $\quad a = \frac{2v}{t} \times \frac{\omega t}{2}$

$$= v\omega$$

But, $\quad v = r\omega$

Therefore, $\boxed{a = r\omega^2}$

or

$$\boxed{a = \frac{v^2}{r}}$$

When $\quad t \to 0, \quad \omega t \to 0$

$=> \quad \theta \to 0 \quad$ since $\theta = \omega t$

From *Fig. 10.3*

when

$$\theta \to 0$$
$$\angle \text{RTS} \to 90°$$

Therefore, Δv is at right angles to v_B (and to v_A). Since the change in velocity is perpendicular to v_B the acceleration is also perpendicular to v_B, *i.e.* the acceleration is along the radius towards the centre of the circle, O. Because the acceleration is directed towards the centre it is called the **centripetal acceleration**. *(For an alternative treatment of this section see Appendix A3.)*

Example

A stone is being swung in a vertical circle on the end of a string 1.5 m long. If the angular speed of the stone (assumed constant) is 10 rad s⁻¹, calculate the centripetal acceleration of the stone.

$$a = r\omega^2$$
$$= 1.5 \times 10^2$$
$$= 150 \text{ m s}^{-2}$$

Ans. 150 m s⁻²

We saw in *Chapter 6* that if a body has an acceleration there must be a resultant external force acting on it (Newton's first law of motion). Since a body travelling in a circular path always has an acceleration towards the centre of the circle there must be a resultant force in that direction also. This force is called the **centripetal force**. Note that, when a body is travelling in a circular path, the centripetal force is the resultant of all the external forces acting on it in the plane of the circle. In all cases the resultant force on a body is given by

$$F = ma$$

Therefore, since centripetal acceleration is given by

$$a = r\omega^2 \quad \text{or} \quad a = \frac{v^2}{r}$$

centripetal force is given by

$$\boxed{F = mr\omega^2}$$

or

$$\boxed{F = \frac{mv^2}{r}}$$

where m is the mass of the body travelling in a circular path of radius r at a constant speed v (angular speed ω).

Example

A body of mass 2.2 kg is moving in a circular path on a horizontal surface, on the end of a string 1.5 m long. If the speed of the body (assumed constant) is 4.0 m s⁻¹ what is the tension in the string?

The only force acting on the body in the plane of the circle is the tension in the string. Therefore,

$$F = \frac{mv^2}{r}$$

$$= \frac{2.2 \times 4^2}{1.5}$$

$$= 23.5 \text{ N}$$

Ans. 23.5 N

10.3 Satellite Motion

A satellite is a heavenly body which orbits another one. Thus, the planets are satellites of the sun, the moon is a satellite of the earth, and so on. Many of these satellites travel in orbits which are approximately circular. In every case, the centripetal force is the gravitational force *(see Chapter 6)* between the orbiting body and the central one. Consider a body of mass *m* moving around another body of mass *M*, in a circular orbit of radius *d*, *Fig. 10.4*. The gravitational force on the smaller body is

$$F = \frac{GMm}{d^2}$$

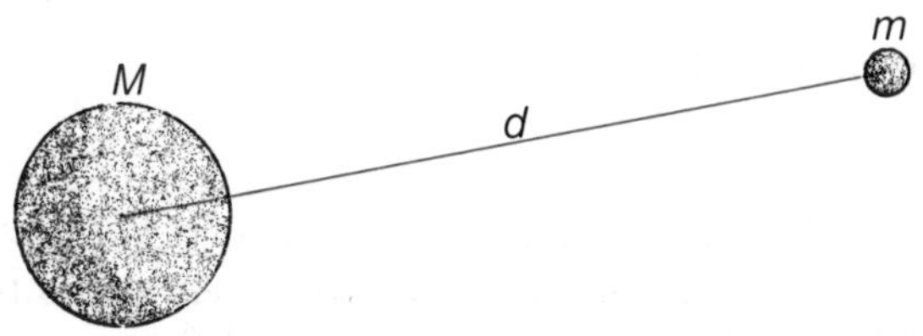

Fig. 10.4

Since this is the resultant force on the smaller body it is the centripetal force and may also be written as

$$F = \frac{mv^2}{d}$$

where *v* is the linear speed of the smaller body.

Combining these two equations we have

$$\frac{GMm}{d^2} = \frac{mv^2}{d}$$

$$\Rightarrow \quad \frac{GM}{d} = v^2$$

or, $$v^2 = \frac{GM}{d} \quad \text{(i)}$$

Since the distance travelled by the smaller body in making one complete orbit is $2\pi d$, the time for one orbit, *i.e.* the **period**, *T*, is given by

$$T = \frac{2\pi d}{v}$$

$$\Rightarrow \quad T^2 = \frac{4\pi^2 d^2}{v^2}$$

Substituting for v^2 from *Eqn. (i)* gives

$$T^2 = 4\pi^2 d^2 \times \frac{d}{GM}$$

$$\Rightarrow \quad \boxed{T^2 = \frac{4\pi^2 d^3}{GM}} \quad \text{(ii)}$$

Example
Calculate the mass of the sun, given that the average radius of the earth's orbit is 1.5 x 10^{11} m.

The period of the earth's orbit is 1 year, so

$$T = (365 \times 24 \times 60 \times 60) \text{ s}$$

$$= 3.15 \times 10^7 \text{ s}$$

Rearranging *Eqn. (ii)* gives

$$M = \frac{4\pi^2 d^3}{GT^2}$$

$$= \frac{4\pi^2 \times (1.5 \times 10^{11})^3}{6.7 \times 10^{-11} \times (3.15 \times 10^7)^2}$$

$$= \frac{4\pi^2 \times 3.375 \times 10^{33}}{6.7 \times 10^{-11} \times 9.92 \times 10^{14}}$$

$$= 2.0 \times 10^{30} \text{ kg}$$

Ans. 2.0×10^{30} kg

From *Eqn. (ii)* we see that the further a planet is from the sun, the greater is its period. Thus, the period of Mercury, the planet nearest the sun (mean distance 5.7×10^7 km), is 88 days; the period of the earth (mean distance 1.5×10^8 km) is 1 year and the period of Pluto, the most distant planet (mean distance 5.8×10^9 km), is 248 years.

Elliptical Orbits

While the orbits of the planets and many satellites are approximately circular, they are in fact, elliptical, with the central body at one focus of the ellipse, *Fig. 10.5*. The ratio of the semi-major axis, a, to the semi-minor axis, b for the planets is given in *Table 10.1*. The ratio of a to b for a circle is exactly 1, so we can see from the table that the orbits of all the planets are very nearly circular.

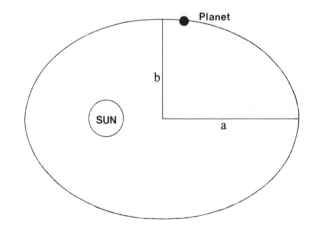

Fig. 10.5 Elliptical orbit

Planet	a/b
Mercury	1.022
Venus	1.000
Earth	1.000
Mars	1.004
Jupiter	1.001
Saturn	1.002
Uranus	1.001
Neptune	1.000
Pluto	1.032

Table 10.1 The ratio, correct to four significant figures, of the semi-major axis to the semi-minor axis for the planets

Geostationary Orbits

When a satellite is in an orbit such that it appears to be stationary over a particular point on the earth's surface, it is said to be in a geostationary, or parking, orbit. If a satellite is to be in a geostationary orbit three conditions must be satisfied: the orbit must be concentric with the centre of the earth; the plane of the orbit must coincide with the equatorial plane, *i.e.* the satellite is always directly above the equator; the period of the satellite must be the same as the period of rotation of the earth, *i.e.* 24 hours. Reference to *Eqn. (ii)* shows that there is only one possible stationary orbit around the earth, *i.e.* since T, G and M are fixed there is only one possible value for d, the radius of the orbit.

Example
Calculate the height of a satellite which is in a stationary orbit above the earth, given that the mass of the earth is 6.0×10^{24} kg and the radius of the earth is 6.4×10^6 m.

The period of the satellite is

$$T = 24 \text{ h}$$

$$= 24 \times 3600 \text{ s}$$

$$= 8.64 \times 10^4 \text{ s}$$

$$T^2 = \frac{4\pi^2 d^3}{GM}$$

Rearranging gives

$$d^3 = \frac{T^2 GM}{4\pi^2}$$

$$= \frac{(8.64 \times 10^4)^2 \times 6.7 \times 10^{-11} \times 6.0 \times 10^{24}}{4\pi^2}$$

$$= 7.60 \times 10^{22}$$

$$\Rightarrow \quad d = 4.24 \times 10^7 \text{ m}$$

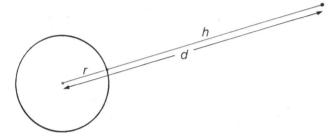

Fig. 10.6

To find the height of the orbit we must subtract the radius of the earth from the radius of the orbit, *Fig. 10.6*. Thus

$$h = 4.24 \times 10^7 - 6.4 \times 10^6$$

$$= 3.6 \times 10^7 \text{ m}$$

$$= 36\,000 \text{ km}$$

Ans. 36 000 km

Fig. 10.7 Communications satellite

Satellites in geostationary orbits make possible continuous transmission of radio and television programmes from one part of the world to another. The first such (commercial) satellite was launched on April 6, 1965. It is called *Early Bird* and is "stationary" over the South Atlantic between Africa and South America. It has an expected lifetime of 10^6 years. A large number of such satellites are now in orbit, providing a world-wide system of telecommunications. Ireland is participating in this system with a £6M earth station at Midleton, Co. Cork. Moreover, Ireland has a share in a satellite which is to be positioned 31° West of Greenwich.

SUMMARY

Angular speed, ω, is the rate of change of angular displacement, θ, with respect to time.

$$\omega = \frac{\theta}{t}$$

Angular speed is related to linear speed, v, by

$$v = r\omega$$

When a body travels in a circular path the direction of its velocity changes continuously so it has an acceleration even though its speed may be constant. This acceleration is called the centripetal acceleration. It is of magnitude

$$a = \frac{v^2}{r} = r\omega^2$$

and is directed towards the centre of the circle. The resultant force causing the acceleration is called the centripetal force.

The period of a satellite in a circular orbit is given by

$$T^2 = \frac{4\pi^2 d^3}{GM}$$

were M is the mass of central body and d is the radius of the orbit. Satellite orbits are elliptical rather than circular. Many, including the orbits of the planets are very nearly circular. A satellite in a geostationary orbit has a period of 24 hours and is placed 36 000 km directly over the equator. Such satellites are used for communications purposes.

Questions 10

DATA: $G = 6.7 \times 10^{-11}$ N m^2 kg^{-2}
$g = 9.8$ m s^{-2}.

Section I

*1. Angular speed is defined as ....

....

*2. Give an expression for the relationship between linear speed and angular speed. Show that the units on both sides of this equation are identical.........................

....

*3. Explain why a body travelling on a circular path at a constant speed has an acceleration.........................

....

*4. The acceleration of a body travelling at a constant speed in a circle is called its..

and its direction at any instant is..........................

*5. What force provides the centripetal acceleration of the moon?...

*6. What force provides the centripetal acceleration of a car travelling around a bend on (i) a level road, (ii) a banked track?...

....

*7. What is a geostationary orbit?..............................

....

....

*8. Why must a satellite in a geostationary orbit be directly over the equator?..

....

....

*9. The period of a satellite in a circular orbit is proportional to ....

*10. The ratio of the periods of two satellites is 4 to 1. What is the ratio of the radii of their orbits?...................

....

Section II

*1. A flywheel has an angular speed of 3000 revolutions per minute (r.p.m.). Express this speed in rad s^{-1}. If the wheel has a radius of 200 mm what is the linear speed of a point on its circumference?

*2. An L.P. rotates on a turntable at 33.3 r.p.m. Express its angular speed in rad s^{-1}. What is the linear speed of a point on the record 10 cm from its centre?

*3. A bicycle is travelling along a road at a constant speed such that the angular speed of its wheels is 15 rad s^{-1}. If the radius of the wheels is 60 cm what is the speed of the bicycle?

*4. A car is travelling around a circular bend at a constant speed of 15 m s^{-1}. If the radius of the bend is 20 m what is the centripetal acceleration of the car?

*5. A particle is travelling in a circular path in such a way that its angular speed is equal in magnitude to its centripetal acceleration. If the radius of the circle is 2.0 m what is the angular speed of the particle?

*6. A body of mass 2.5 kg is travelling at a constant speed of 4 m s^{-1} in a circular path of radius 40 cm. Calculate (i) the angular speed of the body, (ii) the centripetal acceleration of the body, (iii) the magnitude of the resultant force on the body.

*7. A stone of mass 1.2 kg is attached to the end of a string of length 50 cm. It is swung in a vertical circle at a constant angular speed of 5 rad s^{-1}. Calculate (i) the linear speed of the stone, (ii) the magnitude of the resultant force on the stone, (iii) the maximum and minimum tensions in the string.

*8. A body of mass 0.5 kg is whirled in a vertical circle on the end of a string 80 cm long with a constant linear speed of 8 m s^{-1}. Calculate the maximum and minimum tensions in the string.

*9. A particle of mass 2.0 kg rests on the smooth inside surface of a cylinder of radius 40 cm, *Fig. I*. The particle is projected horizontally from the point P with an initial speed u. Determine the minimum value of u for which the particle will follow a complete circle without leaving the surface of the cylinder.

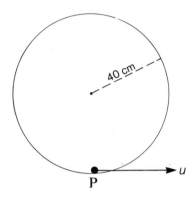

Fig. I

*10. An astronaut is standing in a spacecraft which is orbiting the earth. Draw a diagram to show the forces acting on the astronaut. Explain why he appears to be weightless. Are there any circumstances in which an astronaut could actually be weightless? Explain.

*11. Calculate (i) the angular speed, (ii) the linear speed, of a satellite in a geostationary orbit.

*12. Calculate the period of the moon, given that the radius of its orbit is 3.8×10^8 m and that the mass of the earth is 6.0×10^{24} kg.

*13. The period of a satellite orbiting the earth is the same as that of a satellite orbiting the moon. Determine the relationship between the radii of the two orbits, given that the mass of the earth is 82 times that of the moon.

11 Simple Harmonic Motion

11.1 Simple Harmonic Motion

A particularly important case of linear motion occurs when a particle or body oscillates back and forth along the same path, *e.g.* a weight on the end of a spring or a person on a swing. It can be shown in many such cases that the acceleration of the body is proportional to its displacement. Such motion is called **simple harmonic motion (s.h.m.).**

> The motion of a particle is simple harmonic motion if its acceleration towards a particular point is proportional to its displacement from that point.

In mathematical terms this becomes

$$a \propto -s$$

or

$$a = -ks$$

where k is a constant.

The constant of proportionality, k, is usually represented by ω^2. Thus, the definition becomes

$$a = -\omega^2 s$$

Note: ω^2 simply represents a number. ω does not, in this context, represent angular speed.

Fig. 11.1

The behaviour of a particle executing simple harmonic motion is illustrated in *Fig. 11.1*. The particle P is oscillating to-and-fro about the point O. The further it travels from O the greater is its acceleration towards O. Its acceleration is a maximum when it is at the points A and B since its displacement from O is greatest at these points. When it is at O its acceleration is zero since its displacement from O is now zero. O is therefore the equilibrium position of the particle, *i.e.* the position at which its acceleration is zero (*see p.75*).

Example
When the particle in Fig. 11.1 is 12 cm from O its acceleration towards O is 4 cm s^{-2}. What is its acceleration when it is 14 cm from O?

$$|a| = \omega^2 |s|$$
$$4 = \omega^2 \times 12$$
$$=> \quad \omega^2 = \frac{4}{12}$$
$$|a| = \omega^2 |s|$$
$$= \frac{4 \times 14}{12}$$
$$= 4.7 \text{ cm s}^{-2}$$

Ans. 4.7 cm s^{-2}

The Period of an S.H.M.

Since a particle undergoing simple harmonic motion travels repeatedly over the same path it will pass a particular point at regular time intervals.

> The period of a simple harmonic motion is the time interval between that moment when the particle passes a particular point going in a particular direction, and the moment when it passes that point again going in the same direction.

Since the general derivation of an expression for the period of a simple harmonic motion requires calculus we shall confine ourselves to a particular case. (For a more rigorous treatment see *Appendix A4.*)

Fig. 11.2 shows a point P moving around the circumference of a circle at a constant angular speed, ω. PQ is a line drawn from P perpendicular to the diameter, AB. As P moves around the circumference at constant speed, Q moves back and forth along the diameter. We shall first show that the motion of Q is simple harmonic motion and then obtain an expression for the period.

The acceleration of P is $r\omega^2$ towards O, where r is the radius of the circle (*see p.102*). The acceleration of P may be resolved into two components, one parallel to AB ($r\omega^2 \cos \theta$) and one perpendicular to AB ($r\omega^2 \sin \theta$), *Fig. 11.3*. Since the point Q

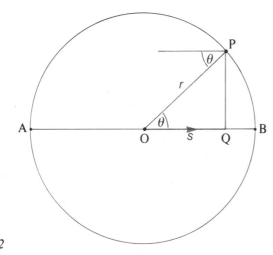

Fig. 11.2

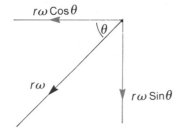

Fig. 11.3

moves only along AB its acceleration is equal to the component parallel to AB, *i.e.*

$$a_Q = r\omega^2 \cos\theta$$

But, $\cos\theta = \dfrac{s}{r}$

$$\therefore \quad a_Q = r\omega^2 \, \frac{s}{r}$$

$$= \omega^2 s$$

But a_Q is directed towards O while s is the displacement of Q *from* O. Therefore

$$a_Q = -\omega^2 s$$

That is, the motion of Q is simple harmonic motion.

The period of Q is equal to the time taken for P to make one complete revolution. Since the angular speed of P is ω

and there are 2π radians in a full circle, the time for one revolution is

$$\frac{2\pi}{\omega}$$

i.e.
$$\boxed{T = \frac{2\pi}{\omega}}$$

Although we have derived this expression for the period of a simple harmonic motion for a particular case it is, in fact, true for all cases of simple harmonic motion. (As noted above, ω would not, in general, represent angular speed. See, for example, next section.)

Fig. 11.4 The motion of a swing is approximately simple harmonic motion

Example

The acceleration of a particle executing simple harmonic motion is 2.5 m s^{-2} when its displacement from its equilibrium position is 20 cm. What is the period of the motion?

$$|a| = \omega^2 |s|$$
$$2.5 = \omega^2 \times 0.2$$
$$\Rightarrow \quad \omega^2 = 12.5$$
$$\Rightarrow \quad \omega = 3.54$$
$$T = \frac{2\pi}{\omega}$$

$$= \frac{2\pi}{3.54}$$

$$= 1.77 \text{ s}$$

Ans. 1.8 s

11.2 The Simple Pendulum

A simple pendulum, *Fig. 11.5* consists of a small, heavy bob suspended by a light inextensible string from a fixed point. When the bob is at position B the forces acting on it are its weight *(mg)* and the tension in the string. If we resolve the weight into two components as shown in the diagram we see that the tension, *T,* is equal to *mg* cos θ, since there is no acceleration parallel to the string. The force pulling the bob back towards O is *mg* sin θ. This force is called the restoring force, *F*.

$$F = mg \sin \theta$$

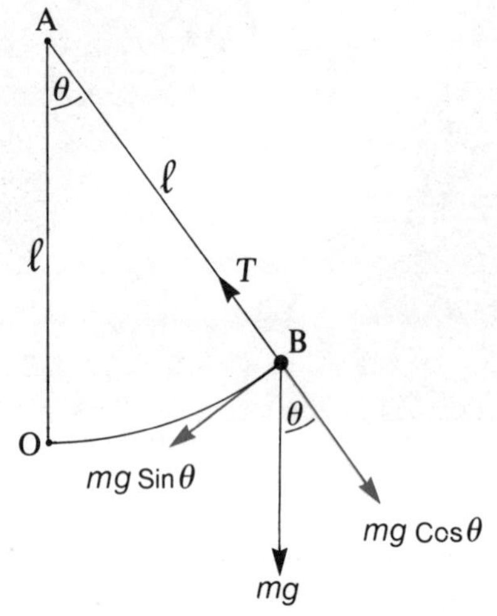

Fig. 11.5 Simple pendulum

in the direction shown in the diagram.

Since, $F = ma$

the acceleration in this direction is

$$a = g \sin \theta$$

If θ is small,

$$g \sin \theta \approx g\theta$$

and is directed towards O, since, when θ is small, the tangent to

the arc at B almost coincides with the chord BO. That is,

$$a = g\theta$$

But, $\theta = \dfrac{|\text{arc OB}|}{l}$

Again, if θ is small, $|\text{arc OB}|$ is approximately equal to the magnitude of the displacement, *s*, of the bob from O. Therefore, the acceleration towards O is

$$a = -\frac{g}{l} s$$

Since *g* and *l* are both constants, g/l is also constant, so *a* is proportional to *s* and in the opposite direction. That is, the motion of the pendulum is simple harmonic motion. The period of the motion is given by

$$T = \frac{2\pi}{\sqrt{\dfrac{g}{l}}}$$

or,

$$\boxed{T = 2\pi \sqrt{\frac{l}{g}}}$$

Note that the motion of a simple pendulum can only be considered to be simple harmonic motion if we can assume sin θ to be equal to θ. For this assumption to be accurate to three significant figures, θ must be less than approximately 5°.

Example

What is the length of a simple pendulum whose period is 1 s?

$$T = 2\pi \sqrt{\frac{l}{g}}$$

$$T^2 = \frac{4\pi^2 l}{g}$$

$$\Rightarrow \quad l = \frac{gT^2}{4\pi^2}$$

$$= \frac{9.8 \times 1}{4\pi^2}$$

$$= 0.25 \text{ m}$$

Ans. 25 cm

EXPERIMENT 11.1

To Determine the Value of g Using a Simple Pendulum

Apparatus: Pendulum, stop clock, retort stand, split cork, metre stick.

Procedure:
1. Place the thread of the pendulum between the two halves of the cork and fix the cork in the clamp of the retort stand. Measure the length of the pendulum, *i.e.* from the point of support to the centre of the bob.
2. Set the pendulum swinging in an arc of not more than 10°. Measure the time for 50 complete oscillations. Repeat for different lengths of the pendulum.
3. The acceleration due to gravity is calculated from

$$T^2 = 4\pi^2 \frac{l}{g}$$

where T is the period of the pendulum and l is its length. Plot a graph of T^2 against l and calculate the value of g from the slope of the graph.

RESULTS

l/m	Time for 50 swings/s	T/s	T^2/s^2

Slope of graph = s^2 m^{-1}.
Acceleration due to gravity = m s^{-2}

Questions

1. What is the advantage of holding the thread in a split cork rather than tying it to the retort stand?
2. Why is the length of the pendulum measured to the centre of the bob?
3. Explain the importance of allowing the pendulum to swing through only a small angle.

SUMMARY

The motion of a body is simple harmonic motion if its acceleration towards a particular point is proportional to its displacement from that point:

$$a = -\omega^2 s$$

The period of such a motion is given by

$$T = \frac{2\pi}{\omega}$$

For small angular displacements of the bob ($\sim 5°$) the motion of a simple pendulum is simple harmonic motion. The period of a simple pendulum of length l is given by

$$T = 2\pi \sqrt{\frac{l}{g}}$$

A simple pendulum may thus be used to determine the value of g experimentally.

Questions 11

DATA: $g = 9.8$ m s^{-2}.

Section I

*1. Define simple harmonic motion...............................
....

*2. In the equation $a = -\omega^2 s$, what does ω^2 represent?
....

*3. What is the unit of ω^2?..
....

*4. What is meant by the period of a simple harmonic motion?..

*5. The period of a simple harmonic motion is given by

....

Show that the units on both sides of this equation are the same.

....

*6. A simple pendulum consists of a............................

bob on the end of a..................................string.

*7. The bob of a simple pendulum passes through the lowest point on its path four times per second. What is the period

of the pendulum?...
*8. The lengths of two simple pendulums are in the ratio of

4 to 1. What is the ratio of their periods?.................

....

Section II

*1. A particle executes simple harmonic motion with a period of 2.0 s. What is its acceleration when its displacement is 20 cm from its rest position?

*2. A particle executing simple harmonic motion passes through a particular point on its path with an acceleration of 4.0 m s^{-2} at time intervals of 2 s and 6 s, alternately. Calculate (i) the period of the motion, (ii) the displacement of the point from the rest position.

*3. Calculate the period of a simple pendulum whose length is 40 cm. What would be the period of the pendulum if its length were increased to 160 cm?

*4. Show that the period of a satellite in a circular orbit is the same as that of a simple pendulum whose length is equal to the radius of the orbit.

*5. Calculate the acceleration due to gravity on the moon given that the period of a pendulum on the moon would be 2.5 times its period on earth.

*6. Derive a relationship between the period of a simple pendulum and its height above the surface of the earth. Sketch a suitable graph to illustrate this relationship.

*7. A simple pendulum of length 60 cm is suspended from the ceiling of a lift. Calculate the period of the pendulum when the lift is (i) at rest, (ii) moving upwards with a constant speed of 10 m s^{-1}, (iii) moving upwards with a constant acceleration of 1.2 m s^{-2}, (iv) moving downwards with a constant acceleration of 1.2 m s^{-2}.

*8. A bucket is filled with water and is attached to a rope tied to the ceiling. The bucket is given a small push so that it swings as a simple pendulum. If the bucket has a small hole in the bottom how will the period of the motion vary with time? Explain your answer. Would there be any difference if the hole were in the side of the bucket rather than the bottom?

REVISION EXERCISES B

DATA: $g = 9.8$ m s^{-2}
$G = 6.7 \times 10^{-11}$ N m^2 kg^{-2}.

Section I

1. *Fig. I* shows the pattern produced by a trolley on a powder mark timer. The time taken by the trolley to travel from the start of one patch to the start of the next is 0.02 s. By taking measurements from the diagram determine the average acceleration of the trolley between A and B.

Fig. I

2. In an experiment to verify the principle of conservation of momentum two trolleys, of masses 220 g and 420 g, respectively, were placed near the centre of a powder mark timer with a compressed spring between them. When the spring was released the patterns on the rail were as shown in *Fig. II*. By taking measurements from the diagram explain how this experiment verified the principle of conservation of momentum. What further procedure should have been followed to verify the principle?

Fig. II

3. In an experiment to verify Newton's second law the acceleration of a trolley was determined for a series of values of the applied force. The results obtained are shown in *Table I*. The acceleration was then determined for trolleys of different mass for a constant value of the applied force. The results obtained are shown in *Table II*.

F/N	0.1	0.2	0.3	0.4	0.5	0.6	0.7
a/cm s^{-2}	9.1	18.0	27.3	36.1	44.8	53.7	63.7

Table I

m/kg	0.2	0.3	0.4	0.5	0.6	0.7	0.8
a/cm s^{-2}	46.0	30.2	22.7	18.6	15.4	13.1	11.4

Table II

Plot suitable graphs from these data and explain how the results of these experiments verify Newton's second law. State how friction might have affected the results of these experiments and explain how the effects of friction might have been minimised.

4. In an experiment to determine the value of g, the acceleration due to gravity, by the free fall method the following values were obtained for the distance fallen by the ball and the time taken.

s/m	1.00	0.90	0.80	0.70	0.60	0.50	0.40	0.30
t/ms	451	430	409	371	342	324	291	244

Using these data plot a suitable graph and hence find a value for the acceleration due to gravity. The experiment was repeated a number of times for each value of the distance and the time recorded in each case. How would the values for the time given in the table have been obtained from the recorded values?

5. The following values were recorded in an experiment to determine the value of the coefficient of dynamic friction for a pair of surfaces. In each case, the mass of the moving body was m and the applied force was F.

m/kg	0.10	0.20	0.30	0.40	0.50	0.60	0.70	0.80
F/N	0.42	0.79	1.26	1.62	1.81	2.42	2.86	2.90

Plot a suitable graph and hence determine the value of the coefficient of dynamic friction. Describe a suitable method for varying the applied force in this experiment.

6. *Fig. III* shows a metre stick in equilibrium under the action of a number of forces. Using the values given on the diagram explain how this arrangement may be said to verify the principle of moments. Take the mass of the metre stick to be 108 g and assume that the centre of gravity of the metre stick lies on the 50 cm mark. If you were carrying out this experiment how would you establish the position of the centre of gravity?

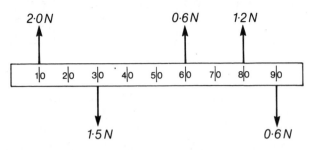

Fig. III

*7. *Fig. IV* shows three identical stretched springs attached to a ring which is in equilibrium. Given that the tension in each spring is proportional to its length, take measurements from the diagram and hence explain how this arrangement verifies the parallelogram law.

Fig. IV

8. In an experiment to determine the value of the acceleration due to gravity using a simple pendulum the following results were obtained.

l/m	1.00	0.90	0.80	0.70	0.60	0.50	0.40	0.30
T/s	2.00	1.92	1.83	1.70	1.53	1.40	1.26	1.12

Use these data to plot a suitable graph and hence find the value of g. Mention two precautions which should be taken when carrying out this experiment to ensure a more accurate result.

Section II

1. Distinguish between vector quantities and scalar quantities. Give three examples of each.

2. Define: Displacement; Velocity; Acceleration. A car starts from rest and reaches a speed of 20 m s^{-1} after 8 s. What is the magnitude of its acceleration and how far does it travel in the 8 s?

3. A ball is thrown vertically upwards from ground level with an initial speed of 20 m s^{-1}. Calculate (i) the greatest height reached, (ii) its speed when 15 m above the ground, (iii) the times at which its height is 12 m.

4. A car is travelling at a speed of 25 m s^{-1} when the brakes are applied. If it comes to rest in a distance of 100 m, what is the magnitude of its acceleration? If the mass of the car is 700 kg what is the average magnitude of the resultant force on the car?

5. A stone is thrown vertically upwards from the top of a building 15 m high with an initial speed of 15 m s^{-1}. Find its greatest height above the ground and the time taken to reach the ground.

6. State the principle of conservation of momentum. A bullet of mass 50 g, travelling at a speed of 400 m s^{-1}, strikes a block of wood of mass 2.5 kg at rest. If the bullet becomes embedded in the block, what is the initial speed of the block?

7. A rifle of mass 5.0 kg fires a bullet of mass 25 g with a muzzle speed of 600 m s^{-1}. Neglecting the effect of the gas expelled, calculate the initial speed of recoil of the rifle.

8. Explain how the principle of conservation of momentum is applied in (i) launching a rocket, (ii) changing the direction in which a spacecraft is travelling in space.

9. State Newton's laws of motion. A car of mass 800 kg, travelling at 30 m s^{-1} is brought to rest in a distance of 150 m when the brakes are applied. Calculate the average magnitude of the force on the car.

10. A force of magnitude 750 N is applied to a car of mass 500 kg for 10 s. Calculate (i) the magnitude of the acceleration, (ii) the speed of the car at the end of the 10 s, (iii) the distance travelled by the car in the 10 s.

11. State Newton's law of gravitation. What do the symbols G and g represent and what is the relationship between them? Describe how you would determine the value of g experimentally.

12. The acceleration due to gravity on the surface of the moon is 1.7 m s^{-2}. Calculate the mass of the moon given that its mean radius is 1.7×10^6 m.

13. Define: Limiting friction; Dynamic friction. Describe an experiment to measure the coefficient of dynamic friction.

14. If the minimum horizontal force required to move a block of wood of mass 1.5 kg along a horizontal surface is 7.5 N what is the coefficient of static friction between the block and the surface?

*15. A body is on the point of sliding down a plane inclined at 30° to the horizontal. What is the coefficient of static friction between the body and the plane?

16. A sleigh is being pulled over a horizontal surface at a constant speed. If the mass of the sleigh is 30 kg and the applied force is 100 N horizontally, what is the coefficient of dynamic friction between the sleigh and the surface?

17. A body of weight 260 N is being dragged along a horizontal surface at a constant speed by a force of 120 N exerted at an angle of 30° to the horizontal. Calculate the coefficient of dynamic friction between the body and the surface.

18. A car of mass 800 kg is being accelerated by a constant force of 2.0 kN. If the magnitude of the acceleration is 1.5 m s^{-2} calculate the magnitude of the force of friction.

19. State the principle of moments and explain how you would verify it experimentally.

20. A see-saw consists of a uniform plank 3.0 m long and supported at its centre. A boy of mass 25 kg sits 50 cm from one end of the plank. How far from the other end must a girl of mass 20 kg sit in order to balance the see-saw in a horizontal position?

21. A spanner, 30 cm long, is being used by a mechanic to tighten a nut. If the magnitude of the force exerted by the mechanic is 500 N, what is the torque on the nut? (Assume the force is applied at the end of the spanner.)

22. A uniform plank of weight 400 N and length 4.6 m rests on two supports which are 50 cm and 10 cm from either end, respectively. Calculate the force exerted by each support on the plank.

23. Define pressure. A solid cylinder of metal of density 8000 kg m^{-3} has a height of 60 cm and a radius of 10 cm. What is the pressure under it when it stands on one end?

24. From the definition of pressure show that the pressure in a fluid of constant density is proportional to the depth. Explain how this may be verified experimentally.

25. State Archimedes' principle and explain how it may be verified experimentally. Show that the law of flotation is a special case of Archimedes' principle.

26. A sphere of metal of density 6000 kg m^{-3} has a radius of 10 cm. It is hanging from a string and is totally immersed in a liquid of density 800 kg m^{-3}. What is the tension in the string?

*27. Define: Work; Energy; Power. Give the unit of each. Show that the loss of potential energy of a freely falling body is equal to the gain in its kinetic energy.

28. A car of mass 800 kg is accelerated from rest to a speed of 20 m s^{-1} over a distance of 100 m. Calculate (i) the kinetic energy gained, (ii) the average magnitude of the resultant force on the car.

29. An elevator on a building site is being used to raise concrete blocks to a height of 6.0 m. If the elevator raises 5 blocks per minute and the weight of each block is 200 N calculate, neglecting friction, the average power output of the elevator motor.

*30. What is the resultant of a horizontal force of 10 N and a vertical force of 15 N?

*31. Two boys are pulling a sledge by means of two ropes attached to the same point on the sledge. If the angle between the ropes is 60° and each boy exerts a force of 150 N, what is the magnitude of the resultant force on the sledge?

*32. A girl can swim at 2.0 m s^{-1} in still water. She heads straight across the river which is flowing at 4.0 m s^{-1} parallel to the bank. What is her resultant velocity? If the river is 12 m wide how long does it take her to reach the opposite bank?

*33. A ship steers due North with a speed of 10 m s^{-1}. A steady current is flowing from the East at a speed of 2.0 m s^{-1}. What is the resultant velocity of the ship?

*34. A car is travelling around a bend of radius 20 m at a constant speed of 20 m s^{-1}. If the mass of the car is 650 kg, what is the centripetal force acting on it? Where does this force act?

*35. A stone of mass 0.25 kg is being whirled in a horizontal circle on the end of a string 1.5 m long at a constant angular speed of 10 rad s^{-1}. Calculate (i) the linear speed of the stone at any instant, (ii) the tension in the string.

*36. If the stone in the previous question were whirled in a vertical circle instead of a horizontal circle, what would be the maximum and minimum tensions in the string?

*37. The moon completes one orbit of the earth every 27.3 days. Given that the mass of the earth is 6.0×10^{24} kg, calculate (i) the mean distance between the earth and the moon, (ii) the average speed of the moon relative to the earth.

*38. Define simple harmonic motion. Show that the motion of a simple pendulum is simple harmonic motion. Describe how a simple pendulum may be used to determine the acceleration due to gravity, g.

12 Wave Motion

12.1 Waves

One type of wave which is familiar to everyone is the water wave, varying in size from tiny ripples on the surface of a pond to the mountainous waves whipped up by an Atlantic storm. We also hear of waves in connection with radio stations — your favourite radio station is said to broadcast on ???? metres on the medium *wave*. We hear of micro*wave* ovens which cook food in a few seconds. What do all of these waves have in common?

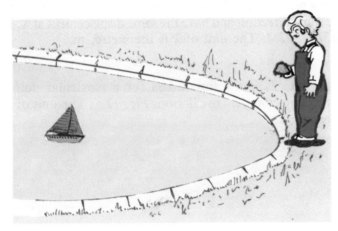

Fig. 12.1

Consider the situation shown in *Fig. 12.1*. A child holds a stone above the surface of the water in a pond. Some distance away a toy sailing boat floats on the water. The stone has a certain amount of potential energy due to its position *(Chapter 8)*. If the child drops the stone it loses its potential energy and a few seconds later the boat starts to bob up and down — it has gained kinetic energy, energy which must have come from the stone. When the stone is dropped a wave spreads out in all directions from the point where the stone hit the water, *Fig. 12.2*. The water does not move from the stone to the boat — it simply moves up and down, eventually coming to rest in its original position. There is no *net* movement of the water, which is the medium through which the wave is moving.

A laboratory experiment which shows wave motion more clearly involves the use of a "Slinky", *Fig. 12.3*. If the free

Fig. 12.2

end of the "Slinky" is moved quickly to one side and back, a **wave pulse** moves along its length and the box moves — energy has been transferred from the person's hand to the box. In this case it is obvious that the medium, *i.e.* the "Slinky", does not move from the hand to the box. Each part of the

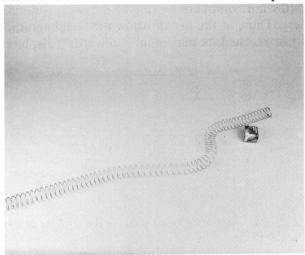

Fig. 12.3 Wave pulse on a Slinky

"Slinky" moves out and back to its original position, *i.e.* there is no net movement. If we look carefully at the "Slinky" as the wave pulse is moving along it we notice that each successive point on it moves in the same way, but at a slightly later time than the previous point. Also, if a continuous wave is passed along the "Slinky", each point moves back and forth along the same path, *i.e.* each point **oscillates** about its rest position.

From the above example we see that a wave may best be

described as a disturbance which moves through a material. The material through which the wave moves is usually called the medium. In moving through a medium, a wave transfers energy from its source to other points in the medium. It is important to realise that no part of the medium itself is transferred from one point to another — only energy is transferred. There is no net movement of the medium; the particles of the medium simply oscillate about their original position and eventually come to rest again in that position. A wave may therefore be defined as follows:

> **A wave is a means of transferring energy through a medium without any net movement of the medium.**

A wave motion can be represented diagrammatically in two ways. *Fig. 12.4* is like a photograph taken of a vibrating medium at a particular time. It shows the displacement, from its rest position, of each point in the medium at a particular instant (the horizontal line represents the rest positions of the particles). Thus, at the instant shown in the diagram, point X has just reached its maximum "upwards" displacement,

point Y is passing through its rest position and point Z has just reached its maximum "downwards" displacement and so on. *Fig. 12.5* shows the displacement of a particular point at different times. In other words, it is really a graph of displacement against time, for a particular point in the medium.

Before proceeding further with this discussion of waves, we must define some of the terms which are commonly used.

The **amplitude**, A, is the maximum displacement of a particular point, *Figs. 12.4 and 12.5*. The unit of A is the **metre**, m.

The **wavelength**, λ, is the distance between any two successive points which are **in phase**, *i.e.* two points which are moving in the same direction and have the same displacement at a given time, *Fig. 12.4*. The unit of λ is the **metre**, m.

The **period**, T, is the time taken for a particular point to undergo one complete oscillation, *Fig. 12.5*. The unit of T is the **second**, s.

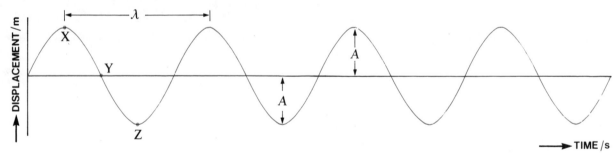

Fig. 12.4 Displacement of different points in the medium at a given time

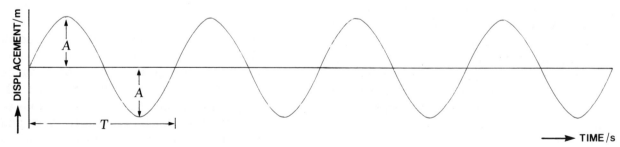

Fig. 12.5 Change in displacement of a given point with time

The **frequency**, f, is the number of complete oscillations a particular point makes in one second. The unit of f is the **hertz**, Hz. If the frequency of a wave is 10 Hz each point makes 10 complete oscillations in one second. Therefore, the time for one oscillation, *i.e.* the period, is one tenth of a second. In general,

$$f \ = \ \frac{1}{T}$$

The **speed**, c, at which a wave travels in a particular medium is the distance travelled by the energy divided by the time taken to travel that distance. The unit of speed is the **metre per second**, m s^{-1}. Since λ is the length of one wave and f is the number of waves produced per second, $f \times \lambda$ is the distance travelled by the wave in one second, *i.e.* the speed, c. That is,

$$c \ = \ f \times \lambda$$

Example
The speed of radiowaves in air is 3.0 $\times$ 10^8 m s^{-1}. Calculate the wavelength of a wave whose frequency is 6.12 $\times$ 10^5 Hz. (This is one of the frequencies on which R.T.E. Radio 2 broadcasts.)

$$c \ = \ f \times \lambda$$
$$3 \times 10^8 \ = \ 6.12 \times 10^5 \times \lambda$$
$$\Rightarrow \quad \lambda \ = \ \frac{3 \times 10^8}{6.12 \times 10^5}$$
$$= \ 490 \text{ m}$$

Ans. 490 m

In the wave motions we have considered so far, *viz.* water waves and waves on a "Slinky", the medium oscillates at right angles to the direction in which the energy is moving. Such waves are called **transverse waves.**

In some cases, the medium oscillates in the same direction as that in which the energy is moving. Such waves are called **longitudinal waves**. These too may be demonstrated with a "Slinky". In this case, the hand is moved to and fro as shown in *Fig. 12.6*. When the hand is moved to the left the first few coils of the "Slinky" are squeezed together and a **compression** is formed. This compression moves down the length of the "Slinky". When the hand is moved to the right the "Slinky" is stretched and a **rarefaction** is formed. This rarefaction follows the compression and the process is repeated. Note that a longitudinal wave may be represented by the diagram of *Fig. 12.5, i.e.* a graph showing change in displacement with time of a particular point in the medium. However, whereas in the case of a transverse wave the displacement is at right angles to the direction in which the wave is travelling, in the case of a longitudinal wave the displacement is parallel to the direction in which the wave is travelling.

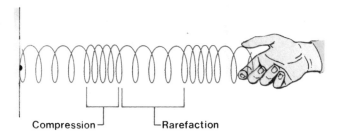

Fig. 12.6 Longitudinal wave

12.2 Diffraction

Diffraction occurs when a wavefront is distorted by an obstacle. We shall look, briefly, at two effects of diffraction.

Fig. 12.7 shows what happens when a wave meets an obstacle — the wave tends to spread into the area behind the obstacle. The extent to which a wave spreads into the area behind an obstacle depends on the wavelength of the wave, the longer the wavelength the greater the extent to which the wave spreads around the obstacle, *Fig. 12.7*.

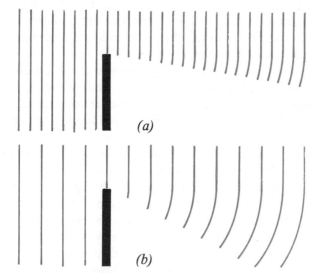

Fig. 12.7 Diffraction at the edge of an obstacle (a) wave of short wavelength, (b) wave of long wavelength

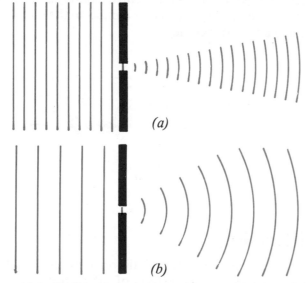

Fig. 12.8 Diffraction at an opening in an obstacle (a) wave of short wavelength, (b) wave of long wavelength

Fig. 12.8 shows the effect of diffraction at an opening in an obstacle. Here, the amount of diffraction depends on the size of the wavelength compared with the width of the opening. The longer the wavelength compared to the width of the opening the greater the extent to which the wave spreads into the area behind the obstacle. Note that the wave spreading out from the opening is similar to that which would be produced by a point source at the opening.

Fig. 12.9 Demonstrating diffraction effects with a ripple tank

12.3 Interference

What happens when two (or more) waves meet? If we send wave pulses from either end of a "Slinky" so that they meet we can observe the result. If the pulses consist of displacements in the same direction they form a larger pulse when they meet, *Fig. 12.10(a)*. If the displacements of the pulses are in opposite directions they tend to cancel each other out when they meet, *Fig. 12.10(b)*. This phenomenon is known as **interference** and these observations may be summarised in a statement which is known as the **Principle of Superposition.**

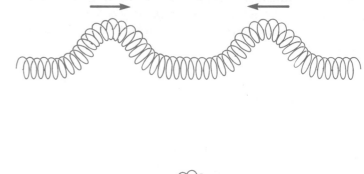

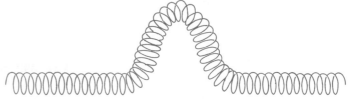

Fig. 12.10(a) Constructive interference

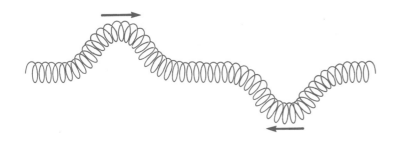

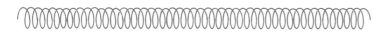

Fig. 12.10(b)   Destructive interference

When two or more wave pulses meet at a point, the resultant displacement at that point is equal to the algebraic sum of the individual displacements.

When two or more pulses add together to give a larger pulse the interference is referred to as **constructive interference**. When the pulses tend to cancel each other out the interference is known as **destructive interference**. In both cases, after they meet, the individual pulses carry on unchanged.

Interference Patterns

A particularly important case of interference occurs when waves from **coherent** sources meet.

Coherent sources are sources which have the same frequency and are in phase with each other.

Fig. 12.11 shows two waves of the same frequency (and hence the same wavelength) spreading out from two coherent sources, A and B. Suppose the red lines represent the "crests" of the waves and the blue lines the "troughs". Then it is clear that along certain lines, *e.g.* those marked X, a crest of one wave always meets a crest of the other; likewise a trough of one wave always meets a trough of the other. Therefore along the lines marked X the disturbance is always a maximum. Points on these lines are called **antinodes**. Conversely, along lines like those marked Y a crest of one wave always meets a trough of the other and *vice versa*. Therefore, along these lines the

disturbance is a minimum. Indeed, if the waves from A and B have the same amplitude, the medium along the lines marked Y will be quite still. Points on these lines are called **nodes**. Thus, while some parts of the medium are oscillating with a large amplitude other parts have a very small amplitude or are at rest. Yet other parts of the medium have a range of amplitudes between these two extremes. That is, an **interference pattern** is produced.

An interference pattern is produced when waves from coherent sources meet.

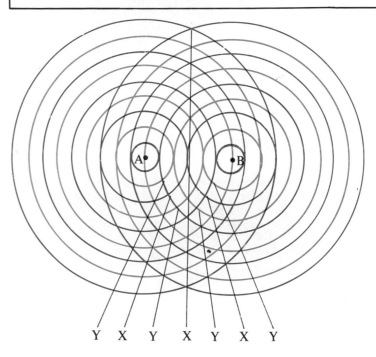

Fig. 12.11 Interference between waves from coherent sources

Stationary Waves

So far we have considered waves in which the disturbance, and the energy, move through the medium form one point to another. Such waves are called progressive waves. In another type of wave motion which is of very great importance in physics, the disturbance does not move through the medium and there is no net transfer of energy. Such waves are called **stationary waves** or **standing waves**.

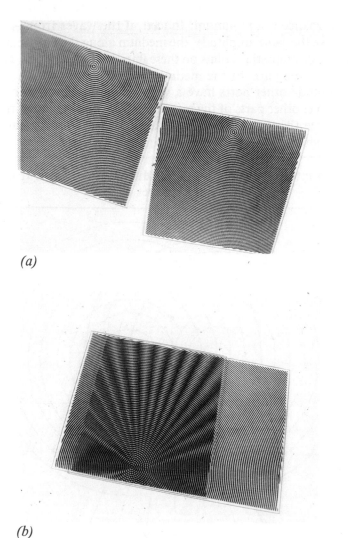

(a)

(b)

Fig. 12.12 Demonstrating an interference pattern using ring plates:
(a) single plates, (b) two plates overlapping

**A stationary wave is produced when two waves of
the same frequency and amplitude and moving in
opposite directions meet.**

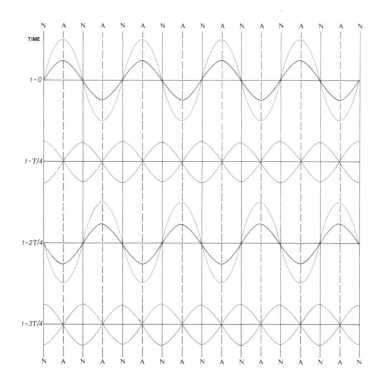

Fig. 12.13 Formation of a stationary wave

A stationary wave is thus a special case of an interference pattern. In the diagram of *Fig. 12.11*, a stationary wave would be set up along the line joining A and B, but nowhere else.

The formation of a stationary wave is illustrated in the series of diagrams of *Fig. 12.13*. Wave A (red line) is moving to the right. Wave B (green line), which has the same frequency as A, is moving to the left. Wave C (blue line) is the actual disturbance in the medium caused by the interference of A and B. The diagrams illustrate the situation at time intervals of $T/4$, where T is the period of A and B.

At $t = 0$ the two waves coincide and so constructive interference occurs at every point in the medium. At $t = T/4$, *i.e.* one quarter of a period later, the waves have moved a quarter of wavelength in opposite directions and so they are now exactly out of step — destructive interference occurs at every point. At $t = T/2$ each wave has moved half a wavelength and so they again coincide. At $t = 3T/4$ destructive interference again occurs at every point. Remembering that the blue line represents the resulting disturbance in the medium at the different times note that the points in the medium along the lines marked N are at rest at all times; these points are nodes. On the other hand, points on the lines marked A undergo the greatest displacement; these points are antinodes.

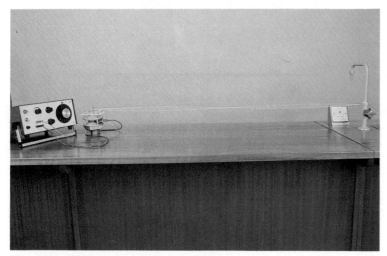

Fig. 12.14 Stationary wave on a string

The properties of stationary waves may be summarised as follows. Note how they differ from the corresponding properties of progressive waves.

1. Not all points in the medium have the same amplitude. Some (the nodes) have zero amplitude, *i.e.* they are permanently at rest. Others (the antinodes) have the maximum amplitude. Other points have amplitudes which lie between these two extremes.
2. The wavelength is twice the distance between two consecutive nodes or antinodes.
3. There is no net transfer of energy through the medium.

Beats

When two waves of almost the same frequency meet a phenomenon known as **beats** occurs. As may be seen from *Fig. 12.15*, the waves arriving at a particular point gradually fall out of step (out of phase) until, at a certain time, they cancel each other out. They are back in step when one wave is exactly one wavelength behind the other. The resultant displacement thus varies from a maximum to a minimum. The beat

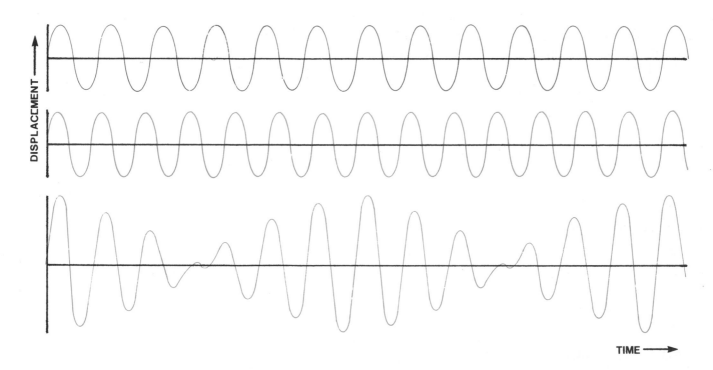

Fig. 12.15 Formation of beats

frequency can be shown to be equal to the difference between the frequencies of the two waves.

12.4 Reflection and Refraction

Waves are reflected from solid surfaces. With water waves or waves on a "Slinky", the reflected wave is difficult to see due to interference between it and the incident wave. However, careful experiment shows that waves obey the laws of plane reflection. Reflection of a single wave pulse on a "Slinky" is easily demonstrated.

When a wave, travelling in a medium in which it has a certain speed, enters another medium in which it has a different speed, its direction changes, *i.e.* it is refracted. (**Note:** There is no change in direction if the direction of the incident wave is perpendicular to the interface between the two media.)

Fig. 12.16 shows a plane wave approaching an interface between two media. Suppose that the speed of the wave is greater in the first medium. When A enters the second medium, it slows down and B, which is still in the first medium is now travelling faster than A. As a result, when B enters the second medium, the wavefront is no longer parallel to the original wavefront, *i.e.* the wave is travelling in a different direction.

Let c_1 be the speed of the wave in the first medium and c_2 be its speed in the second medium. Let t be the time taken for B to reach C. Then,

$$t = \frac{|BC|}{c_1} \quad \text{(i)}$$

In the same time A will travel some distance, say $|AD|$. Then,

$$t = \frac{|AD|}{c_2} \quad \text{(ii)}$$

Combining *Equations (i)* and *(ii)* gives

$$\frac{|BC|}{c_1} = \frac{|AD|}{c_2} \quad \text{(iii)}$$

Since

$$c_2 < c_1$$

$$|AD| < |BC|$$

i.e. the new wavefront is not parallel to the old wavefront.

The relative refractive index (*see p.23*) of the two media,

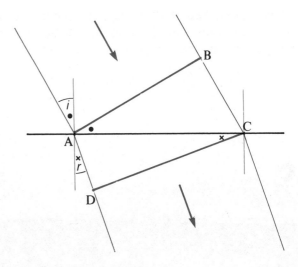

Fig. 12.16 Refraction of a wave

$_1n_2$, is defined by the equation

$$_1n_2 = \frac{\sin i}{\sin r}$$

From *Fig. 12.16*

$$\sin i = \frac{|BC|}{|AC|}$$

and

$$\sin r = \frac{|AD|}{|AC|}$$

$$\therefore \ _1n_2 = \frac{|BC|}{|AC|} \times \frac{|AC|}{|AD|}$$

$$= \frac{|BC|}{|AD|}$$

But, from *Eqn. (iii)*,

$$\frac{|BC|}{|AD|} = \frac{c_1}{c_2}$$

$$\therefore \quad \boxed{_1n_2 = \frac{c_1}{c_2}}$$

12.5 Resonance

All bodies are capable of vibrating at certain fixed frequencies called their natural frequencies. A simple example is a child's swing which, like a simple pendulum (*p.110*), has a period given by

$$T = 2\pi \sqrt{\frac{l}{g}}$$

where *l* is the length of the pendulum. In other words, the period of a simple pendulum depends only on its length, since *g* is constant. It follows that the frequency, *f*, also depends only on the length since $f = 1/T$. Therefore, if a pendulum, or a swing, of a fixed length is set in motion it will swing freely with a certain frequency. Even though the amplitude will gradually decrease as energy is "lost" due to friction the frequency will remain constant.

Now, if it is desired to keep the swing going or to make it go higher, *i.e.* to give it a greater amplitude, it must be pushed "at the right time." "At the right time" means, in fact, giving it a series of pushes which has the same frequency as the natural frequency of the swing. This phenomenon is known as **resonance** and the swing is said to resonate. Given such a series of pushes the amplitude would, in theory, increase indefinitely. (In practice, the amplitude will stop increasing when the rate at which the energy is being supplied by the pusher is equal to the rate at which it is being "lost" due to friction.)

Resonance may be demonstrated in the laboratory using an apparatus known as Barton's pendulums, Fig. 12.17. As we

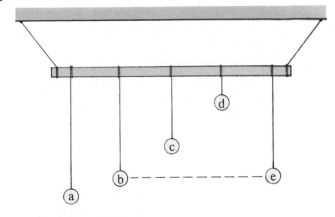

Fig. 12.17 Barton's pendulums

noted above, a simple pendulum has only one natural frequency, which depends on its length. When pendulum e is set in motion, all of the pendulums start to move but the only one to gain an appreciable amplitude is the one which has the same length as e, *i.e.* pendulum b. Energy is transferred easily between the two bodies which have the same natural frequency, *i.e.* they resonate. Resonance may thus be defined as follows.

> **Resonance is the transfer of energy between two bodies which have the same natural frequency.**

Examples of resonance occur in all areas of physics and engineering and are of great practical importance. For example, in the design of machines, care must be taken to ensure that resonance cannot occur between different parts. Similar considerations apply to structures like buildings, bridges, etc. Much of the destruction caused by the 1985 earthquake in Mexico city occured because of resonance — many of the buildings and the soil on which they were built had the same natural frequency (about 2 Hz) as the shock wave from the earthquake which had its epicentre 225 miles away. Other examples include the well-known one of a glass being shattered by a singer singing a high note. Resonance also occurs in electrical circuits, *e.g.* tuned circuits in radios and televisions.

12.6 Polarisation

If a "Slinky" is moved from left to right only the resulting wave is said to be plane polarised — the oscillation is in one plane only, *viz.* the horizontal plane. An unpolarised wave is one in which vibrations occur in all planes at the same time. It is not possible to demonstrate this with a "Slinky" but an approximation can be obtained by moving the "Slinky" in a circle. Such a wave may be plane polarised by passing it through a slit which is approximately the same width as the diameter of the "Slinky". Only the vibration parallel to the slit gets through and the resulting wave on the other side of the slit is plane polarised. If two slits are placed at right angles to each other then no vibration can pass through.

It should be clear from the above discussion that only transverse waves can be polarised — longitudinal waves cannot. A longitudinal wave on the "Slinky" will pass through a slit unaffected. If we can show that a particular wave can be polarised then we may conclude that we are dealing with a transverse wave.

The lenses of sunglasses are sometimes made from a

polarising material. These glasses work in two ways. Firstly, they reduce the brightness of the light from all sources since they only transmit light waves which are vibrating in one plane, the remainder of the light being absorbed. Secondly, they reduce "glare," *i.e.* bright light reflected from a smooth surface, *e.g.* water. The "glare" is reduced because the light is partly polarised by reflection from the surface. *(See also Chapter 13.)*

The waves used to carry television signals are also polarised. Some are polarised vertically and are received by "vertical" aerials; other are polarised horizontally and are received by "horizontal" aerials. Although all aerials will receive both types of wave the response in a horizontal aerial to vertically polarised waves is very weak. Similarly, horizontally polarised waves produce a very weak response in a vertical aerial. Using polarised waves thus prevents the wrong signals being picked up, even if the frequencies are the same.

Fig. 12.18 Television aerials

12.7 The Doppler Effect

To a person standing near the track, the pitch of the note from a train's whistle seems to drop suddenly as the train passes. The same effect is noticed with a police-car or ambulance siren, a low-flying aircraft, *etc.,* and is known as the **Doppler effect.** The change in pitch in each case is due to a change in the frequency of the sound wave *(see p.149)*. The examples quoted refer to sound waves but the effect may be observed with any type of wave.

> **The Doppler effect is the apparent change in the frequency of a wave due to the motion of the source of the wave.**

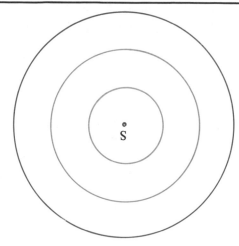

S

Fig. 12.19 Wave source at rest

Fig. 12.19 shows a wave source, S, at rest; the waves spread out equally in all directions. (The lines represent successive wavefronts, *e.g.* the crests of the waves. The red line represents the crest of the first wave produced; the green line represents the crest of the second and the blue line represents the crest of the third.) In Fig. 12.20, S is moving at constant speed towards A. It is clear from this diagram that, from the point of view of a person standing at A, the wavefronts appear to be closer together than they were when the source was stationary. The wavelength of the wave is equal to the distance between two crests so, to the person at A the wavelength seems shorter. Since the speed of the wave does not change this means that the frequency appears to be higher. (Remember: The speed of a wave is given by $c = f\lambda$.) On the other hand, to a person standing at B the wavefronts seem further apart than when the source was stationary, *i.e.* the wavelength seems longer and therefore the frequency seems lower. Thus, to a person standing on the line joining A and B, the frequency of the wave seems to drop as the source passes him/her.

The Doppler effect is used by astronomers to determine the speeds of distant galaxies by measuring the frequencies of the light emitted by them. Garda "speed traps" use the Doppler effect with microwaves to determine the speed of cars and a similar method is used to track satellites. In medicine the Doppler effect is used to investigate the performance of the

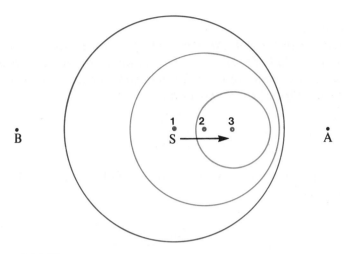

Fig. 12.20 Wave source moving

heart. Movements of the heart cause changes in the frequency of the reflected waves and these changes can then be used to calculate the speed at which the heart muscle is contracting and relaxing.

Change in Frequency

Referring to Fig. 12.20, suppose the speed of the source is v in a medium where the speed of the wave is c. Let f_s and λ_s be the frequency and wavelength of the wave emitted by the source. Let f_o and λ_o be the frequency and wavelength as they appear to an observer at A. Let T be the period of the source, *i.e. T* is the time taken to produce one complete wave. In a time T the source moves a distance vT towards A. Therefore, the distance between successive wavefronts in this direction is $\lambda_s - vT$. But, the distance between successive wavefronts is the wavelength as seen by the observer at A. That is,

$$\lambda_o = \lambda_s - vT$$

But,

$$T = \frac{1}{f_s}$$

$$\Rightarrow \quad \lambda_o = \lambda_s - \frac{v}{f_s}$$

But,

$$\lambda_o = \frac{c}{f_o} \quad \text{and} \quad \lambda_s = \frac{c}{f_s}$$

$$\therefore \quad \frac{c}{f_o} = \frac{c}{f_s} - \frac{v}{f_s}$$

$$= \frac{1}{f_s}(c - v)$$

Dividing both sides by c gives

$$\frac{1}{f_o} = \frac{1}{f_s}\left(1 - \frac{v}{c}\right)$$

or,

$$f_o = \frac{f_s}{(1 - v/c)} \qquad (iv)$$

If $\quad 0 < v < c$

$$1 - v/c < 1$$

$$\therefore \quad f_o > f_s$$

If the increase in frequency is Δf, then

$$\Delta f = f_o - f_s$$

$$= \frac{f_s}{1 - v/c} - f_s$$

$$= f_s\left(\frac{1}{1 - v/c} - 1\right)$$

$$= f_s\left(\frac{v/c}{1 - v/c}\right)$$

If $\quad v \ll c$

then $\quad \boxed{\Delta f \approx f_s\,(v/c)}$

From *Eqn. (iv)*, if $\quad v \longrightarrow c$

$$f_o \longrightarrow \infty$$

and $\quad \lambda_o \longrightarrow 0$

In other words, the wavefronts coincide and the resulting build-up of energy impedes the passage of the source. This is what happens when an aircraft breaks the "sound barrier."

When the observer is at B the source is moving away from him/her and a similar analysis shows that the apparent frequency is given by

$$f_o = \frac{f_s}{(1 + v/c)} \qquad \text{(v)}$$

If

$$0 < v < c$$

$$1 + v/c > 1$$

and

$$f_o < f_s$$

Again, as in the case of the source moving towards the observer we can show that the change in the frequency, Δf, is given by

$$\Delta f \approx f_s \frac{v}{c}$$

when $v \ll c$

Eqns. (iv) and *(v)* give us the apparent frequency when the source of the waves is moving towards the observer and when the source is moving away from the observer, respectively. These equations differ only in the sign of v in the denominator. As we did in the case of the mirror/lens formula, we can reduce these to one equation by the use of a sign convention. In this case we shall make the value of v positive if the source is moving *away* from the observer and negative when the source is moving *towards* the observer. In both situations the apparent frequency is then given by

$$f_o = \frac{f_s}{(1 + v/c)}$$

or,

$$\boxed{f_o = \frac{cf_s}{c + v}}$$

Example

A car is travelling at 20 m s^{-1} when its horn emits a note of frequency 100 Hz. What is the frequency of the note heard by a person standing near the road some distance in front of the car? (Speed of sound in air = 340 m s^{-1}.)

$$f_o = \frac{cf_s}{c + v}$$

Since the source is moving towards the observer the value of v is negative.

$$f_o = \frac{340 \times 100}{340 - 20}$$

$$= \frac{34000}{320}$$

$$= 106.25 \text{ Hz}$$

Ans. 106 Hz

Example

Microwaves are reflected from a car moving at 30 m s^{-1}. If the frequency of the waves striking the car is 1.5 x 10^{10} Hz what is the change in frequency of the reflected waves? (Speed of microwaves in air = 3.0 x 10^{8} m s^{-1}.)

$$\Delta f = f_s(v/c)$$

$$= \frac{1.5 \times 10^{10} \times 30}{3 \times 10^8}$$

$$= 1.5 \times 10^3 \text{ Hz}$$

Ans. 1.5×10^3 Hz

SUMMARY

A wave is a disturbance which moves through a medium.

In a progressive wave energy is transferred from one point to another.

A stationary wave is produced when two progressive waves of the same frequency and amplitude and travelling in opposite directions meet. In this case there is no net transfer of energy. The wavelength of a stationary wave is twice the distance between two consecutive nodes or antinodes.

Waves may also be classified as transverse (the medium vibrates at right angles to the direction in which the wave is travelling) or longitudinal (the medium vibrates parallel to the direction in which the energy is travelling).

Waves undergo reflection and refraction. They also show diffraction effects — bending around obstacles and spreading out on passing through an opening in an obstacle. The longer the wavelength the more noticeable are the diffraction effects.

When two or more waves meet at a point interference occurs, the resultant displacement being equal to the algebraic sum of the individual displacements. If the waves come from coherent sources an interference pattern is produced. Standing waves are a special case of an interference pattern.

Resonance is the transfer of energy between bodies which have the same natural frequency.

A polarised wave is one which vibrates in one plane only. Longitudinal waves cannot be polarised.

The Doppler effect is the apparent change in the frequency of a wave due to the motion of the source of the wave. When the source is moving towards the observer the apparent frequency is higher than the actual frequency and when the source is moving away the reverse is the case.

If the speed of the source, v, is very much less than the speed of the wave, c, the difference between the frequency of the wave emitted by the source, f_s, and the apparent frequency, f_o, is given by

$$\Delta f = f_s \frac{v}{c}$$

In general, the apparent frequency, f_o, is given by

$$f_o = \frac{cf_s}{c + v}$$

where v is taken to be positive when the source is moving away from the observer and negative when it is moving towards the observer.

Questions 12

DATA: Speed of light in vacuum $= 3.0 \times 10^8$ m s^{-1}.

Section I

1. What is a wave?.. ....

2. What is the relationship between the hertz and the second? ....

3. The speed of a wave is equal to its......................... multiplied by its..

4. A longitudinal wave consists of a series of............... .... and ....

5. The ability of a wave to spread around an obstacle is an effect of...

6. What condition must be satisfied in order for an interference pattern to be observed?......................... ....

7. What type of interference pattern is produced when two waves of the same frequency and amplitude and travelling in opposite directions meet?...................... ....

8. Give two differences between standing waves and progressive waves..............................

....

....

9. How are beats produced?..................................

....

....

*10. What is the relationship between the relative refractive index of two media and the speeds of a wave in the two media?...............................

....

11. What is meant by the natural frequency of a body?

....

....

12. What is meant by resonance?...........................

....

....

13. When a wave is confined to vibrating in one plane only it is said to be....................................

....

14. What is the Doppler effect?..............................

....

....

15. The frequency of the light from a distant star is observed to increase and decrease alternately. What deductions might be made from this observation?

....

Section II

1. Explain the terms: wavelength; frequency; amplitude; period. What is the relationship between (i) the frequency and the period, (ii) the frequency, wavelength and the speed?

2. Explain the following: Diffraction effects; Interference pattern; Coherent sources; Stationary waves; Resonance.

3. Explain, with the aid of a diagram, why waves change direction when they pass from one medium into another of different density.

*4. Given that light is a wave motion, calculate the speed of light in glass of refractive index 1.5.

5. Visible light has wavelengths ranging from 3.7×10^{-7} m to 7.0×10^{-7} m. What is the frequency range of visible light?

6. The frequency of the waves produced in a ripple tank is 50 Hz. If the average distance between successive crests in the water is 6.0 mm, what is the speed of the waves in the water?

7. R.T.E. Radio 2 broadcasts in the Dublin area in the medium wave band with a wavelength of 235 m and in the VHF band at a frequency of 93.5 MHz. Calculate the corresponding frequency in the MW band and the corresponding wavelength in the VHF band.

8. A wave of frequency 2000 Hz is reflected back along its own path so that a standing wave is set up. If the distance between consecutive nodes on the standing wave is 8.5 cm, what is the speed of the wave?

*9. A train is travelling at a speed of 30 m s^{-1} when its whistle emits a note of frequency 500 Hz. If the speed of sound in air is 330 m s^{-1}, what is the frequency of the note heard by a person standing near the track when the train is (i) approaching, (ii) receding?

*10. A car, travelling at a constant speed of 20 m s^{-1}, passes a person standing at the side of the road. As the car passes, the frequency of the note from its horn, as heard by the pedestrian, falls from 430 Hz to 380 Hz. Calculate the actual frequency of the note and the speed of sound in air.

13 Wave Nature of Light

13.1 The Corpuscular Theory

In Chapter 1 we learned that light is a form of energy. Using the methods of geometrical optics we have been able to learn a good deal about the behaviour of light in certain circumstances. However, while we have been able to say what light does in certain cases, we have not attempted to explain how or why it does it.

The question "What is Light", has exercised the mind of man since the beginning of recorded history and probably before. In the sixth century B.C. Pythagoras thought that when a person looked at an object invisible rays went out from his eyes and when they touched the object he had the sensation of sight. This strange idea was accepted for over two hundred years until another Greek, Epicurus, realised that objects were seen when light is reflected from them into the eye.

Newton is credited with being the first person to put forward a theory which attempted to explain the nature of light, *i.e.* to answer the question "What is light?" According to this theory, now known as the **corpuscular theory**, light is to be imagined as being made up of tiny particles which travel in straight lines from the source of the light. Newton suggested that, although these particles have mass they are too small to be seen individually but, when they strike our retina, they cause the sensation of sight.

Newton's corpuscular theory provides a satisfactory explanation for reflection. It predicts that when light is reflected, the angle of incidence should be equal to the angle of reflection. This is easily shown experimentally to be the case, as we have already seen.

The theory also provides an explanation for refraction. In this case, it predicts that light should have a higher speed in a denser medium. In Newton's time it was not clear that light had a speed other than infinity; it was certainly not possible to measure the speed of light in a medium like glass or water. So Newton had no way of checking experimentally if his explanation of refraction was satisfactory.

We now know that light does in fact travel more slowly in a denser medium, so the theory is not satisfactory in respect of refraction. Neither does it offer any explanation for diffraction effects or interference, both of which were known, though perhaps not recognised as such, in Newton's time.

In view of the above, it is not surprising to learn that the corpuscular theory is no longer used. As we have seen, it is totally inadequate in explaining most of the phenomena related to light, its only success being in explaining reflection. However, probably because Newton was so well known, the corpuscular theory was widely accepted for more than a century, not being finally abandoned until the beginning of the nineteeneth century.

It is important to realise that scientific theories are merely "models" designed by scientists to help us understand the world around us. As such, theories can never, strictly speaking, be "right" or "wrong", their value can only be judged by their usefulness. The corpuscular theory was abandoned because it was limited in its application to a very few specific cases and therefore not very useful. Other theories, as we shall see, have much wider application and are thus more useful in unlocking the secrets of Nature.

13.2 The Wave Theory

A Dutchman, Christiaan Huygens, proposed that the behaviour of light could best be explained in terms of a wave motion. The fact that his theory was largely ignored while Newton's was accepted was probably partly due, as already stated, to the prestige enjoyed by Newton and partly to Huygens' inability to design an experiment which demonstrated conclusively that light exhibited interference and diffraction effects.

In 1802 Thomas Young, an Englishman, designed an experiment which demonstrated both interference and diffraction of light. *Fig. 13.1* shows the method used by Young.

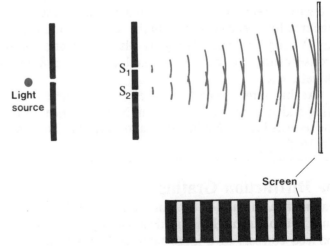

Fig. 13.1 Young's experiment

The two slits, S_1 and S_2 act as coherent sources. (They are coherent because different parts of the same wave pass through both at the same time.) The waves spread out from S_1 and S_2 due to diffraction and produce an interference pattern in the space where they overlap. This pattern, which consists of a series of bright and dark lines (antinodes and nodes respectively), may be observed on a screen placed anywhere in this space.

Not only did this experiment show conclusively that light has a wave nature but it actually provided a method of measuring the wavelength of the light. Using this arrangement Young was able to show that the wavelength of light was of the order of 10^{-7} m.

Christiaan Huygens (1629 — 1695)

Huygens was born in The Hague, the son of a Dutch government official. He started his academic life as a mathematician but later became interested in astronomy and physics.

Using improved telescopes, which he built himself, Huygens made a number of important discoveries, including the Orion Nebula and one satellite and the rings of Saturn. He also attempted to estimate the distance to the stars. His greatest invention was the first mechanical clock based on the motion of the pendulum. In 1690 he proposed that light was a wave motion rather than particles, as had been proposed by Newton. However, Newton's reputation ensured that Huygens' ideas were ignored for over a century until the issue was finally settled by Young in 1802. Huygens also did work on the conservation of momentum and the conservation of kinetic energy.

As his fame spread, Huygens spent some time in England and at the court of Louis XIV in France. In 1681 he returned to the Netherlands and he died in The Hague in 1695.

The Diffraction Grating

As a practical method of determining the wavelength of light, Young's experiment has the disadvantage that the pattern produced consists of a large number of bright lines (called fringes) which are relatively faint and very near together. It is therefore very difficult to measure accurately the distance between them. The diffraction grating reduces the number of bright lines seen to perhaps 5 or 7 which are much brighter and much further apart.

A diffraction grating consists of a large number of parallel lines ruled on a sheet of transparent material. Light cannot pass through lines but can pass through the spaces between them. These spaces then act like a series of parallel slits. The principle is the same as in Young's experiment, except that now we have several thousand slits instead of only two. This has the effect of making the bright lines produced much narrower (sharper). Also, since the slits are very much nearer together (the separation of the slits is typically of the order of 10^{-6} m), the bright lines are much further apart.

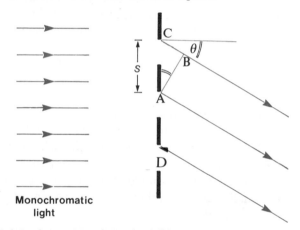

Fig. 13.2 Light passing through a diffraction grating

In order to use the diffraction grating to determine the wavelength of light we must derive an expression for the wavelength in terms of the position of the bright lines in the interference pattern. *Fig. 13.2* shows part of a diffraction grating with a beam of monochromatic light (light of one wavelength) falling normally on it. (This diagram is not drawn to scale. The distance between the slits, s, is only of the order of 10^{-6} m.) At the slits, diffraction occurs and the light spreads out from each slit as in Young's experiment (*see Fig. 13.1*) but to a greater extent since the slits are narrower. Consider the light travelling in one particular direction from the grating, *e.g.* making an angle θ with the normal to the grating, *Fig. 13.2*. As in Young's experiment the waves leaving all the slits are in phase since they were all parts of the same wave originally. However, since they have different distances to travel, they gradually fall out of phase. For the waves from two adjacent slits, *e.g.* C and A, to enter an observer's eye (or the telescope of a spectrometer, *see Experiment 13.1 on p.134*) in phase the "path difference" between the waves must be a whole number of wavelengths, *e.g.* one wavelength or two wavelengths, *etc.*, *Fig. 13.3*. In other words, for a bright line to be observed at an angle θ to the normal the path difference, $|CB|$, must equal a whole number of wavelengths. That is

$$|CB| = n\lambda \quad \text{where } n = 0, 1, 2, \text{ etc.}$$

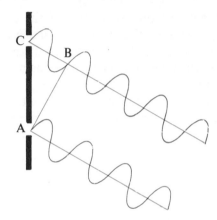

Fig. 13.3

From the diagram of *Fig. 13.2* we see that

$$\frac{|CB|}{|CA|} = \sin \theta$$

$$=> \qquad |CB| = |CA| \sin \theta$$

$$= s \sin \theta$$

where s is the distance between the slits.

Thus, the necessary condition for a bright line to be formed at an angle θ to the normal is

$$\boxed{n\lambda = s \sin \theta}$$

While we have derived this formula for two adjacent slits it holds for the waves from any pair of slits, *e.g.* C and D. You should check that this is so by repeating the above derivation for waves from slits C and D. For the first bright line on either side of the normal n has the value 1; for the second bright line $n = 2$, *etc.* n is called the **order** of the bright line.

While we have referred to bright *lines* in the interference pattern the shape of the image formed depends on the shape of the source. If the source is a rectangular slit, *e.g.* the slit in the collimator of a spectrometer, the images will be lines. If the source is a laser the images will be dots, *etc.* If the light used is not monochromatic but contains a number of wavelengths a series of bright lines will be formed for each value of *n. (See Dispersion, p.136).*

Example

Light of wavelength 6.0×10^{-7} m is passed through a diffraction grating which has 500 lines per mm ruled on it. At what angles to the normal will bright lines be formed?

$$s = \frac{1}{500}$$

$$= 2.0 \times 10^{-3} \text{ mm}$$

$$= 2.0 \times 10^{-6} \text{ m}$$

$$n\lambda = s \sin \theta$$

$$=> \qquad \sin \theta = \frac{n\lambda}{s}$$

$$= \frac{n \times 6 \times 10^{-7}}{2 \times 10^{-6}}$$

$$= n \times 3 \times 10^{-1}$$

$$= 0.3 \times n$$

(i) $\qquad n = 0$

$=> \qquad \sin \theta = 0$

$=> \qquad \theta = 0$

(ii) $\qquad n = 1$

$=> \qquad \sin \theta = 0.3$

$=> \qquad \theta = 17° \ 27'$

(iii) $\qquad n = 2$

$=> \qquad \sin \theta = 0.6$

$=> \qquad \theta = 36° \ 52'$

(iv) $\qquad n = 3$

$=> \qquad \sin \theta = 0.9$

$=> \qquad \theta = 64° \ 9'$

(v) $\qquad n = 4$

$\qquad => \qquad \sin \theta = 1.2$

Since no angle can have a sin greater than 1 a fourth order image does not exist.

Ans. Bright lines are formed on the normal and at angles of $17° \, 27'$, $36° \, 52'$ and $64° \, 9'$ on either side of the normal, *i.e.* 7 bright lines in all.

Thomas Young (1773 — 1829)

Young, born in Somerset, England, was an infant prodigy. He could read at the age of two and by the age of four had read the Bible twice. As he grew up he learned a dozen languages and could play a variety of musical instruments. At university he studied medicine, obtaining his degree in 1796.

Young discovered how the lens of the eye changes shape to enable us to focus on objects at different distances. He also discovered the reason for astigmatism (blurred vision resulting from light rays entering the eye not being brought to a common focus). He was the first to suggest the idea of there being only three primary colours. In 1807 he was the first to use the word energy in its present sense. However, he is best remembered for demonstrating conclusively the wave nature of light. He did this in 1802 by showing firstly that light was diffracted at a narrow opening and, secondly, that when light waves from two narrow slits overlapped an interference pattern was produced. In 1817 he proposed that light waves were transverse rather than longitudinal.

Young's work met with considerable opposition in England, mainly because it contradicted Newton's particle theory. However, the results of his experiments could not be ignored and, with the support of a number of French physicists, the wave theory came to be accepted.

In addition to his work as a physicist, Young was an acknowledged expert on the ancient language of the Egyptians. He also contributed to developments in biology. He died in London in 1829.

EXPERIMENT 13.1

To Measure the Wavelength of Light

Apparatus: Spectrometer, sodium lamp, diffraction grating.

Procedure:

1. Adjust the telescope of the spectrometer for parallel light by focussing it on a distant object, *e.g.* a tree. Adjust the position of the eyepiece until the image of the crosswires coincides (no parallax) with the image of the distant object. Refocus the telescope if necessary.
2. Place the lamp in front of the slit of the collimator and adjust the position of the slit until a sharp image of it is seen in the telescope. Adjust the width of the slit until the image seen in the telescope is as narrow as possible without being too faint.
3. Place the diffraction grating on the table of the spectrometer so that it is at right angles to the collimator.

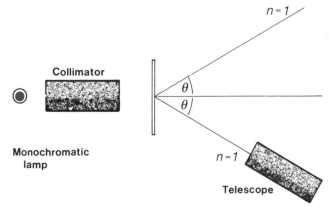

Fig. 13.4 Using a spectrometer to measure the wavelength of light

4. Rotate the telescope to the left, counting the number of bright lines from the centre, until the cross wires are on the last bright line. Note the order, n, of this line and the reading, θ_L, on the scale. Moving the telescope to the right, repeat the measurements for each of the bright lines. Angles to the right of the central line are θ_R.
5. The difference between θ_L and θ_R for each value of n is 2θ (see Fig. 13.4). Read the number of lines per mm from the grating and calculate s.
6. Calculate a value for λ from $n\lambda = s \sin \theta$ for each value of n. Calculate an average value of λ.

RESULTS

$s = $ m.

n	$\theta_L/°$	$\theta_R/°$	$\theta/°$	$\lambda/$m

Wavelength of sodium light =   m.

Questions

1. Why is the telescope first focussed on a distant object?
2. What is the effect of the width of the slit on the accuracy of the result?
3. If the table of the spectrometer is not level what is the effect on the images seen in the telescope?
4. What determines the number of images seen in the telescope?

Polarisation

Having established that light is a wave motion we must now ask if the waves are transverse or longitudinal.

Polaroid is a material which consists of long molecules arranged parallel to each other (the molecules may be imagined to form a series of parallel slits). When light is passed through two sheets of polaroid onto a screen, *Fig. 13.5*, it is found that, as one of the sheets is rotated, the brightness of the light on the screen varies from zero in one position to a maximum when the sheet is turned through a further 90°. We can imagine that when the "slits" in the polaroid are parallel to each other the light gets through but when the "slits" are at right angles to each other the light cannot get through. From this we may conclude that light can be polarised and hence that light is a transverse wave motion (*see p.125*).

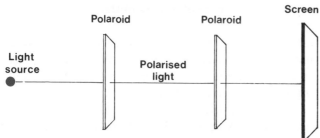

Fig. 13.5 Demonstrating polarisation of light

Polaroid is used in sun-glasses, as explained in *Chapter 12*, to reduce the brightness of the light and to cut out the "glare" produced when light is reflected from a horizontal surface, *e.g.* a pool of water. This light is polarised in a horizontal plane and so cannot pass through the polaroid which is arranged so that it transmits only light waves which are vibrating in a vertical plane.

13.3 The Speed of Light

The first person to deduce that light has a finite speed was a Danish astronomer, Ole Romer (1644 -1710). While observing the planet Jupiter at the Royal Observatory in Paris, he noticed that the period of one of Jupiter's moons appeared to vary from time to time. Further investigation showed that as the earth moved away from Jupiter, the period seemed to increase and that as the earth approached Jupiter, the period seemed to decrease. Since it seemed very unlikely that the speed of the moon changed in this way Romer concluded that as the earth moved away from Jupiter the light had further to travel and hence the period seemed to increase because it took light from the moon longer to reach the earth. In 1676 Romer announced that, according to his calculations, the speed of light was 2.25×10^8 m s^{-1}. This is about 25% smaller than the correct value but it was rather good for a first attempt!

The first successful terrestrial method of determining the speed of light was devised by a French physicist, Armand Fizeau (1819 — 1896). The value obtained by Fizeau was nonetheless 10 times too small. In 1850 another French physicist, Jean Foucault (1819 — 1868), devised a method with which he succeeded in determining the speed of light, not only in air but also in water. His value for the speed in air was within 1% of the presently accepted value and he succeeded in showing that the speed was less in water. This discovery was important because it provided further evidence that light is a wave motion rather than particles as predicted by Newton (*see p.131 and p.124*).

Foucault's method, with some improvements, was used by an American physicist, Albert A. Michelson (1852 — 1931). Michelson made a number of measurements of the speed of light between 1878 and his death in 1931. (The results of his last experiment were not in fact published until some years after his death.) The experiment described here is one he carried out in 1927.

Michelson first measured accurately the distance between

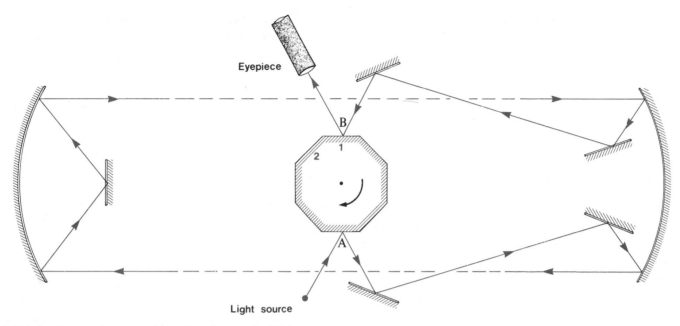

Eyepiece

Light source

Fig. 13.6 Michelson's experiment to determine the speed of light

the peaks of two mountains, Mt. San Antonio and Mt. Wilson, in California. He then placed a concave mirror on each mountain with a small octagonal mirror between them, *Fig. 13.6*. The octagonal mirror was made of steel and could be rotated at very high speeds. The light followed the path shown from the source to the eyepiece. When the mirror is rotating the light will enter the eyepiece only if side 2 moves to position B in the time it takes light to travel from A to B. Michelson found that in order to satisfy this condition, the mirror had to be rotating at 528 rev s^{-1}. This meant that the time for one revolution was 1/528 s and the time taken for side 2 to move to position B was 1/(528 × 8) s or 0.24 ms. This then was the time taken for the light to travel from A to B. The speed was then found by dividing the total distance travelled by the time taken.

The speed of light has been determined a number of times by different methods and with increasing accuracy since the 1930's. In 1982 it was decided to *define* the speed of light in a vacuum to be 2.997 924 58 × 10^8 m s^{-1}. This means that the unit of length, the metre, can now be defined in terms of the speed of light. Thus, one metre is the distance travelled by light in a vacuum in 1/299 792 458 s.

The speed of light in a vacuum is a fundamental constant in physics. We have already met it in Einstein's famous equation, $E = mc^2$ (*see p.91*) and it also occurs in a number of other areas, particularly in atomic physics.

13.4 Dispersion

In 1666 Newton discovered that when a narrow beam of sunlight (white light) entered a prism a wide beam containing all the colours of the rainbow emerged at the other side, *Fig. 13.7*. Newton correctly interpreted this as evidence that white light was a mixture of the other colours. (This contradicted the generally accepted belief at the time that white light was indivisible, *i.e.* could not be broken up into something simpler.) He confirmed his theory by using a second prism to bring the colours together again, forming white light, *Fig. 13.8*. The phenomenon discovered by Newton is known as **dispersion.**

> **Dispersion is the breaking up of white light into its constituent colours.**

Dispersion is caused by the fact that, while all colours have the same speed in air, each has a different speed in glass (and other media), *e.g.* red light travels faster in glass than does violet light. Since refractive index depends on speed *(see p. 124)* the glass has a different refractive index for each colour, *i.e.* each colour is bent by a different amount on entering and leaving the glass.

The range of colours produced when white light is passed

Fig. 13.7 Dispersion by a prism

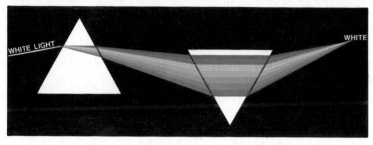

Fig. 13.8 Recombination of colours to give white light

through a prism is called the **visible spectrum**. Using a prism on its own as shown in *Fig. 13.7* the colours overlap. A pure spectrum, *i.e.* one in which the colours do not overlap, may be produced by using two converging lenses with the prism as shown in *Fig. 13.9*.

Primary and Secondary Colours

Over one hundred years after Newton's experiments with prisms, Thomas Young put forward the theory that only three colours were necessary to produce white light and that any shade of colour could be produced by mixing these colours in suitable proportions. This theory is now known as the three colour theory. The three colours are red, blue and green and they are now called the **primary colours** of light.

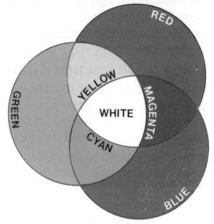

Fig. 13.10 Mixing primary colours

When two primary colours are mixed a **secondary colour** is produced. Thus, red and green give yellow; green and blue give cyan and red and blue give magenta, *Fig. 13.10*. A primary colour and a secondary colour which together give white light are called **complementary colours**, *e.g.* blue and yellow are complementary colours, *Fig. 13.10*. The other complementary colours are red and cyan and green and magenta.

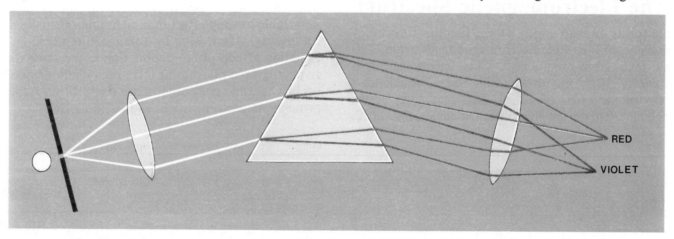

Fig. 13.9 Production of a pure spectrum

Primary Colours	Secondary Colours	Complementary Colours
Red Blue Green	Cyan Yellow Magenta	Red, Cyan Blue, Yellow Green, Magenta

Colour photography and colour television both involve applications of the three colour theory. For example, for colour television each scene is filmed in each of the three primary colours. The final picture on the screen is then produced by recombining the three colours in their original proportions.

Dispersion Using a Diffraction Grating

When light of one colour (monochromatic light) is passed through a diffraction since grating the position of the image formed depends on the wavelength of the light used since $n\lambda = s \sin \theta$ where λ is the wavelength and θ is the angular displacement of the image from the normal to the grating (*see p.133*). When white light is passed through a grating a visible spectrum is produced for each value of n, *Fig. 13.11*. In other words, the images of different colours are formed in different positions. Therefore, we see that different colours have different wavelengths. From the photograph of *Fig. 13.11* we see that red has the longest wavelength (the red image is formed furthest from the centre) and violet has the shortest wavelength (the violet image is formed nearest the centre). The range of wavelengths in the visible spectrum is found by experiment to be from 4×10^{-7} m (violet) to 7×10^{-7}m (red). The other colours have wavelengths which lie between these two extremes.

13.5 The Electromagnetic Spectrum

The visible spectrum with which we are familiar forms only a very small part of the whole electromagnetic spectrum. This spectrum extends from gamma rays, which have wavelengths of the order of 10^{-12} m, up to radio waves, which have wavelengths of the order of 10^3 m. Different parts of the electromagnetic spectrum are given different names, *Fig. 13.12*,

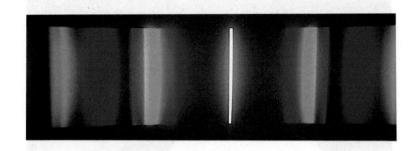

Fig. 13.11 Dispersion by a diffraction grating. Only part of the second order spectrum on either side is shown

but all the radiations have a common source (oscillating electric charges) and all travel at the same speed in a vacuum (3.0×10^8 m s^{-1}). They differ only in their wavelength and frequency.

The radiation to which the human eye responds is called **visible** radiation or, simply, **light**. As we saw earlier, light has wavelengths between 4×10^{-7} m s^{-1} and 7×10^{-7} m s^{-1}. Light is emitted as a result of the movement of electrons in atoms when, for example, a substance is heated. We shall learn more about this process in *Chapter 26*.

Infra-Red Radiation

Infra-red radiation is the name given to radiation which has wavelengths just longer than light, *Fig. 13.12*. The infra-red region extends up to about 10^{-3} m. Like light, infra-red radiation is emitted from hot bodies, although those emitting mainly infra-red are generally at lower temperatures than those emitting light. Also like light, infra-red results from movements of electrons within atoms. Infra-red radiation produces a more noticeable heating effect than light and is usually detected by its heating effect, *e.g.* by using a thermometer. Infra-red radiation affects photographic film in the same way as light. In photographs taken using infra-red radiation hot object show up bright while cold objects are dark, *Fig. 13.13*. A similar principle is used in "night sights" on rifles.

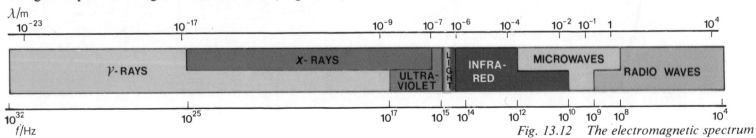

Fig. 13.12   The electromagnetic spectrum

Fig. 13.13 Infra-red photograph.
(*Courtesy of Dr. E. O'Mongáin. U.C.D.*)

Ultra-Violet Radiation

At the other end of the visible spectrum from infra-red is ultra-violet radiation. Ultra-violet radiation also arises from electron movements within atoms of hot bodies, although in this case the bodies are generally at higher temperatures than those which emit mainly light. Wavelengths of ultra-violet radiation range down to about 10^{-8} m.

Ultra-violet radiation causes certain substances, *e.g.* Vaseline, to **fluoresce**. This means that the Vaseline absorbs the ultra-violet radiation and re-emits the energy as light. Fluorescent substances are sometimes included in washing powders to make clothes appear brighter. Particles of the fluorescent substance remain in the material after washing and emit white light when exposed to ultra-violet radiation. The effect is most noticeable in a disco where there is a high proportion of ultra-violet radiation. A similar idea is used as a security measure by some financial institutions, *e.g.* building societies. When a customer opens an account he/she signs a card with a special ink which is invisible in white light but fluoresces under ultra-violet radiation. When the customer wishes to make a withdrawal the signature on the withdrawal form can be compared with the signature on the card using an ultra-violet lamp.

Ultra-violet radiation is present in the sun's radiation and although much of it is absorbed by the earth's atmosphere, some reaches the surface of the earth where it produces sun-tan and sun-burn in people exposed to it. Large amounts of ultra-violet radiation may initiate skin cancer. Ultra-violet radiation does not pass through ordinary glass, so it is not possible to get a sun-tan in a greenhouse! Ultra-violet lamps use a special type of glass (quartz glass) which is transparent to ultra-violet radiation.

Heinrich Hertz (1857 — 1894)

Hertz was born in Hamburg and studied physics at the University of Berlin under, among others, Gustav Kirchhoff (*see p.224*). In 1888 Hertz discovered that an oscillating electric current produced electromagnetic radiation. (This was as had been predicted by the Scottish physicist and mathematician, James Clerk Maxwell, 20 years earlier.) Hertz determined the wavelength of these waves to be 66 cm and, when Marconi used them as a form of "wireless" communication, they came to be known as radio waves. In the course of these experiments Hertz noticed that ultra-violet radiation falling on the negative terminal of his "radio transmitter" improved its efficiency. Hertz did not pursue the investigation of this phenomenon but it was, in fact, the first observation of what has come to be known as the photoelectric effect (*see p.288*)

Hertz died in Bonn at the age of 36 from chronic blood poisoning. As a tribute to his work his name is used for the unit of frequency.

13.6 Spectra

As we have seen in the previous section when white light from a lamp is passed through a diffraction grating, a spectrum is obtained. This is an example of an **emission** spectrum — it results from the light emitted from a particular body. There are three types of emission spectrum. If the source of the light is a hot solid, *e.g.* the filament in a lamp, or a hot liquid, the spectrum is a **continuous** spectrum, *Fig. 13.14*. In a continuous spectrum all the wavelengths in a given range are present. If the source is a monatomic gas, *e.g.* neon, a **line** spectrum is obtained, *Fig. 13.15*. Remember that the position of the image produced by a diffraction grating depends on the wavelength of the light used ($n\lambda = s \sin \theta$). Therefore, each line in the spectrum corresponds to a different wavelength. If the source

Fig. 13.14 Continuous spectrum (emission)

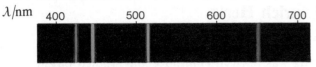

Fig. 13.15(a) Line spectrum (emission)

is a polyatomic gas, *e.g.* oxygen, a **band** spectrum is obtained, *Fig. 13.16*. This type of spectrum consists of a series of lines arranged in groups or bands.

Since the wavelengths of the radiation emitted in a given case are characteristic of the source (*see p.291*), spectra can be used to identify unknown substances or to establish the

Fig. 13.16(a) Band spectrum (emission)

presence of a particular element in a sample. One advantage of this method of analysis is that only very small quantities are required. It is also useful in cases where the source is inaccessible. For example, helium was discovered in the sun's atmosphere from an examination of the sun's spectrum.

Spectra may also be produced when light is absorbed by a substance. Such spectra are called **absorption** spectra and, like emission spectra, they may be divided into three types — continuous, line and band. For example, when white light is directed onto a red solid and the reflected light is passed through a diffraction grating a continuous absorption spectrum is obtained. Such a spectrum might look like the spectrum in *Fig. 13.14* with the blue, green and yellow parts missing, these colours having been absorbed by the red solid. When white light is passed through a monatomic gas a line absorption spectrum is produced, *Fig. 13.17*. The dark lines correspond to wavelengths which have been absorbed by the gas. These

Fig. 13.15(b) Spectrum of iron   (Courtesy of Dr. S. O'Connor, U.C.D.)

Fig. 13.16(b) Spectrum of air   (Courtesy of Dr. S. O'Connor, U.C.D.)

Fig. 13.17 Line spectrum (absorption)

wavelengths would be precisely the same as those emitted by the same gas if it were heated. (Note that, for an absorption spectrum to be observed, the absorbing substance must be at a lower temperature than the source of the white light. For an explanation of how emission and absorption spectra are produced see *Chapter 26*.)

Fraunhofer Lines

One particularly important absorption spectrum is that of the sun. A close examination of the sun's spectrum shows the familiar "rainbow" of Newton's experiments to be crossed by a large number of dark lines. These lines were discovered by an English scientist, William Wollaton (1766 - 1828) in 1802. From 1814 the German physicist, Joseph Von Fraunhofer (1787 — 1826), investigated the lines, determining their position and hence their wavelength. In recognition of this work the lines are now known as Fraunhofer lines. While Fraunhofer discovered several hundred lines, some 25 000 are known today. They are due to absorption of certain wavelengths of the sun's radiation by the effectively cooler gases, mainly hydrogen and helium, in the sun's atmosphere. In fact, helium was discovered in the sun's atmosphere in this way before it was known on earth. (The word helium comes from the Greek word *helios*, meaning the sun.)

SUMMARY

The first theory of the nature of light was the corpuscular theory put forward by Newton. This held that light consisted of tiny particles travelling in straight lines. While

this theory explained reflection, it did not explain refraction, interference or diffraction effects.

The wave theory, proposed by Huygens, also explained

reflection and refraction and predicted that light should show interference and diffraction effects. When these were demonstrated experimentally by Young in 1802 the wave theory was generally accepted. An improved version of Young's experiment involves the use of a diffraction grating (a piece of transparent material with a very large number of lines ruled on it) to determine the wavelength of light. Using a diffraction grating, the wavelength, λ, is given by the formula

$$n\lambda = s \sin \theta$$

where s is the distance between the lines on the grating, θ is the angular displacement of the images from the normal to the grating and n has the values 0, 1, 2, 3, *etc.* n is called the order of the image.

Light can be polarised, showing it to be a transverse, rather than a longitudinal, wave motion.

The speed of light may be determined by reflecting light off a many-sided mirror rotating at high speed. The speed of light in a vacuum is a fundamental constant in physics and has the value 3.0×10^8 m s^{-1}.

Dispersion is the breaking up of white light into its constituent colours, each of which has a different wavelength and travels at a different speed in glass and other transparent media. (All colours travel at the same speed in a vacuum.) The wavelength range of the visible spectrum is from 4×10^{-7} m to 7×10^{-7} m.

The primary colours are red, blue and green. When mixed together these give white light. The mixing of two primary colours gives a secondary colour *(see Fig. 13.10, p.137)*. A primary colour and a secondary colour which together give white light are called complementary colours.

Infra-red radiation has longer wavelengths than light and has a more noticeable heating effect. Ultra-violet radiation has shorter wavelengths than light; it causes fluorescence in certain substances and does not pass through ordinary glass. It also causes sun-tan and sun-burn. Like light, infra-red and ultra-violet radiation are emitted from hot bodies; the hotter the body the shorter the average wavelength of the radiation emitted.

Emission spectra are formed when light is emitted from a hot substance; absorption spectra result when certain wavelengths are absorbed by cooler bodies. Spectra from solids and liquids are continuous; line spectra arise from monatomic gases; band spectra arise from polyatomic gases.

Fraunhofer lines are dark lines on the sun's spectrum caused by absorption in the sun's atmosphere.

Questions 13

DATA: Speed of light in vacuum (air) = 3.0×10^8 m s^{-1}.

Section I

1. State two contributions made by Thomas Young to the development of physics..

2. What is a diffraction grating?............................... ....

3. Would it be possible to use a number of light sources to produce an interference pattern? Explain.............. ....

4. What is the distance, in metres, between the lines on a grating which has 300 lines per mm?......................

5. In the formula $n\lambda = s \sin \theta$, n represents .... The value of n is always ....

6. How do we know that light is a transverse wave?.....

7. What is meant by dispersion?...............................

8. Name the primary colours and the secondary colours.

9. What are complementary colours? Name the three pairs of complementary colours..................................

*10. For what is the American physicist, A. A. Michelson chiefly remembered?...

*11. Why is it important that the value of the speed of light be known accurately?...

12. Give two properties which are common to all the types of radiation which make up the electromagnetic spectrum ....

13. The radiations on either side of light in the spectrum are called.. ....

14. Will a thermometer, placed beyond the violet end of the visible spectrum, show a rise in temperature? Explain.

....

....

15. Name the three types of emission spectrum..............

....

16. What are Fraunhofer lines?................................

....

Section II

1. Light is a form of energy which is transmitted by a transverse wave motion. What evidence is there to support this statement?

2. A beam of white light passes through a red filter and a diffraction grating and onto a screen. What would be the effect, if any, on the separation of the images of:

   (i) moving the screen nearer the grating;
   (ii) reducing the separation of the lines on the grating;
   (iii) moving the light source nearer the grating;
   (iv) replacing the red filter with a blue one?

3. Monochromatic light is passed through a diffraction grating having 500 lines per mm. The angle between the first order image and the zero order is found to be $15°$. What is the wavelength of the light?

4. Monochromatic light of wavelength 4.76×10^{-7} m is passed through a diffraction grating which has 6000 lines per cm and is placed on the table of a spectrometer. When the telescope is focussed on the zero order image the reading on the scale is $0°$. Calculate the angular positions of the first, second and third order images.

5. What is the difference between diffraction and dispersion? In what way does the spectrum of white light produced by a prism differ from that produced by a diffraction grating?

6. State the adjustments which must be made to a spectrometer before it is used with a diffraction grating to measure the wavelength of light. In such an experiment, the diffraction grating has 500 lines per mm and a total of five bright lines are observed. These are at angles of $50° 4'$, $68° 25'$, $85° 32'$, $102° 44'$ and $121° 6'$, respectively. Calculate the wavelength of the light.

7. A parallel beam of light of wavelength 590 nm falls normally on a diffraction grating which has 500 lines per mm. What is the highest order image which may be seen?

8. A helium lamp emits light of wavelength 590 nm and 670 nm. This light is allowed to fall normally on a diffraction grating which has 600 lines per mm. Calculate the angular separation of the first order images.

9. A monochromatic lamp, a metre stick and a diffraction grating are arranged as shown in *Fig. I*. The lamp is just above the 50 cm mark on the metre stick. The distance from the metre stick to the grating is 2.4 m and the grating has 300 lines per mm. Given that the wavelength of the light is 650 nm calculate the positions on the metre stick where the first order images would be seen.

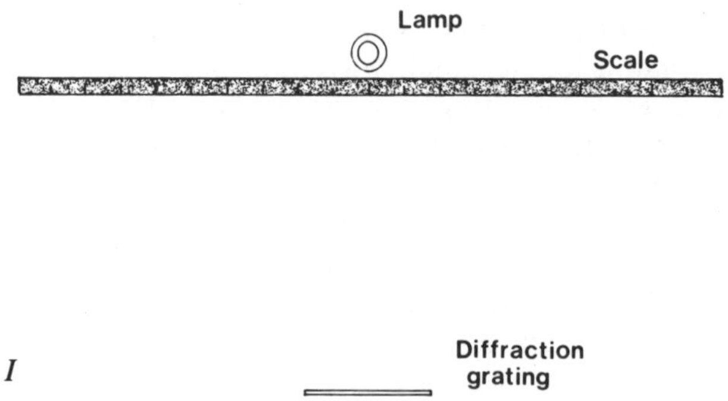

Fig. I

10. Give an expression for the refractive index of glass in terms of the speed of light in vacuum and in the glass. A beam of light has a wavelength λ as it travels through air. It enters a block of glass of refractive index n. Establish an expression, in terms of λ and n, for the change in the wavelength of the light as it enters the block.

*11. The refractive index of a certain type of glass is 1.45. What is the wavelength in the glass of electromagnetic radiation of frequency 4.53×10^{14} Hz?

*12. In the experiment described on *p136* for determining the speed of light the eight-sided mirror rotated at a speed of 528 rev s^{-1}. Taking the speed of light to be 3.0×10^8 m s^{-1}, what was the total distance travelled by the light?

*13. In one experiment to measure the speed of light Michelson used a 32-sided mirror. If the total distance travelled by the light was 3 km, what was the minimum number of revolutions made by the mirror in one second?

14 Sound

14.1 Wave Nature of Sound

Sound, like light is a form of energy. All sounds have as their source something which is vibrating, *e.g.* a guitar string. If you place your fingers on your throat, you can feel your vocal chords vibrate as you speak. Likewise, if you lightly touch a drum which has just been struck you will feel the vibration.

Since sounds are caused by vibrating or oscillating bodies we might immediately suspect that sound is carried by a wave motion. How can we verify that this is the case? If sound is carried by a wave we should be able to show interference and diffraction effects.

Fig. 14.1 Interference of sound waves using a tuning fork

To demonstrate interference we need a pair of coherent sources, *e.g.* a tuning fork. Strike a tuning fork and rotate it slowly beside your ear, *Fig. 14.1*. You will hear the loudness of the sound increasing and decreasing in a regular way — a typical interference effect. Interference may also be demonstrated using two identical loudspeakers connected to the same signal generator, *Fig. 14.2*. (A signal generator produces an electrical signal of a particular frequency. A loudspeaker converts an electrical signal to a sound wave of the same frequency.) A person walking along the line XY will hear the loudness of the sound increasing and decreasing as he/she moves along the line. If one of the loudspeakers is

disconnected the effect disappears — the loudness of the sound along the line is uniform.

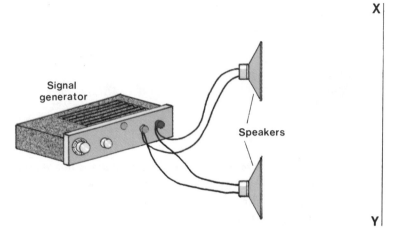

Fig. 14.2 Interference of sound waves using loudspeakers

No special equipment is required to demonstrate diffraction effects with sound. It is well known that sound travels around corners — there is no "sound shadow" behind a building for example.

That sound is carried by a longitudinal, rather than a transverse, wave may be demonstrated by placing a lighted candle in front of a loudspeaker which is connected to a signal generator which is producing a low-frequency signal. The flame waves to and fro in the direction in which the sound is travelling. Indeed, it is not even necessary to have the candle — it is usually possible to see the diaphragm of the loudspeaker move in and out.

Transmission of Sound

To understand how a sound wave travels through a medium, *e.g.* air, consider the situation shown in *Fig. 14.3*. When the drum is set vibrating the drumskin moves from right to left and back. When the drumskin moves to the right the air molecules near to it are squeezed together; a compression is formed which moves away from the drum. When the drumskin moves to the left the air molecules are free to spread out and a rarefaction is formed. When the wave reaches our ear, a compression causes the eardrum to move in and a rarefaction allows it to spring out. Thus, the eardrum vibrates at the same frequency as the drum which made the sound. The movement of the eardrum is converted into an electrical signal of the same frequency. This signal is then carried to the brain via the auditory nerve, giving the sensation of hearing.

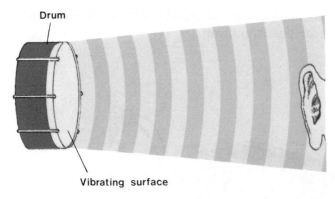

Fig. 14.3 Sound waves are longitudinal (This diagram is not to scale)

From the previous paragraph it should be clear that sound requires a medium through which to travel. (This was first established by Robert Boyle around 1660, *see p.163.*). Sound cannot travel through a vacuum since there are no molecules to vibrate. This can be demonstrated using the apparatus illustrated in *Fig. 14.4*. Initially, the bell is set ringing with the bell-jar full of air. The sound can be heard clearly. When the air is removed from the bell-jar with a vacuum pump the sound can no longer be heard, although the bell can be seen to continue to work. This shows that light can travel through a vacuum (we can see the bell) but sound cannot; sound requires a medium in which to travel.

Fig. 14.4 Sound does not travel in a vacuum

Refraction and Reflection of Sound

It is common experience that sounds can be heard clearly over greater distances on a cold night than during the day. This is

due to refraction of the sound waves. At night, the temperature of the ground falls faster than that of the air, thus the temperature of the air increases with height above the ground. Sound travels more slowly in colder air so the sound is continuously refracted towards the ground, *Fig. 14.5*. (*Cf. the formation of a mirage, p.30*)

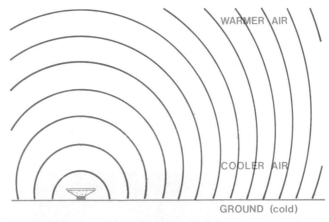

Fig. 14.5 Refraction of sound waves

Refraction of sound in the laboratory may be demonstrated using the arrangement shown in *Fig. 14.6*. Carbon dioxide is more dense than air and, since sound travels more slowly in a denser gas, the balloon acts as a converging lens, bringing the sound to a focus. (The microphone converts the sound wave to an electrical signal of the same frequency. An oscilloscope is a device, rather like a television, which displays an electrical signal as a trace on a screen. The wave pattern on the screen is thus a representation of the original sound wave.)

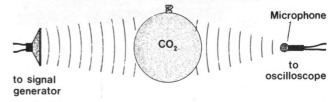

Fig. 14.6

When sound strikes a surface it is reflected in the same way as light. Everyday examples of this are echoes. An echo is simply sound reflected from, for example, a tall building or a cliff. Reflection of sound in the laboratory may be investigated using the arrangement shown in *Fig. 14.7*. Sound from a small loudspeaker travels down a tube and is reflected from a wall or screen. The position of the second tube which gives a maximum trace on the oscilloscope is determined. The angles of incidence and reflection are measured and, as in the case of light, are found to be equal.

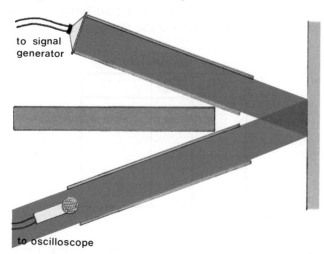

Fig. 14.7 Reflection of sound waves

14.2 Harmonics

We learned in *Chapter 12* that all bodies have certain natural frequencies at which they can vibrate. Resonance occurs when a body is set vibrating at one of its natural frequencies by another body already vibrating at that frequency (*see p.125*). Resonance involving sound waves is easily demonstrated using two tuning forks of the same natural frequency. When one is set vibrating the other, placed close to it, starts to vibrate as well. Energy is transferred from one to the other by means of sound waves.

While we shall deal here with harmonics and standing waves in relation to sound only, it is worth noting that similar principles may be applied to problems encountered in the transmission of light through optical fibres and in radar systems.

Stationary Waves on a String

When a string, or other body, vibrates at one of its natural frequencies a stationary wave exists in it. The simplest stationary wave which can exist in a string fixed at both ends (*e.g.* a guitar string) is one which has a node at either end and an antinode in the middle, *Fig. 14.8(a)*. When the string is vibrating in this way it is said to vibrate in its fundamental mode. This mode of vibration is also called the first **harmonic.**

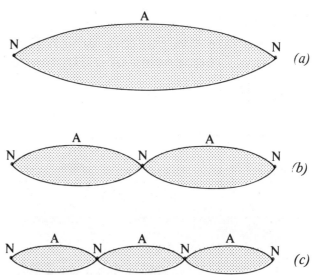

Fig. 14.8 Stationary waves on a string

From the diagram we see that the wavelength of the wave is equal to twice the length of the string. Unlike a swing or a simple pendulum, a stretched string has more than one natural frequency. For example, a string can also vibrate in such a way that there is a node at both ends and one in the middle, *Fig. 14.8(b)*. This is the second simplest mode of vibration and is called the second harmonic or first **overtone.** In this case the wavelength is equal to the length of the string. Thus the wavelength of the second harmonic is half the wavelength of the fundamental. Since the speed of the wave does not change this means that the frequency of the second harmonic is twice the frequency of the fundamental ($c = f\lambda$, so, if the wavelength is halved the frequency is doubled). The third harmonic (second overtone) is shown in *Fig. 14.8(c)*. Again from the diagram we can see that the frequency of the third harmonic is three times the frequency of the fundamental. In theory, an infinite number of harmonics is possible, each harmonic being a multiple of the fundamental.

> **Harmonics are frequencies which are multiples of a fundamental frequency.**

A stretched string can vibrate with any of these frequencies or, most likely, with a number of them at the same time. It is partly due to the presence of these overtones that a note of a given frequency sounds different when played on different instruments.

EXPERIMENT 14.1

To Investigate the Factors which Determine the Natural Frequencies of a Stretched String

Method 1:

Apparatus: Sonometer, selection of tuning forks, (weights), small block of wood.

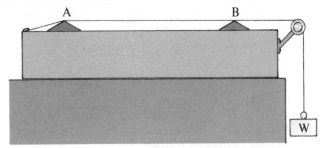

Fig. 14.9 Sonometer

Procedure:

1. Set the tension at some intermediate value. (In some versions of the apparatus the tension is adjusted by means of a key, in others the wire is passed over a pulley and weights are hung from the end of it.)
2. Strike a tuning fork on the block of wood (held in the hand) and place it on the box of the sonometer. Adjust the length of the wire by moving the bridge until the wire resonates with the tuning fork. (Resonance may be detected more easily by placing a small piece of paper near the centre of the wire.) Measure the length of the wire between the bridges.
3. Repeat for tuning forks of different frequency and plot a graph of frequency against $1/l$.

RESULTS

f/Hz	l/m	$\frac{1}{l}$/m^{-1}

4. Keeping the length fixed adjust the tension until resonance is obtained between the wire and a tuning fork as described in Step 2 above. Record the tension (weight).
5. Repeat with tuning forks of different frequency and plot a graph of frequency against the square root of the tension.

RESULTS

f/Hz	T/N	$\sqrt{T}$/N$^{\frac{1}{2}}$

Questions

1. Why should the piece of paper be placed near the middle of the wire to detect resonance?
2. What factor, other than the two investigated, affects the natural frequency of the wire?

Method 2:

Apparatus: Sonometer, signal generator, two leads, (weights), horseshoe magnet. (*For the theory of this method see p.256*)

Procedure:

1. Set the tension at some intermediate value. Connect the ends of the wire to the low resistance terminals of the generator (these are sometimes marked "speaker"). Ensure that the wire does not get warm by turning down the amplitude control on the generator if necessary and/or connecting a suitable resistor in series with the wire.
2. Place the magnet half way between the bridges with one pole on either side of the wire.
3. Adjust the frequency of the generator until the wire resonates *(see Step 2 above)*. Record the frequency and the length.
4. Repeat for different lengths of wire and plot a graph of frequency against $1/l$.

RESULTS

f/Hz	l/m	$\frac{1}{l}$/m^{-1}

5. Keeping the length fixed repeat for different tensions. Record the length and the tension in each case and plot a graph of frequency against the square root of the tension.

RESULTS

f/Hz	T/N	$\sqrt{T}$/N^{½}

Questions

1. Why should the magnet be placed near the middle of the wire?
2. Why is it important that the wire should not become warm?

In the previous experiment we saw that the natural frequency of a wire depends on its length, l, and its tension, T. It may also be shown that the natural frequency depends on the mass per unit length, μ, of the wire. *Fig. 14.10* shows how the frequency depends on each of these factors when the remaining two are kept constant.

From these graphs we can see that

$$f \propto \frac{1}{l}$$

$$f \propto \sqrt{T}$$

$$f \propto \frac{1}{\sqrt{\mu}}$$

Combining these results we have

$$f \propto \frac{1}{l}\sqrt{\frac{T}{\mu}}$$

or,

$$f = \frac{k}{l}\sqrt{\frac{T}{\mu}} \qquad (i)$$

where k is a constant.

We saw earlier that if a wire can vibrate with a certain frequency, say f, then it will also vibrate with frequencies $2f$, $3f$, etc. We would therefore expect that k in *Eqn. (i)* can have a series of values, all of which are multiples of a single basic value. This is in fact the case. It may be shown experimentally that

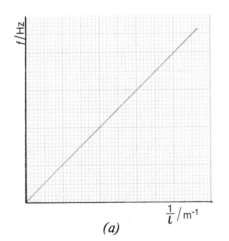

(a)

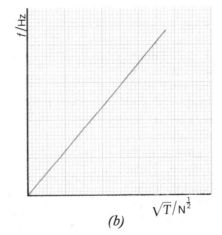

(b)

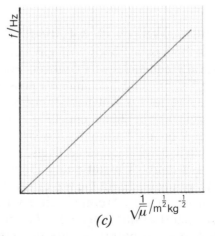

(c)

Fig. 14.10 Relationship between frequency and (a) length, (b) tension, (c) mass per unit length

$$k = \frac{n}{2}$$

where $n = 1, 2, 3$, *etc.*

Thus, we have that the natural frequencies of a wire are given by

$$f = \frac{n}{2l} \sqrt{\frac{T}{\mu}}$$

When $n = 1$, f is the fundamental frequency or first harmonic. $n = 2$ gives the second harmonic, and so on.

Example
What is the fundamental frequency of a stretched wire whose effective length is 1.00 m and whose mass per unit length is 1.25×10^{-4} kg m^{-1}, if the tension in the wire is (i) 20 N, (ii) 80 N?

(i)
$$f = \frac{n}{2l} \sqrt{\frac{T}{\mu}}$$

$$= \frac{1}{2} \sqrt{\frac{20}{1.25 \times 10^{-4}}}$$

$$= 200 \text{ Hz}$$

(ii) Since $f \propto \sqrt{T}$, if the tension is increased by a factor of 4, f is increased by a factor of 2.

$$\therefore \quad f = 400 \text{ Hz}$$

Ans. 200 Hz; 400 Hz

Stationary Waves in a Closed Pipe
Like solid bodies, a "body" of gas, *e.g.* air in a pipe, also has natural frequencies of vibration. For a closed pipe, *i.e.* one closed at one end, the simplest possible stationary wave which may be set up in it has a node at the closed end and an antinode at the open end, *Fig. 14.11(a)*. This is the fundamental or first harmonic. *Fig. 14.11(b)* and *(c)* show the third and fifth harmonics, respectively. (**Note:** It is not possible to draw a picture of a stationary longitudinal wave. These diagrams simply represent the waves.) As in the case of strings a column of air may vibrate with a number of its natural frequencies at the same time. Note that, in a closed pipe, only the odd-numbered harmonics are present.

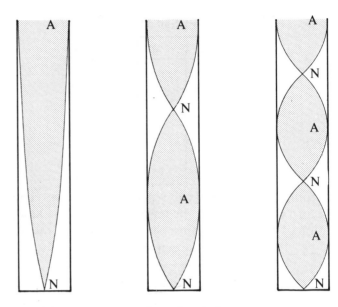

Fig. 14.11 *Stationary waves in a closed pipe*

Stationary Waves in an Open Pipe
A pipe which is open at both ends must have an antinode at both ends. Therefore, the simplest possible stationary wave is as shown in *Fig. 14.12(a)*. The second and third harmonics are shown in *Fig. 14.12(b)* and *(c)*, respectively. Note that, unlike a closed pipe, all the harmonics may be present in an open pipe.

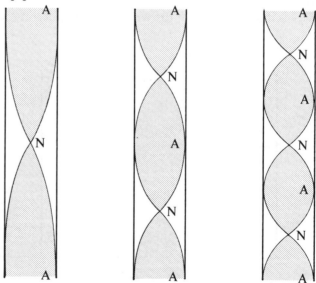

Fig. 14.12 *Stationary waves in an open pipe*

It should be noted that in the case of both open and closed pipes the antinodes at the open ends do not, in fact, coincide with the end of the pipe but are slightly outside the pipe. It may be shown experimentally that the distance from the end of the pipe to the position of the antinode is approximately equal to 0.3 times the diameter of the pipe.

14.3 Musical Notes

Practically everyone can recognise the difference between a noise and a musical note. (There might, however, be some disagreement between people of different generations!) Physically, a **noise** is a wave made up of a large number of unrelated frequencies superimposed on each other. A **note** is a wave made up of either a single frequency or a relatively small number of superimposed frequencies which are multiples of some fundamental frequency, *i.e.* harmonics, *Fig. 14.13*.

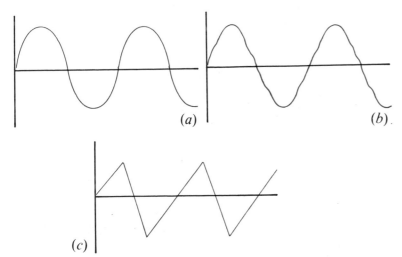

Fig. 14.13 A note of a certain frequency from (a) a tuning fork (no overtones), (b) a piano, (c) a violin

Two notes may differ in one or more of three ways. These are **loudness, pitch** and **quality** (or timbre).

Loudness depends on the amount of energy carried by the wave. This in turn depends on the amplitude of the wave. Loudness also depends on the frequency of the wave and on the sensitivity of the listener. The human ear is most sensitive to notes which have a frequency of around 1000 Hz.

Pitch depends mainly on the frequency of the wave — the higher the frequency the higher the pitch. Middle C on the physical scale has a frequency of 256 Hz. The human ear responds to frequencies in the range 20 Hz to 20 kHz,

approximately. These values are referred to as the limits of audibility. The upper limit decreases with age and few adults can hear sounds of frequency above about 16 kHz.

Quality is what distinguishes a note played on one instrument from a note of the same frequency and loudness played on a different instrument. The quality of a note depends on a number of factors. One of these is the number of harmonics present — the more harmonics the richer and fuller the note.

Musical Instruments

The piano, guitar and violin are examples of musical instruments which depend on the vibrations in stretched strings to produce musical notes. Vibrations in the strings are started in different ways, depending on the instrument. For example, in a piano a note is produced when a string is struck by a small hammer. In the case of the guitar the vibrations start when a string is plucked. For the violin vibrations are set up by drawing the bow over the strings. The piano contains a complete set of strings of fixed length and tension. In other words, each string produces a single note. The guitar and violin have only a few strings of variable length. The effective length may be varied by the player placing his/her fingers at different points on the strings. Each string may thus be used to give a variety of notes.

The organ, trombone and tin whistle are examples, among many, of wind instruments, *i.e.* instruments which produce notes from vibrating air columns. Referring to *Figs. 14.11* and *14.12* we can see that the frequency of the note emitted from such instruments depends on the length of the air column — the shorter the length the shorter the wavelength and therefore the higher the frequency. In an organ, *Fig. 14.14*, a set of pipes of different length is used. In other wind instruments the effective length of the pipe is changed, *e.g.* in the case of the tin whistle the effective length is changed by the player placing his/her fingers over holes in the side of the instrument. An instrument consisting of an open pipe, *e.g.* the trombone, gives a richer note than one consisting of a closed pipe, *e.g.* the tin whistle, since only the odd harmonics are present in the closed pipe.

14.4 Speed of Sound

The first recorded attempt at measuring the speed of sound was made by a French physicist, Pierre Gassendi, in 1635. He measured the time between seeing the puff of smoke from a

Fig. 14.14 Pipe organ

cannon and hearing the "bang". He assumed, correctly, that the time taken by light to travel a short distance was negligible and, dividing the distance between the cannon and himself by the time, he arrived at a value of 478 m s^{-1} for the speed of sound in air.

This value was soon recognised to be too high. In 1750, a more careful experiment in Paris gave a value of 332 m s^{-1} for the speed of sound at 0 °C. This value is accurate to within 1%. Approximately 50 years later, at the beginning of the nineteenth century, the speed of sound in water and in an iron pipe were determined and found to be approximately four times and sixteen times, respectively, the speed of sound in air.

The fact that the speed of a sound wave depends on the nature of the medium through which it is travelling is made use of in the search for oil and natural gas. A controlled explosion is set off at a particular point in the area under investigation. The speeds of the resulting shock waves in different directions are then determined. A knowledge of the speeds helps geologists to establish the nature of the rock formations in the area, thus enabling them to predict the locations of possible oil and/or gas deposits.

EXPERIMENT 14.2
To Measure the Speed of Sound in Air
Method 1:

Apparatus: Resonance tube, large graduated cylinder, selection of tuning forks, small block of wood, retort stand.

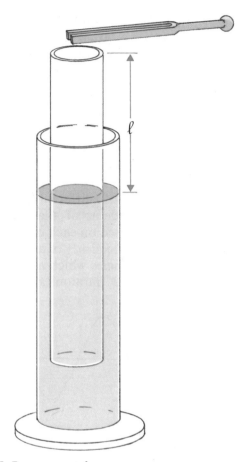

Fig. 14.15 Resonance tube

Procedure:
1. Fill the graduated cylinder with water. Clamp the resonance tube lightly so that it is resting on the bottom of the graduated cylinder.
2. Take the tuning fork of highest frequency and strike it on the block of wood (held in the hand). Holding the vibrating tuning fork over the resonance tube, slowly raise the tube until it resonates with the tuning fork.
3. Measure the distance, l, from the top of the water in the graduated cylinder to the top of the resonance tube. Measure the diameter, d, of the resonance tube. The wavelength, λ, of the sound is equal to $4(l + 0.3d)$ and the speed is calculated from $c = f\lambda$, where f is the frequency of the tuning fork.
4. Repeat with tuning forks of different frequency and calculate an average value for the speed of sound in air.

RESULTS

Diameter of tube =   m.

f/Hz	l/m	λ/m	c/m s^{-1}

Questions

1. Explain how the expression for the wavelength, $\lambda = 4(l + 0.3d)$, is obtained.
2. What are the practical factors which determine the length of the resonance tube?
3. Suggest a method of determining the wavelength if the diameter of the tube could not be measured.

Method 2:

Apparatus: Signal generator, loudspeaker, microphone, oscilloscope, reflector (large board or wall), metre stick.

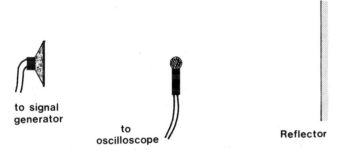

to signal generator

to oscilloscope

Reflector

Fig. 14.16

Procedure:

1. Connect the loudspeaker to the low resistance terminals of the generator (these are sometimes marked "speaker"). Set the frequency of the generator to 1 kHz.
2. Place the loudspeaker 1 — 2 m from the reflector and parallel to it.
3. Connect the microphone to the input of the oscilloscope. Switch on the signal generator and adjust the time base (time/div switch) of the oscilloscope until several waves are seen on the screen.
4. Move the microphone slowly between the loudspeaker and the reflector and mark the positions of the microphone for which the trace on the screen is a minimum. Measure the average distance, d, between these positions.
5. The speed of the sound is calculated from $c = 2df$, where f is the frequency of the signal from the generator. Repeat for different values of the frequency and calculate an average value for c.

RESULTS

f/Hz	d/m	λ/m	c/m s^{-1}

Questions

1. Explain the principle on which this experiment is based.
2. What is the effect of adjusting the distance between the loudspeaker and the reflector? Explain.
3. What are the practical limits on the range of frequencies which can be used in this experiment.

Speed of Sound in Media Other than Air

We noted above that the speed of sound in iron was sixteen times its speed in air. The methods given above for determining the speed of sound in air are not readily adaptable to determining its speed in other media so an alternative method is required. The method usually adopted uses an apparatus known as Kundt's tube, *Fig. 14.17*. This consists of a long glass tube which is stopped at one end by a plunger, P, and has a moveable piston attached to a long metal (or wooden) rod at the other end. The bottom of the tube is covered with a layer of fine, dry dust. The rod is clamped at its mid-point and when it is rubbed with a rosined cloth it vibrates longitudinally in its fundamental mode, *i.e.* with an antinode at either end and a node at the centre. The wavelength of the sound in the rod is thus twice the length, l, of the rod. A sound wave travels down the tube and is reflected at P. P is then adjusted until a stationary wave is set up in the tube, *i.e.* the air in the tube resonates with the rod. The dust in the tube then moves away from antinodes and settles at the nodes. The wavelength of the sound in the gas in the tube is then twice the average distance, d, between two successive piles of dust. The speed of sound in the gas is then given by

151

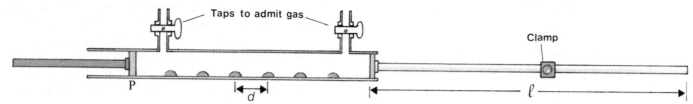

Fig. 14.17 Kundt's tube

$$c_g = f\lambda_g = 2df$$

and the speed of sound in the rod is given by

$$c_r = f\lambda_r = 2lf$$

where f is the frequency of the sound in both the rod and the gas.

Combining these two equations gives

$$\frac{c_g}{c_r} = \frac{2df}{2lf}$$

$$\boxed{\frac{c_g}{c_r} = \frac{d}{l}}$$

Thus, if c_g is known c_r may be calculated, or *vice versa*. Since the speed of sound in air may be determined by other methods, the experiment is normally first carried out with the tube full of air. This allows c_r to be found. The experiment may then be repeated with other gases in the tube or with rods of other materials.

EXPERIMENT 14.3
To Measure the Speed of Sound in a Gas/Solid

Apparatus: Kundt's tube, lycopodium powder, retort stand, a cloth, some rosin, metre stick.

Procedure:
1. Make sure that the tube is dry. Sprinkle some of the powder evenly along the bottom of the tube.
2. Place the piston in the end of the tube and clamp the rod firmly at its mid point, *Fig. 14.17*.
3. Put some rosin on the cloth and rub the rod hard with it. The rod should emit a high pitched note. Adjust the position of the plunger (while rubbing the rod) until the powder gathers in piles as shown in the diagram.
4. Find the average distance, d, between two successive piles

of dust by measuring the distance between the first and, say, the eleventh. Find also the length of the rod. The speed of sound in the rod may be calculated from $c_a/c_r = d/l$, where c_a is the speed of sound in air, c_r is the speed of sound in the rod and l is the length of the rod.

RESULTS

Distance from first pile to eleventh pile...	=	m
Average distance between two piles	=	m
Length of rod	=	m
Speed of sound in rod	=	m s^{-1}.

5. To measure the speed of sound in gas, fill the tube with the gas using the taps provided and repeat procedure given above. The speed of sound in the rod is now known and so the speed of sound in the gas may be calculated.

Questions

1. Explain why the dust gathers into piles.
2. What is the purpose of putting rosin on the cloth used to rub the rod?
3. What assumption is made in the determination of the wavelength of the vibration in the rod?

14.5 Ultrasonics

We noted earlier that vibrations between approximately 20 Hz and 20 kHz stimulate the human ear, giving the sensation of hearing. These audible vibrations are called sounds. Vibrations above 20 kHz are called ultrasonics. The ears of certain animals and birds are receptive to ultrasonic vibrations up to certain frequencies. A dog's ears are sensitive to frequencies up to about 35 kHz, hence the use of "silent" dog whistles which emit notes of frequencies between 20 and 35 kHz. Some bats are sensitive to frequencies up to 80 kHz, while whales are known to respond to frequencies as high as 200 kHz.

Sonar

A ship may determine its distance from the sea-bed, a cliff or an iceberg by sending out a high frequency wave and measuring the time taken for it to return. This method of determining distances is known as sonar and it was first used during World War I to detect submarines. A similar method is used for mapping the sea-bed, finding sunken wrecks and detecting the presence of whales and shoals of fish.

Uses in Medicine

The applications of ultrasonics in medicine are many and varied. Ultrasonic waves are used in place of X-rays where the use of the latter might be dangerous, *e.g.* in examining the brain or in examining the uterus of a pregnant woman.

In dentistry, teeth may be cleaned using ultrasonic waves — small particles, in the form of a slurry, are vibrated against the teeth at frequencies of 20 — 30 kHz.

Medical instruments may be cleaned by immersing them in a solvent irradiated by an ultrasonic beam of frequency from 20 kHz to 1 MHz. This method of cleaning is much more thorough than traditional methods.

14.6 Intensity of Sound

The intensity of a sound is a measure of the amount of energy passing a particular point. **Sound intensity**, I, is defined as follows.

> **The sound intensity at a point is the rate at which energy is crossing unit area perpendicular to the direction in which the sound is travelling.**

The unit of sound intensity is the watt per metre squared, W m^{-2}.

Example

A loudspeaker, fixed to a wall, is producing sound at a rate of 0.2 mW, i.e. the loudspeaker is emitting 0.2 mJ of energy per second. Assuming that the sound spreads out equally in all directions and ignoring reflections from walls, etc., calculate the sound intensity at a point 1.2 m from the loudspeaker.

Imagine the loudspeaker surrounded by a hemisphere of radius 1.2 m, *Fig. 14.18*. Then, in each second, 0.2 mJ of energy cross the surface of the hemisphere. The area of the hemisphere is

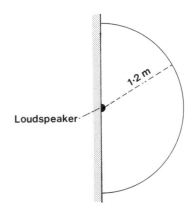

Fig. 14.18

$$A = \tfrac{1}{2} \times 4\pi(1.2)^2$$
$$= 9.05 \text{ m}^2$$

Therefore, the amount of energy crossing unit area in one second, *i.e.* the sound intensity, is

$$I = \frac{0.2 \times 10^{-3}}{9.05}$$
$$= 2.21 \times 10^{-5} \text{ W m}^{-2}$$

Ans. 2.2×10^{-5} W m^{-2}

The amounts of energy emitted by most common sources of sound are very small. For example, the intensity of normal conversation is about 10^{-6} W m^{-2} or 1 μW m^{-2}. The human ear can respond to sound intensities between 10^{-12} W m^{-2} and 1 W m^{-2}. The lower value is called the **threshold of audibility** or the **threshold of hearing**. Intensities above 1 W m^{-2} cause pain and may permanently damage the ear.

The ear responds to relative increases in sound intensity rather than actual increases. In other words, the loudness of a particular sound depends on what other sound the person has just heard. If you were sitting alone in a quiet house at night and someone tapped on the window it would seem very loud! If the same thing happened during the day when the radio was on or there was traffic passing you probably would not hear it. (The same is true of other nervous responses. If someone were to stick a pin in your arm while you were reading a book it would be very painful. However, if your finger was being held in very hot water when the pin was stuck in your arm you would not feel it at all.) Taking a quantitative example, a change in intensity from 1 μW m^{-2} to 2 μW m^{-2} would be easily noticed, while a change from 10 μW m^{-2} to 11 μW m^{-2}

153

would go undetected; to obtain the same apparent increase from $10\ \mu W\ m^{-2}$ the actual intensity would have to be increased to $20\ \mu W\ m^{-2}$. In other words, the apparent increase in intensity, the increase in loudness, depends on the ratio of the intensities rather than the difference between them.

The relative increase in intensity, *i.e.* the ratio of one intensity to another, is measured in bels, B, named after Alexander Graham Bell (1847 — 1922), the inventor of the telephone. If the intensity of one sound is 10 times the intensity of another then the difference in their intensities is said to be one **bel**.

> **The relative increase in intensity from I_1 to I_2 is 1 B if $I_2 = 10\ I_1$.**

If $I_2 = 100\ I_1$, *i.e.* $10 \times 10\ I_1$, the relative increase in intensity is 2 B. In general, if $I_2 = 10^n\ I_1$ the relative increase in intensity is n B. If

$$\frac{I_2}{I_1} = 10^n$$

$$\log_{10} \frac{I_2}{I_1} = n$$

In other words,

> $$\text{No. of bels} = \log_{10} \frac{I_2}{I_1}$$

In practice, the bel is a rather large unit and so changes in sound intensity are usually measured in **decibels**, dB. $1\ B = 10\ dB$. Therefore,

> $$\text{No. of decibels} = 10 \log \frac{I_2}{I_1}$$

Example

The sound intensity in a room increases from $10^{-8}\ W\ m^{-2}$ to $10^{-4}\ W\ m^{-2}$ when a vacuum cleaner is switched on. What is the relative increase in sound intensity in dB?

$$\text{No. of bels} = \log \frac{I_2}{I_1}$$

$$= \log \frac{10^{-4}}{10^{-8}}$$

$$= \log 10^4$$

$$= 4$$

$$= 40\ dB$$

Ans. 40 dB

The bel scale, being a logarithmic scale, has the advantage that a large range of intensities can be represented by fairly small numbers. We saw earlier that audible sounds vary in intensity from $10^{-12}\ W\ m^{-2}$ to $1\ W\ m^{-2}$, *i.e.* by a factor of 10^{12}. This corresponds to a change of 12 B or 120 dB.

Intensity Level

In the last section we saw how apparent changes or differences in sound intensity could be expressed. We did not say how the apparent intensity of a particular sound could be determined. This is like saying that one piece of string is ten times as long as another without being able to say what length either piece is. In the case of length we choose a particular length as our basic unit and call it a metre. Then a piece of string which is four times as long as our basic unit is said to be 4 m long. For apparent sound intensity we take as our basic unit the threshold of audibility at a frequency of 1 kHz. This is taken to have the value $10^{-12}\ W\ m^{-2}$. The intensity level of a sound is then expressed relative to this value. Using the bel scale, the intensity level of a sound whose actual intensity is I is given by

> $$\text{Intensity Level} = \log \frac{I}{10^{-12}}$$

Example

The sound intensity near a busy street is $4.0 \times 10^{-5}\ W\ m^{-2}$. What is the intensity level in decibels?

$$\text{Intensity Level} = \log \frac{I}{10^{-12}}$$

$$= \log \frac{4 \times 10^{-5}}{10^{-12}}$$

$$= \log (4 \times 10^7)$$

$$= \log (4) + 7$$

$$= 7.60\ B$$

$$= 76.0\ dB$$

Ans. 76 dB

Summary

Sound is a form of energy which is transmitted by a longitudinal wave motion in a material medium — sound cannot travel through a vacuum. Interference of sound waves may be demonstrated using a tuning fork or two matched loudspeakers.

A stretched string and a column of air have a number of natural frequencies of vibration. The lowest of these is called the fundamental and the others, which are whole number multiples of the fundamental, are called overtones. All frequencies which are whole number multiples of the fundamental are called harmonics. In an open pipe only the odd harmonics are present so the first overtone is the third harmonic.

The factors which determine the natural frequencies of a stretched string are its length, l, tension, T and mass per unit length, μ. The natural frequencies of a stretched string are given by

$$f = \frac{n}{2l} \sqrt{\frac{T}{\mu}}$$

where n has the values 1, 2, 3, *etc*. This relationship may be verified using a sonometer and a range of tuning forks. The frequency of the wire is determined by obtaining resonance between it and a tuning fork of known frequency.

The three characteristic properties of musical notes are: loudness, pitch and quality. These depend primarily on the amplitude and frequency, the frequency and the number of overtones, respectively.

The speed of sound in air may be found by setting up a stationary wave between a loudspeaker and a wall and finding the distance between successive nodes using a microphone and an oscilloscope. Speed is given by $c = f\lambda$. The speed of sound in air may also be found using a resonance tube. The speed of sound in a gas or a solid may be found using Kundt's tube:

$$\frac{c_{\text{rod}}}{c_{\text{gas}}} = \frac{l}{d}$$

where l is the length of the rod and d is the distance between piles of dust in the tube.

The human ear can respond to vibrations with frequencies in the approximate range 20 Hz to 20 kHz. Vibrations with higher frequencies are called ultrasonics and are audible to certain animals. They are used in sonar and in medicine.

The intensity of a sound is the amount of energy crossing unit area (normal to the direction in which the sound is travelling) per second. The unit of sound intensity is the W m^{-2}.

The apparent increase in intensity (the increase in loudness) depends on the ratio of the intensities. If the new intensity is ten times the old intensity then the relative increase is 1 B.

$$\text{No. of bels} = \log_{10} \frac{I_2}{I_1}$$

A decibel is one tenth of a bel, *i.e.* 1 B = 10 dB. Intensity level, in bels, of a sound whose actual intensity is I, is given by

$$\text{Intensity Level} = \log_{10} \frac{I}{10^{-12}}$$

10^{-12} W m^{-2} is the threshold of audibility (hearing) at a frequency of 1 kHz. It is the lowest intensity to which the human ear will respond.

Questions 14

DATA: Speed of sound in air $= 340$ m s^{-1}
Speed of sound in water $= 1500$ m s^{-1}.

Section I

1. In order to show that sound is a wave motion it is necessary to demonstrate.....................................
and/or ....

2. Sound waves are..
....

3. Give three differences between sound waves and light waves. ....
....
....

4. What are harmonics?..
....
....

5. Stationary waves are produced when......................
....
....

6. What happens when two tuning forks of almost the same frequency are sounded together?..........................
....

7. State the factors which determine the natural frequencies of a wire...
....

8. In what way do the harmonics in a closed pipe differ from those in an open pipe of the same length?..............
....

9. Give the three characteristics of a musical note. In each case give the physical property of the wave on which the characteristic depends..
....
....

10. Why does the pitch of the note from an ambulance siren change as the ambulance passes?..........................
....
....

11. Why is there normally a delay between seeing a flash of lightning and hearing the thunder? What determines the duration of this delay?.......................................
....
....

12. What are ultrasonics?...
....
....

*13. The sound intensity at a point is defined as.............
....
....

*14. The unit of sound intensity is................................
....

*15. The unit of intensity level is the..........................
or the...

Section II

1. How would you verify experimentally that sound is a wave motion?

2. Two matched loudspeakers are connected to a signal generator and are placed facing each other 1.00 m apart. A microphone, connected to an oscilloscope, is moved along a line between the loudspeakers. Describe how the trace on the oscilloscope screen changes as the microphone is moved and explain what is happening.

3. If, in the previous question, the distances moved by the microphone between successive maxima are 8.2 cm, 8.7 cm, 8.2 cm, 8.6 cm and 8.3 cm, respectively, calculate the wavelength of the sound. If the signal generator is set at a frequency of 2000 Hz, calculate the speed of the sound in the air.

4. Calculate the refractive index of water for sound waves. Compare the value obtained with that for light waves. Will a sound ray entering water from air be bent towards the normal or away from it?

5. A tube of length 20 cm, closed at one end, vibrates at its fundamental frequency. What is the frequency of the note emitted? What is the frequency of the first overtone from such a pipe?

6. If the pipe in the previous question were open at both ends, what would be the frequencies of the fundamental and the first overtone?

7. A fishing trawler sends out an ultrasonic wave of frequency 30 kHz and 24 ms later it detects the reflected wave. If the wave is reflected from a shoal of fish, calculate the depth of the shoal.

8. In a Kundt's tube experiment, the distance between the first and tenth piles of dust was 79.2 cm and the length of the metal rod was 1.32 m. What was the ratio of the speed of sound in the gas to the speed of sound in the rod?

9. The fundamental frequency of a sonometer wire is 240 Hz. Calculate its fundamental frequency when (i) its length is doubled, the tension being kept the same, (ii) the tension is doubled, the length being kept the same. State the third factor on which the fundamental frequency depends and give the relationship between this quantity and the fundamental frequency.

10. A sonometer wire has a cross-sectional area of 0.10 mm^2 and an effective length of 80 cm. The tension in the wire is 24 N and it is made of steel of density 7.8×10^3 kg m^{-3}.

Calculate (i) the mass per unit length of the wire, (ii) the fundamental frequency of vibration.

11. The fundamental frequency of a wire was determined for a series of values of its length, the tension being kept constant. The following results were obtained.

l/m	0.2	0.3	0.4	0.5	0.6	0.7	0.8	0.9
f/Hz	540	364	268	218	184	154	138	120

Draw a graph showing that the frequency is inversely proportional to the length and, from the graph, determine the ratio of the tension to the mass per unit length of the wire.

12. Describe how you would determine the frequency of an unmarked tuning fork using a sonometer and a number of tuning forks of known frequency.

*13. The sound intensity in the average home is about 10^{-9} W m^{-2}. What is the intensity level in decibels? If the sound intensity is doubled what is the change in the intensity level?

*14. A small loudspeaker emits sound uniformly in all directions at a rate of 0.12 mW. What is the sound intensity at a point (i) 20 cm, (ii) 40 cm, from the loudspeaker? What is the intensity level at each of these points?

*15. The intensity level in a factory is 85 dB. What is the sound intensity? If, when a large machine is switched off, the sound intensity drops by a factor of 5, what is the new intensity level?

15 Temperature

15.1 Temperature Scales

Everyone has a general idea of what is meant by the terms "hot" and "cold". Most people would agree that boiling water is very hot and freezing water is very cold. However, water which might feel cool to a person in an overheated room would probably feel quite warm to a person who has spent some time outside on a frosty night (*cf. p.153*). For scientific work we obviously need something more definite than "feeling hot" or "feeling cold". The physical quantity associated with "hotness" is called temperature.

> **The temperature of a body is a measure of how hot it is.**

In order to measure temperature we need a unit in which to measure it. The first standard temperature scale was defined by the Swedish astronomer, Anders Celsius in 1742 and, in a slightly modified form, is the scale most widely used today. On the **Celsius scale** the freezing point of pure water is given the value of 0 degrees Celsius (0 °C) and the boiling point of water is given the value of 100 degrees Celsius (100 °C). These two temperatures are called the lower and higher **fixed points**, respectively.

Measuring Temperature

To establish a temperature scale we need:

1. Two fixed points. These are two temperatures which are easily reproducible, *e.g.* the freezing point and the boiling point of water.
2. A body which has a property which changes continuously with changing temperature, *e.g.* the volume of a fixed mass of gas at constant pressure. Such a property is called a **thermometric property**.
3. A scale, *e.g.* 0° to 100°.

A commonly used thermometric property is the length of a column of liquid in a glass tube. To use this property to measure temperature the tube is put in a mixture of ice and water, *Fig. 15.1*, and the length of the column of liquid is measured. This length corresponds to 0 °C. The tube is then put in steam above boiling water, *Fig. 15.2*, and the length of the liquid column is again measured. This length corresponds

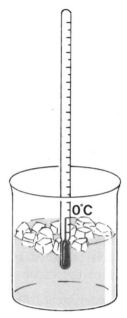

Fig. 15.1 Establishing the lower fixed point

to 100 °C and the difference between the two lengths corresponds to 100 Celsius degrees. Thus, by definition, one Celsius degree corresponds to one hundredth of the difference between these two lengths. The following example should help to clarify how an unknown temperature may be determined.

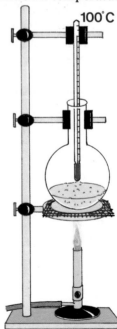

Fig. 15.2 Establishing the higher fixed point

Example

The length of a column of liquid in freezing water is 2 cm and in boiling water is 22 cm. When the tube is placed in a beaker of warm water the length of the column is 7 cm. What is the temperature of the warm water?

$$100 \text{ Celsius degrees} = 22 - 2$$
$$= 20 \text{ cm}$$
$$1 \text{ Celsius degree} = 0.2 \text{ cm}$$
$$1 \text{ cm} = 5 \text{ Celsius degrees}$$

The increase in the length in the warm water is

$$7 - 2$$
$$= 5 \text{ cm}$$
$$= 25 \text{ Celsius degrees}$$

Since 2 cm corresponds to 0 °C, 7 cm corresponds to 25 °C and the temperature of the warm water is therefore 25 °C.

Ans. 25 °C.

We can generalise the calculation of the previous example as follows. Let

$$l_0 = \text{length of column at f.p.}$$
$$l_{100} = \text{length of column at b.p.}$$
$$l_\theta = \text{length at unknown temperature, } \theta.$$

Then θ is given by

$$\theta = \frac{l_\theta - l_0}{l_{100} - l_0} \times 100 \qquad \text{(i)}$$

The liquids most commonly used to measure temperature in this way are mercury and alcohol. As well as the length of a column of liquid other thermometric properties which can be used include: the resistance of a wire or thermistor *(Chapter 21)*; the e.m.f. of a thermocouple *(p.160)*; the colour of certain substances; pV (pressure × volume) for a fixed mass of gas *(p.164)*.

Eqn. (i) may be further generalised to define a Celsius scale of temperature based on any thermometric property. Letting Y represent the thermometric property and replacing *l* in *Eqn. (i)* with the appropriate value of Y we have

$$\theta = \frac{Y_\theta - Y_0}{Y_{100} - Y_0} \times 100 \qquad \text{(ii)}$$

Anders Celsius (1701 — 1744)
Celsius was born into a famous scientific family in Uppsala, Sweden, in 1701. In 1743 he was appointed professor of astronomy at Uppsala University.

Celsius studied the Aurora Borealis and was the first to associate it with changes in the earth's magnetic field. He also attempted to determine the magnitude of stars by measuring the brightness of their light. In 1742 he devised the scale of temperature for which he is now remembered. He first gave the value 0° to the boiling point of water and 100° to the freezing point but this was reversed the following year to give us the scale we use today.

15.2 Thermometers

There are many different types of thermometer available, the type used depending on the particular situation. The most common is probably the mercury-in-glass thermometer, *Fig. 15.3*. It consists of a sealed capillary tube with a bulb full of mercury at one end. When the bulb is placed in contact with a hot body, the mercury expands up the capillary tube. This type of thermometer can be used to measure temperatures between —39 °C and 357 °C, the freezing point and boiling point, respectively, of mercury. (The upper limit can be extended considerably by filling the tube with a gas such as nitrogen. This increases the pressure of the mercury and so increases its boiling point.) The mercury thermometer is relatively cheap and easy to use although it is not particularly accurate.

A similar type of thermometer uses alcohol instead of mercury. Compared with the mercury thermometer the alcohol thermometer is much more sensitive, *i.e.* it can detect much smaller changes in temperature. It can also measure lower temperatures since the freezing point of alcohol is —115 °C.

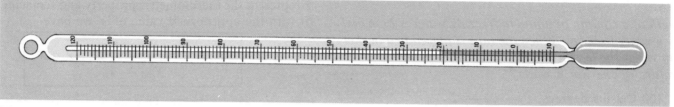

Fig. 15.3 Mercury-in-glass thermometer

However, since the boiling point of alcohol is 78 °C, it is not suitable for measuring higher temperatures.

The **platinum resistance thermometer** consists of a fine platinum wire wound on a strip of mica and placed in a silica tube, *Fig. 15.4*. As the name suggests, the thermometric property on which this thermometer is based is the electrical resistance of a wire *(p.217)*. The platinum resistance thermometer can measure temperatures between — 200° C and 1200 °C, so it has a much wider range than the mercury thermometer. It is very accurate but it cannot be used to measure rapidly changing temperatures because of its high heat capacity *(see Chapter 17)* and because the silica tube is a poor conductor. A more modern version of the platinum resistance thermometer consists of a thin layer of platinum on a base of alumina (aluminium oxide) and protected by a ceramic coating. This type of thermometer can measure temperatures in the range —50 °C to 500 °C.

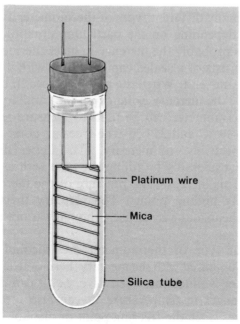

Fig. 15.4 Platinum resistance thermometer

Another type of electrical resistance thermometer is based on the **thermistor** *(p.229)*. Thermistor thermometers are more sensitive than platinum thermometers, especially at lower temperatures. Another type of electrical thermometer is based on the **thermocouple**, *Fig. 15.5*. This consists of two wires made from different metals, *e.g.* copper and iron, and joined together at both ends. When the two junctions are kept at different temperatures, an e.m.f. *(see Chapter 19)* is generated in the circuit and a current flows around it. This type of thermometer can measure temperatures between —250 °C and 1500 °C. It is not as accurate as the resistance thermometers but, because of its low heat capacity, it can be used to measure rapidly changing temperatures and the temperature at a point.

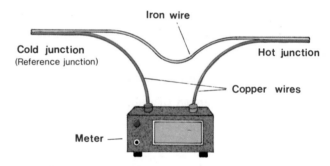

Fig. 15.5 Thermocouple thermometer

Fig. 15.6 shows a colour thermometer as used in a freezer. At room temperature all parts of the thermometer are blue. As the temperature falls the colours of different parts of the thermometer change as shown in *Fig. 15.6 (a), (b)* and *(c)*. Another type of "colour" thermometer uses the radiation emitted from a hot body to establish its temperature. We noted in *Chapter 13* that the hotter a body the lower the average wavelength of the radiation emitted from it. Thus the wavelength of the emitted radiation may be used as a thermometric property. This is the only type of thermometer which can be used to measure temperatures above about 1500 °C.

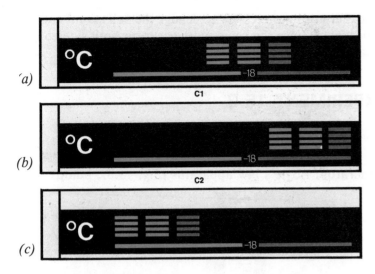

(a)

C1

(b)

C2

(c)

Fig. 15.6 Colour thermometer

Before a thermometer can be used it must first be calibrated. This is done by finding the value of the particular thermometric property at two fixed points (the points chosen depend on the range of temperatures to be measured). The difference between the two values is then divided into a number of equal parts as described earlier for the mercury-in-glass thermometer. A scale in degrees Celsius may then be marked on the instrument or the temperature may be calculated using *Eqn. (ii) (p.159)*.

EXPERIMENT 15.1

To Calibrate and use a Thermocouple Thermometer

Apparatus: Thermocouple, 2 beakers, round-bottomed flask, some ice, retort stand, bunsen burner, galvanometer.

Procedure:
1. Place some ice and cold water in the beakers. Connect the galvanometer to the thermocouple and place one junction of the thermocouple in each of the beakers. Note the reading, E_0, on the galvanometer. This reading should be zero.
2. Boil some water in the flask. Place one junction of the thermocouple in the steam above the boiling water, keeping the other junction (the reference junction) in the melting ice in the beaker, and note the reading, E_{100}, on the galvanometer.

3. Keeping the reference junction in the melting ice, warm some water in the other beaker and place the other junction in it. Note the reading, E_θ, on the galvanometer. The temperature, θ, of the water is then calculated from

$$\theta = \frac{E_\theta - E_0}{E_{100} - E_0} \times 100$$

4. Measure the temperature of the water in the beaker using a mercury thermometer.

RESULTS
Reading in ice .... =

Reading in steam .... =

Reading in warm water .... =

Temperature of water (thermocouple) .... = °C.

Temperature of water (mercury) .... = °C.

Questions
1. Comment on the difference or similarity of the temperature reading obtained from the thermocouple and that obtained from the mercury thermometer.

(**Note:** An experiment to calibrate and use a resistance thermometer is described in *Chapter 21*.)

15.3 Boyle's Law
Around 1660, Robert Boyle investigated the relationship between the volume of a gas and its pressure. The exact relationship was established by the French physicist, Edmé Mariotte (1620 — 1684), in 1676. This relationship is now known as Boyle's law.

> **For a given mass of gas at a constant temperature, the pressure is inversely proportional to the volume.**

In symbols,

$$p \propto \frac{1}{V}$$

or, $\quad p = k \dfrac{1}{V}$

where k is a constant.

Rearranging gives

$$pV = k$$

Since pressure is inversely proportional to volume a graph of pressure against the inverse of the volume for a fixed mass of gas at constant temperature is a straight line through the origin, *Fig. 15.7*.

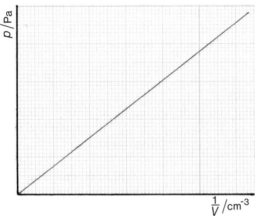

Fig. 15.7 Pressure against 1/volume

Example

A certain quantity of gas has a volume of 200 cm³ at a pressure of 100 kPa. If the pressure is increased to 120 kPa, the temperature being kept constant, what is the new volume of the gas?

Since pV is constant,

$$p_1 V_1 = p_2 V_2$$
$$100 \times 200 = 120 \times V$$

where V is the new volume of the gas.

i.e. $\quad 120V = 20\ 000$

$\qquad\quad V = 166.7 \text{ cm}^3$

$\qquad\qquad\qquad\qquad\qquad\quad$ Ans. 167 cm³

Note that it is not necessary to change the units to m³ and Pa, respectively, since the same units occur on both sides of the equation.

EXPERIMENT 15.2

To Verify Boyle's Law

Apparatus: Boyle's law apparatus *(Fig. 15.8)*, pump.

Fig. 15.8 Boyle's law apparatus

Procedure:

1. Open the tap. If the pressure gauge does not read atmospheric pressure use a Fortin barometer to find the atmospheric pressure. This must then be added to the pressure readings from the gauge.
2. Connect the pump to the apparatus and pump up the oil until the pressure gauge reaches its maximum reading. Close the tap.
3. Wait a few minutes and then read the pressure and the volume. Open the tap slightly until the pressure falls by about 0.5×10^5 Pa. Close the tap, wait a few minutes and then read the new pressure and volume. Repeat until the pressure reaches atmospheric pressure and then plot a graph of pressure against the inverse of the volume.

RESULTS

Atmospheric pressure =   kPa

p/kPa	V/cm^3	$\frac{1}{V}$/cm^{-3}

Questions

1. Why is the pressure plotted against the inverse of the volume rather than the volume?
2. Why should you wait a few minutes before taking the reading in each case?

Method 2:

Apparatus: J-tube, metre stick, mercury.

Procedure

1. Pour some mercury into the tube, trapping a volume of air in the closed end.
2. Measure the height of the air column, H. This is proportional to the volume of air in the tube.
3. Measure the difference in heights, h, of the mercury columns in the two sides of the tube. This gives a measure of the difference in pressure between the air in the closed end and the atmosphere.
4. Record the atmospheric pressure, A, from a Fortin barometer. The pressure of the air in the closed end of the tube is equal to $h + A$.
5. Pour some more mercury into the tube and repeat from Step 2.
6. Repeat a number of times and plot a graph of pressure against the inverse of the volume.

RESULTS

Atmospheric pressure =   cm of Hg

p/cm	V/cm	$\frac{1}{V}$/cm^{-1}

Questions

1. Why is it not necessary to know the actual volume of air in the closed end of the tube?
2. What precautions should be taken in handling mercury and why?

Note to teachers:

Mercury is a hazardous substance and should be handled with care. It is recommended that this method of verifying Boyle's law should be done as a teacher demonstration rather than a student experiment.

Robert Boyle (1627 — 1691)

Robert Boyle was born in Lismore, Co. Waterford, the fourteenth child of the Earl of Cork. He was a child prodigy who could speak Latin and Greek by the age of eight; at fourteen he was in Italy studying the works of Galileo.

Like Galileo, Boyle was a convinced experimentalist. Using his own design of vacuum pump, he showed that a piece of lead and a feather fall at the same speed in a vacuum. He also showed that sound cannot travel in a vacuum. He was the first chemist to collect a gas (he is sometimes referred to as the Father of Chemistry) and in 1662 he investigated the relationship between the pressure and the volume of a gas at constant pressure. This relationship is now known as Boyle's Law.

In later life Boyle devoted himself to the study of religion, learning Hebrew and Aramaic to extend his Biblical studies. He was a founder member of the Royal Society and in 1680 was elected its president, although he declined the honour because of the form of the oath. Boyle died in London on December 30, 1691.

The Ideal Gas

Boyle's law is obeyed exactly only when the density of the gas is low. Thus, gases like hydrogen, helium and oxygen obey the law quite well at normal temperatures and pressures but not at high pressures or low temperatures, when their density is also high.

The **ideal gas** is defined as a gas which obeys Boyle's law exactly at all temperatures and pressures. No such gas exists but the behaviour of real gases at low pressures approaches that of the ideal gas.

15.4 The Ideal Gas Temperature Scale

As we have seen, a variety of physical properties may be used to define a Celsius temperature scale. Unfortunately, different thermometers may give different temperatures for the same degree of "hotness". For example, a resistance thermometer placed in a beaker of warm water might give a reading of 35.8 °C, while a mercury thermometer placed in the same beaker might give a reading of 35.4 °C. This does not mean that one, or both, of the thermometers is giving the wrong temperature. Each is giving the correct temperature as defined by that particular thermometric property. The different values arise from the fact that each property varies slightly differently with changing "hotness". In other words, if we were to plot a graph of length (of mercury column) against resistance we would not get a straight line.

Since thermometers based on different thermometric properties give different temperatures it is obvious that one property must be chosen as the standard. In 1848 the Scottish mathematician and physicist, William Thompson (later, Lord Kelvin), proposed that the product pV (pressure multiplied by volume) of the ideal gas be taken as the standard thermometric property. In other words, temperature is defined to be proportional to pV for the ideal gas. That is

$$\boxed{T \propto pV}$$

In other words, if pressure multiplied by volume is the same for two equal quantities of the gas then, by definition, the temperatures of the two quantities are the same. If pV for one quantity is twice that for the other quantity, then their temperatures are also in the ratio of two to one. In general,

$$\boxed{\frac{T_1}{T_2} = \frac{p_1 V_1}{p_2 V_2}} \qquad \text{(i)}$$

The unit of temperature on the gas scale is the **kelvin**, K, and the scale is sometimes referred to as the Kelvin scale or the Absolute scale.

To establish a temperature scale we need two fixed points. For the Kelvin scale, the **triple point** of water is chosen as the upper fixed point. The triple point of water is the unique temperature at which ice, water and water vapour can exist in contact with each other. This temperature is given the value 273.16 K. This value was chosen so that the kelvin and the Celsius degree have the same value. The lower fixed point on the Kelvin scale is the temperature at which the pressure becomes zero. This temperature is given the value 0 K and is called the absolute zero of temperature. Thus, from *Eqn. (i)* we see that temperature on the ideal gas scale may be defined by the equation

$$\boxed{\frac{T}{273.16} = \frac{pV_{\text{T}}}{pV_{\text{tp}}}} \qquad \text{(ii)}$$

where pV_{T} is the value of pV at temperature T and pV_{tp} is its value at the triple point.

As mentioned above the Kelvin scale is designed so that a kelvin has the same value as a Celsius degree. Thus, 0 °C = 273 K, 100 °C = 373 K, *etc.*, *Fig. 15.9*. Thus, to convert degrees Celsius to kelvins simply add 273.

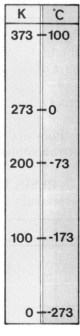

K	°C
373	100
273	0
200	-73
100	-173
0	-273

Fig. 15.9

Note that the unit of temperature on the Kelvin scale is the kelvin, not the degree kelvin. The kelvin is the SI unit of temperature; the Celsius degree, while it may be used with the SI system, is not part of it. Consequently, in calculations involving other quantities the kelvin, and not the Celsius degree, must be used.

Standard Temperature and Pressure

Since the volume of a gas depends on both its temperature and pressure it is sometimes useful to have a standard temperature and pressure at which volumes, *etc.*, may be compared. Standard temperature is taken as 273 K (0 °C). Standard pressure is taken as 1.01×10^5 Pa (76 cm of mercury). Note that pressure is sometimes measured in bars or millibars, 1 bar being equal to 1.01×10^5 Pa. The bar is not an SI unit.

The Constant Volume Gas Thermometer

One practical type of thermometer based on the ideal gas scale is the constant volume gas thermometer. In its simplest form, *Fig. 15.10*, this thermometer consists of a glass bulb containing a fixed mass of gas. A fine capillary tube leads to a U-tube containing mercury. The outer arm of the U-tube can be raised or lowered as necessary to maintain the level of the mercury in the inner tube at the reference mark.

When the bulb is placed in contact with a hot body the gas expands, pushing the mercury down from the mark. The outer arm of the tube is then raised to bring the mercury level back to the mark, thus increasing the pressure of the gas and keeping the volume constant. The hotter the body in question the higher the pressure of the gas, *i.e.* the pressure of the gas is the thermometric property on which the constant volume gas thermometer is based. The pressure is determined as follows. The pressure of the gas is equal to the pressure of the mercury at the reference mark. This in turn is equal to the pressure of the mercury at N (two points at the same horizontal level have the same pressure, *p.79*). The pressure at N is the pressure due to a column of mercury of height h ($\rho g h$, where ρ is the density of mercury), plus atmospheric pressure. Thus, by measuring the height h and finding the value of atmospheric pressure from a barometer the pressure of the gas in the bulb may be determined. The temperature of the gas, and hence of the body with which it is in contact, is found by determining

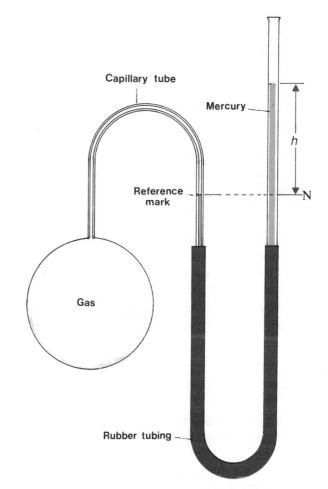

Fig. 15.10 Simple constant volume gas thermometer

the pressure at the triple point of water and at the temperature in question and substituting into *Eqn. (ii)*. Since the volume of the gas remains constant *Eqn. (ii)* becomes

$$\frac{T}{273.16} = \frac{p_T}{p_{tp}}$$

Since the behaviour of the real gas in the thermometer only approximates to that of the ideal gas a slight correction must be made to the reading given by the thermometer. Corrections must also be made to allow for the fact that the gas in the capillary tube is not at the same temperature as the gas in the bulb and for the fact that the bulb itself expands with increasing temperature.

165

This type of thermometer can be made to measure temperature very accurately and has a very wide range, being able to measure temperatures between 3 K and 1200 K. It is also very sensitive. However, it is much too cumbersome for everyday use and is used only as a standard by which other thermometers are calibrated.

William Thompson (Lord Kelvin) (1824 — 1907)

Kelvin was born in Belfast, the son of an eminent Scottish mathematician. He was a child prodigy who attended his father's lectures when he was eight and wrote his first paper on mathematics while still in his teens. He was professor of physics at Glasgow University for over 50 years. A renowned lecturer, he was one of the first to teach physics in the laboratory.

Kelvin's main interest was in the areas of heating and temperature. It was he who was chiefly responsible for promoting the work of James Joule (*p.235*). Investigating the expansion of gases with temperature, he suggested, in 1848, that no temperature below —273 °C existed. He called this temperature absolute zero and suggested a temperature scale with its zero at this point. This scale is now known as the gas scale or the Kelvin scale and the unit of temperature on it is called the kelvin in his honour.

As well as his work on heating and temperature, Kelvin made a number of important contributions in the area of electricity. He was knighted for his achievements in 1866 and was made Baron Kelvin in 1892. He died in Ayr, Scotland, in 1907 and is buried in Westminster Abbey, beside Newton.

SUMMARY

Temperature is a measure of the "hotness" of a body. The requirements for establishing a temperature scale are: two fixed points; a physical quantity which varies continuously with changing temperature (such quantities are called thermometric properties); a scale. *e.g.* 0 ° to 100°. Examples of thermometric properties are the length of a column of liquid, the volume or pressure of a gas, the resistance of a wire or thermistor, the e.m.f. of a thermocouple and the colour of certain substances. For a Celsius temperature scale the two fixed points chosen are the freezing point and boiling point of water and these are given the values 0 °C and 100 °C, respectively. Temperature, θ, on a Celsius scale is then defined by

$$\theta = \frac{Y_\theta - Y_0}{Y_{100} - Y_0} \times 100$$

where Y_θ is the value of the chosen thermometric property at the temperature θ and Y_{100} and Y_0 are its values at the higher and lower fixed points (boiling point and freezing point of water) respectively.

The most common types of thermometer are the mercury-in-glass thermometer (cheap, narrow but useful range); the platinum resistance thermometer (wide range, very accurate but cannot measure rapidly changing temperatures); the thermocouple thermometer (wide range, not very accurate but can measure rapidly changing temperatures).

Boyle's law states that, for a given mass of gas at constant temperature, pressure is inversely proportional to volume ($pV = k$). The ideal gas is one which obeys Boyle's law exactly.

On the ideal gas scale, temperature is defined by $T \propto pV$. The fixed points on this scale are absolute zero (0 K) and the triple point of water, which is given the value 273.16 K. Thus, temperature on this scale is given by

$$\frac{T}{273.16} = \frac{pV_T}{pV_{tp}}$$

where pV_T is the value of pV at temperature T and pV_{tp}

is its value at the triple point. The unit of temperature on this scale is the kelvin, K. The kelvin and the Celsius degree have the same value so temperatures are converted from degrees Celsius to kelvins by adding 273. Thus 0 °C = 273 K, 100 °C = 373 K, *etc.*

The constant volume gas thermometer uses the pressure of a fixed mass of gas at constant volume as the thermometric property. It has a wide range and is very accurate but it is cumbersome to use and so is used only as a standard by which other thermometers are calibrated.

Questions 15

Section I

1. Define temperature...

....

2. What is meant by a thermometric property? Give three examples..

....

....

3. What are the two fixed points on a Celsius scale?.....

....

....

4. Give an expression which defines a Celsius scale of temperature for a resistance thermometer................

....

5. State Boyle's law...

....

....

6. In addition to his work on gases Robert Boyle is remembered for ....

....

....

7. The ideal gas is one which ....

8. Which thermometric property is used to define temperature on the Kelvin scale?...........................

9. The triple point of water is ....

....

*10. Why is a capillary tube, rather than an ordinary tube, used in a constant volume gas thermometer?....

....

....

*11. Why is the constant volume gas thermometer used as a standard thermometer?................................

....

*12. The bulb of a gas thermometer is usually made of a special type of glass. What special properties do you think this glass should have?..........................

....

Section II

1. State the principles underlying the establishment of a temperature scale. A Celsius temperature scale for a thermocouple thermometer may be defined by the equation

$$\theta = \frac{A - B}{C - B} \times 100.$$

What do the letters A, B and C represent?

2. The length of a column of mercury in a glass tube is 2.5 cm at the freezing point of water and 28.5 cm at the boiling point. What is the temperature, in degrees Celsius, when the length of the column is 14 cm? State two properties which the glass tube should have if this arrangement is to be used as a thermometer.

3. The resistance of a piece of wire is to be used to measure the temperature of a beaker of water. If the resistance of the wire in the water, in melting ice and in boiling water is, respectively 7.5 Ω, 5.0 Ω and 9.0 Ω, calculate the temperature of the water in (i) °C, (ii) K.

4. The volume of air in a Boyle's law apparatus was found to be 40 cm^3 on a day when the atmospheric pressure was 1.02×10^5 Pa. A few days later the volume was found to be 42 cm^3. What was the atmospheric pressure on the second day?

5. The product pV for a fixed mass of gas is found to be 10 Pa m^3 at 27 °C. What is the value of pV for the gas at 87 °C? What assumption must be made in this calculation and in what circumstances is the assumption justified?

*6. At the triple point of water the pressure reading on a constant volume gas thermometer is 1.0×10^5 Pa. What is the temperature of the bulb when the pressure reading is 1.2×10^5 Pa?

*7. The readings on a constant volume gas thermometer at the boiling point and freezing point of water are, respectively, p_{100} and p_0. Determine the ratio of p_{100} to p_0.

8. A certain quantity of gas has a volume V at a pressure p and a temperature of 27 °C. The gas is then compressed until its volume is reduced by half. If the pressure of the gas is now $2.5p$ what is its temperature (assuming ideal behaviour)?

9. The speed of sound in a gas is given by $c = \sqrt{(\gamma p / \rho)}$, where p and ρ are the pressure and density, respectively, of the gas and γ is a constant. Show that the speed of sound in a gas depends on the temperature of the gas and is independent of changes in the pressure.

16 Kinetic Theory

16.1 Brownian Motion

The modern atomic theory of matter was first proposed by an English schoolmaster, John Dalton (1766 — 1844), in 1803. (The idea that matter was made up of atoms was due originally to the ancient Greeks. However, Dalton was the first to base his theory on experimental observations.) In 1811 the Italian scientist, Amedeo Avogadro (1776 — 1856), extended Dalton's ideas by proposing that atoms of the same element, or of different elements, could join together to form molecules, a molecule being the smallest particle of a substance which can exist on its own. The first direct evidence for the existence of molecules (though not recognised as such at the time) was discovered in 1827 by a Scottish botanist, Robert Brown (1773 — 1858), who noticed that fine pollen grains suspended in water were in a state of rapid random motion. This phenomenon, which is now known as **Brownian motion**, can be observed for many kinds of small particle suspended in different fluids (liquids and gases). Although Brown failed to understand the significance of his discovery, Brownian motion is now considered to be due to collisions between molecules and the suspended particles. Brownian motion thus provides direct evidence for the existence of molecules and hence of atoms.

One method of demonstrating Brownian motion is illustrated in *Fig. 16.1*. Some smoke is enclosed in a small glass cell. The cell is illuminated from the side and viewed from above through a microscope. The smoke particles are then seen to move around in a random manner.

The Avogadro Constant

The masses of atoms and molecules are very small, the mass of a hydrogen atom, for example, being 1.67×10^{-27} kg. The kilogram is thus an inconvenient unit for the measurement of such masses. In practice, therefore, atomic masses and molecular masses are measured relative to the mass of a carbon atom which is given a mass of 12 "units". The unit of mass on this scale is called the **unified atomic mass unit**, u. (The name of this unit is sometimes abbreviated to a.m.u.) 1 u is approximately equal to 1.66×10^{-27} kg. Thus, the mass of

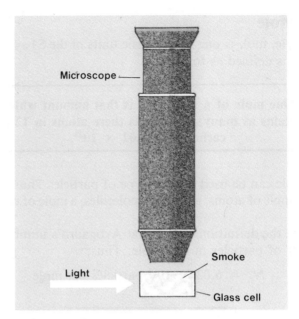

Fig. 16.1 Observing Brownian motion

a carbon atom is 12 u or 1.99×10^{-27} kg. On this scale the mass of a hydrogen atom is 1 u, *i.e.* the mass of a hydrogen atom is one twelfth of the mass of a carbon atom. Likewise, the mass of an oxygen atom is 16 u, *i.e.* its mass is 4/3 times that of a carbon atom. It follows that the mass of a water molecule, which is made up of two hydrogen atoms and one oxygen atom, is 18 u. Since masses on this scale are measured relative to the mass of the carbon atom, the terms "relative atomic mass" and "relative molecular mass" are also used. Thus, the relative atomic mass of hydrogen is 1, the relative molecular mass of water is 18, *etc.*

Since the mass of a carbon atom is 12 times the mass of a hydrogen atom, it follows that 1 g of hydrogen will contain 12 times as many atoms as 1 g of carbon. In other words, 12 g of carbon will contain the same number of atoms as 1 g of hydrogen. Likewise, there is the same number of molecules of water in 18 g of water as there are atoms in 12 g of carbon. In general, the relative atomic mass of any element expressed in grams contains the same number of atoms as there are in 12 g of carbon; the relative molecular mass of any substance contains the same number of molecules as there are atoms in 12 g of carbon. This number is known as **Avogadro's number** or the **Avogadro Constant**, N_A. Its value is found by experiment to be 6.02×10^{23}.

The Mole

The mole, mol, is one of the basic units of the SI system (*cf. p.6*). It is defined as follows.

> **One mole of a substance is that amount which contains as many particles as there atoms in 12 g of carbon, *i.e.* 6.02 × 10²³.**

The mole can be used for any type of particle. Thus, we can have a mole of atoms, a mole of molecules, a mole of electrons, *etc*.

From the definition we see that Avogadro's number is the number of particles in one mole. Thus,

$$N_A = 6.02 \times 10^{23} \text{ particles per mole}$$

or

$$N_A = 6.02 \times 10^{23} \text{ mol}^{-1}$$

EXPERIMENT 16.1

To Estimate the Diameter of a Molecule

Apparatus: Large shallow tray, burette, small beaker, talcum powder, 0.01% solution of oleic acid in alcohol*.

Procedure:
1. Clean the burette and the tray with alcohol.
2. Pour approximately 30 cm³ of the solution into the burette and then open the tap until the liquid starts to flow from the burette. Close the tap and note the reading on the burette.
3. Open the tap slowly and count 50 drops into the beaker. Note the new reading on the burette and hence calculate the volume of one drop of the solution. Calculate the volume of acid in one drop of the solution.
4. Pour water into the tray to a depth of about 1 cm. Dust a little talcum powder onto the surface of the water. Place the tray under the burette and allow one drop of the solution to fall onto the surface of the water.
5. Estimate the area of the patch of acid on the surface of the water. This may be done by treating the patch as a circle and measuring its radius or by having a sheet of graph paper

on the bottom of the tray and counting the number of squares covered by the patch.
6. Repeat a number of times and find an average value for the area. The patch of acid is assumed to be one molecule deep so the diameter of a molecule is found by dividing the volume of acid in one drop by the area of the patch of acid.

RESULTS

Area/cm²

Average area =   cm².

Volume of 50 drops of solution..........	=	cm³
Volume of one drop of solution.........	=	cm³
Volume of acid in one drop	=	cm³
Area of patch	=	cm²
Diameter of molecule	=	cm

* This solution is made up as follows. Pour some ethanol into a clean 100 cm³ volumetric flask. Add 1 cm³ of oleic acid and shake the flask well. Add ethanol to make the solution up to 100 cm³ and shake the flask. Take 1 cm³ of this solution and make it up to 100 cm³ with ethanol in a second clean volumetric flask. The solution is now one part in 10 000 oleic acid, *i.e.* a 0.01% solution of oleic acid.

16.2 The Kinetic Theory of Gases

The kinetic theory of gases attempts to explain the behaviour of gases in terms of the motion of their molecules. It is based on a number of assumptions which are justified by the fact that predictions based on the theory agree with experimental results. The basic assumptions are:

1. Even a small volume of gas contains a very large number of molecules which move at high speeds in a random manner.

2. Forces between molecules are negligible except during collisions.
3. The time occupied by collisions is negligible compared with the time spent between collisions.
4. The volume of the molecules is negligible compared with the total volume occupied by the gas.
5. On average, molecules neither gain nor lose energy in collisions.

Working from these assumptions we shall now derive an expression for the pressure of a gas. (**Note:** The same symbol, p, is used for both momentum and pressure. Both quantities appear in this section, although they never appear together in the same equation. You should take extra care to ensure that you understand what the symbols in each equation represent.)

The Kinetic Theory Equation

Consider a cubical box, *Fig. 16.2(a)*, of side l. Suppose it contains N molecules, each of mass m. Let the velocity of one of these molecules be c_1. This velocity may then be resolved into three components, u_1, v_1 and w_1 parallel to the x, y and z axes, respectively, *Fig. 16.2(b)*. First, consider the motion

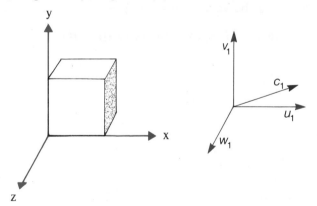

Fig. 16.2(a)   *Fig. 16.2(b)*

of the molecule parallel to the x-axis, *i.e.* perpendicular to the shaded side of the box. When the molecule strikes this side its velocity changes from $+u_1$ to $-u_1$. Its momentum therefore changes from $+mu_1$ to $-mu_1$. That is, the change in momentum, Δp_1, of the molecule due to the collision is

$$\Delta p_1 = 2mu_1$$

The time taken for the molecule to travel to the opposite side and back again is

$$t = \frac{2l}{u_1}$$

Each time the molecule strikes the side its momentum changes by Δp_1. That is, in time t its momentum changes by Δp_1. Therefore, the average rate of change of its momentum is

$$\frac{\Delta p_1}{t} = 2mu_l \times \frac{u_1}{2l}$$
$$= \frac{mu_1^2}{l}$$

But, from Newton's second law, force equals rate of change of momentum. Therefore, the average force on the wall due to this molecule is

$$F_1 = \frac{mu_1^2}{l}$$

The total force on the wall due to all N molecules is

$$F = F_1 + F_2 + F_3 + \ldots + F_N$$
$$= \frac{m}{l}(u_1^2 + u_2^2 + u_3^2 + \ldots + u_N^2) \quad \text{(i)}$$

Let $\bar{u}^2$ be the mean value of $u_1^2, u_2^2, u_3^2, \ldots u_N^2$

Then,

$$\bar{u}^2 = \frac{u_1^2 + u_2^2 + u_3^2 + \ldots + u_N^2}{N}$$
$$\Rightarrow \quad N\bar{u}^2 = u_1^2 + u_2^2 + u_3^2 + \ldots + u_N^2$$

Substituting in *Eqn. (i)* gives

$$F = \frac{mN\bar{u}^2}{l} \quad \text{(ii)}$$

Since there is no net movement of the gas the pressure must be the same at all points in the gas, including at the side. (If there were a difference in pressure between two points there would be a resultant force on the gas from the point of higher pressure towards the point of lower pressure and so the gas would move in that direction.) The pressure, p, at the shaded side, and therefore throughout the gas, is given by

$$p = \frac{F}{A}$$

where A is the area of the side.

$$A = l^2$$

$$\Rightarrow \quad p = \frac{F}{l^2}$$

$$= \frac{mN\overline{u^2}}{l^3} \quad \text{(from Eqn. (ii))}$$

$$= \frac{mN\overline{u^2}}{V} \tag{iii}$$

where V is the volume of the box and hence of the gas.

From *Fig. 16.2(b)* we see that

$$\overline{c^2} = \overline{u^2} + \overline{v^2} + \overline{w^2}$$

(Compare the addition of *two* vectors, p.95)

Therefore, $\overline{c^2} = \overline{u^2} + \overline{v^2} + \overline{w^2}$

Also, since there is no net movement of the gas in a particular direction,

$$\overline{u^2} = \overline{v^2} = \overline{w^2}$$

$$\Rightarrow \quad \overline{c^2} = 3\,\overline{u^2}$$

$$\Rightarrow \quad \overline{u^2} = \tfrac{1}{3}\overline{c^2}$$

Substituting for u^2 in *Eqn. (iii)* gives

$$p = \frac{1}{3}\frac{mN\overline{c^2}}{V} \tag{iv}$$

This is the **kinetic theory equation.** It relates the pressure of a gas to its volume and the number, mass and speed of its molecules.

The total mass of the gas is mN and its volume is V. Therefore its density, ρ, is given by

$$\rho = \frac{mN}{V}$$

Therefore, *Eqn. (iv)* may also be written as

$$p = \tfrac{1}{3}\rho\overline{c^2} \tag{v}$$

In these equations $\overline{c^2}$ is the mean of the squares of the velocities of the molecules. The square root of $\overline{c^2}$ is called the **root-mean-square (r.m.s.) speed** of the molecules. Note that the r.m.s. speed is not equal to the average speed, *i.e.*

$$\sqrt{\overline{c^2}} \neq \overline{c}$$

or $\quad \overline{c^2} \neq (\overline{c})^2$

For example, the average of 3 and 5 is 4 while the r.m.s. value of 3 and 5 is the square root of $(9 + 25)/2$, *i.e.* 4.12.

The Kinetic Theory and Temperature

We saw in the previous chapter that temperature on the Kelvin scale is defined by

$$T \propto pV$$

or, $\quad pV = kT \tag{a}$

where k is a constant.

From the kinetic theory equation *(Eqn. (iv))*

$$pV = \tfrac{1}{3}\,mN\overline{c^2}$$

$$= \tfrac{2}{3}N(\tfrac{1}{2}m\overline{c^2})$$

$$= \tfrac{2}{3}N\,\overline{E}_k \tag{b}$$

where $\overline{E}_k$ is the average kinetic energy of the molecules of the gas. Combining *Eqns. (a)* and *(b)* we have

$$\tfrac{2}{3} N \, \bar{E}_4 = kT$$

$$=> \quad \bar{E}_k = \frac{3k}{2N} \, T$$

Since, for a given quantity of gas, N is constant, we have

$$\boxed{\bar{E}_k \propto T} \qquad \text{(vi)}$$

That is, the average kinetic energy of the molecules of a gas is proportional to its absolute, or Kelvin, temperature. This, of course, is only strictly true for the ideal gas.

Example
Calculate the r.m.s. speed of the molecules of hydrogen at standard temperature and pressure (s.t.p.). Density of hydrogen at s.t.p. is 0.09 kg m^{-3}. If the temperature is increased to 546 K, the pressure being kept constant, what is the new r.m.s. speed of the molecules?

From *Eqn. (v)*

$$\sqrt{\bar{c^2}} = \sqrt{\frac{3p}{\rho}}$$

$$= \frac{3 \times 1.0 \times 10^5}{0.09}$$

$$= 1.83 \times 10^3 \text{ m s}^{-1}$$

Standard temperature is 273 K, so the temperature is increased by a factor of 2. From *Eqn. (vi)*, if the temperature increases by a factor of 2 the r.m.s. speed increases by a factor of $\sqrt{2}$. So the new r.m.s. speed is

$$\sqrt{\bar{c^2}} = 1.83 \times 10^3 \times 1.41$$

$$= 2.58 \times 10^3 \text{ m s}^{-1}$$

$$\text{Ans. } 1.8 \times 10^3 \text{ m s}^{-1}; \quad 2.6 \times 10 \text{ m s}^{-1}$$

Avogadro's Law
We have derived the kinetic theory equation from a number of assumptions. If this equation is valid it must agree with the known experimental laws of gases. One such law is Avogadro's law which, although it cannot be verified directly by experiment, leads to conclusions which can. Avogadro's law may be stated as follows.

> **Equal volumes of all gases at the same temperature and pressure contain equal numbers of molecules.**

From the kinetic theory equation we showed that

$$pV = \tfrac{2}{3} N \, \bar{E}_k$$

$$=> \quad V = \frac{2\bar{E}_k}{3p} \, N$$

Assuming ideal gas behaviour, if temperature is constant $\bar{E}_k$ is constant. If p is also constant, then

$$V = KN$$

where K is a constant. That is,

$$V \propto N$$

That is, the volume of a gas is proportional to the number of molecules in it. Therefore, gases which have the same volume at a given temperature and pressure also have equal numbers of molecules.

Thus, in this case, the kinetic theory leads to a conclusion which is consistent with experimental results. It is this, and similar evidence in relation to other experimental laws of gases, which leads us to believe that the assumptions of the kinetic theory are substantially correct.

SUMMARY

Brownian motion, the motion of small particles suspended in a fluid, indicates that fluids are made up of molecules which are in continuous random motion.

Atomic and molecular masses are measured relative to the mass of a carbon atom which is given the value 12 u. The atomic or molecular mass of a substance expressed in grams contains Avogadro's number (constant) of atoms or molecules, respectively. Avogadro's number, N_A, has the value 6.02×10^{23} mol^{-1}. One mole, mol, of a substance contains Avogadro's number of particles.

The kinetic theory of gases explains the behaviour of gases in terms of the motion of their molecules. It is based on a number of assumptions (*see p.170*) which are justified by the fact that they lead to conclusions which are verifiable experimentally. The kinetic theory equation is an expression, based on these assumptions, for the pressure of a gas in terms of its volume and the mass,

number and speeds of its molecules. The kinetic theory equation may be written as

$$p = \frac{1}{3} \frac{mN\overline{c^2}}{V}$$

or

$$p = \frac{1}{3} \rho \overline{c^2}$$

For the ideal gas, temperature, on the absolute or Kelvin scale, is proportional to the mean kinetic energy of the molecules (approximately true for real gases).

Avogadro's law states that equal volumes of all gases at the same temperature and pressure contain equal numbers of molecules.

Questions 16

DATA: $N_A = 6.0 \times 10^{23}$ mol^{-1}.

Section I

*1. What is Brownian motion?..................................

....

*2. One unified atomic mass unit is equivalent to ....

....

*3. Avogadro's number (constant) is equal to the number

of ....

in...

*4. What is a mole?...

....

*5. What does each of the symbols in the kinetic theory

equation represent?...

....

....

*6. What is meant by the root-mean-square speed of the

molecules of a gas?..

....

*7. State the relationship between the density of a gas and the root-mean-square speed of its molecules at constant

pressure...

....

*8. According to the kinetic theory, what is the relationship between the temperature of a gas and the kinetic energy

of its molecules?...

....

*9. In what circumstances does this relationship apply to real

gases?..

....

*10. State Avogadro's law...

....

Section II

*1. Calculate the number of water molecules in a beaker containing 45 cm³ of water, given that the relative molecular mass of water is 18. (Density of water = 1.0 × 10³ kg m⁻³.)

*2. Given that one mole of gas occupies 22.4 litres at s.t.p. calculate the number of hydrogen molecules in 1 cm³ of the gas at s.t.p.

*3. A sample of oxygen contains 4.8 × 10²² molecules at s.t.p. Calculate (i) the number of moles in the sample, (ii) the mass of the sample. (Relative molecular mass of oxygen = 32; 1 u = 1.7 × 10⁻²⁷ kg.)

*4. A school laboratory vacuum pump can reduce the pressure in a sealed container to 8.0 × 10⁻³ Pa. How many air molecules remain in a container of volume 0.10 m³ at this pressure and a temperature of 273 K? (1 mol of air at s.t.p. occupies 22.4 litres.)

*5. Calculate (i) the mean, (ii) the root-mean-square, of the following numbers: 4, 6, 7, 6, 3, 5, 5, 8.

*6. The r.m.s. speed of the molecules of a gas is 1.2 × 10³ m s⁻¹ at 20 °C. What is the r.m.s. speed of the molecules at 100 °C?

*7. The r.m.s. speed of the molecules of a gas is 1.4 × 10³ m s⁻¹ when the pressure is 75 cm of Hg. What is the density of the gas? (Density of Hg = 1.36 × 10⁴ kg m⁻³; g = 9.8 m s⁻².)

*8. The r.m.s. speeds of the molecules of two gases at a certain temperature are 0.46 km s⁻¹ and 1.8 km s⁻¹. Calculate the ratio of the densities of the two gases.

17 Heating and Internal Energy

17.1 Specific Heat Capacity

In order to raise the temperature of a body, *i.e.* to make it hotter, we must supply energy to it. There are two ways in which this energy may be supplied. It may be supplied by **heating** the body, *i.e.* by placing it near, or in contact with, a hotter body. Alternatively, the energy may be supplied by **doing work** on the body, *e.g.* by rubbing it with another body. In either case, the **internal energy** of the body is increased.

> **The internal energy of a body is the sum of the potential and kinetic energies of its molecules.**

Note that the internal energy does not include the kinetic or potential energy that the body as a whole may have due to its velocity, position, *etc.*

The addition of a given amount of energy does not always result in the same rise in temperature. For example, a half-full beaker of water will reach 50 °C more quickly than a full beaker if both start from the same temperature and energy is supplied to both at the same rate, *Fig. 17.1*. Similarly, a beaker of oil will reach a given temperature more quickly than a similar beaker of water, *Fig. 17.2*. Obviously, the temperature of a body depends not only on the amount of energy supplied to it, but also on its size and nature. In order to be able to compare the abilities of different materials to absorb energy we define the quantity, **specific heat capacity (s.h.c.),** *c*.

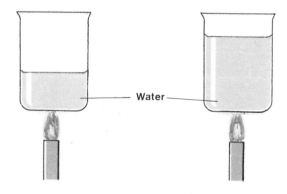

Fig. 17.1

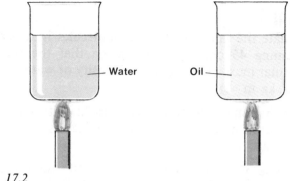

Fig. 17.2

> **The s.h.c. of a substance is the amount of energy which will change the temperature of 1 kg of the substance by 1 K.**

In other words,

$$\text{s.h.c.} = \frac{\text{Energy added or removed}}{\text{Mass} \times \text{Change in Temp.}}$$

In symbols this becomes

$$c = \frac{\Delta E}{m \times \Delta \theta}$$

The unit of *c* is the **joule per kilogram kelvin**, J kg^{-1} K^{-1}.

Example
What is the s.h.c. of copper, given that the temperature of 200 g of copper is increased from 15 °C to 25 °C by the addition of 780 J?

$$c = \frac{\Delta E}{m \times \Delta \theta}$$

$$= \frac{780}{0.2 \times 10}$$

$$= 390 \text{ J kg}^{-1} \text{ K}^{-1}$$

Ans. 390 J kg^{-1} K^{-1}

The s.h.c.'s of some common substances are shown in *Table 17.1*. From the table it can be seen that water has a very high s.h.c. This is one of the reasons why water is used as a coolant, *e.g.* in car engines — relatively small quantities of water can carry away large amounts of energy. It also helps to explain why the sea is colder than the land on a summer day and, conversely, why land cools more quickly than the sea on a summer night.

Storage heaters make use of the high s.h.c. of special bricks. Electrical energy is used to warm the bricks at night when electricity is cheap. The bricks can store enough energy at relatively low temperatures (~ 100 °C) to keep a room warm during the day. Some solar heating systems use large tanks of water to store energy. Energy from the sun warms the water during the day. In the evening this energy is delivered to the house by pumping the warm water through radiators.

Substance	c/kJ kg^{-1} K^{-1}
Water	4.2
Ice	2.1
Paraffin	2.1
Wood	1.7
Aluminium	0.91
Copper	0.39
Tin	0.23
Mercury	0.14
Lead	0.13

Table 17.1 Specific Heat Capacities

EXPERIMENT 17.1

To Measure the s.h.c. of Water

Apparatus: Calorimeter, battery or low voltage power supply, heating coil, joulemeter, thermometer, large beaker with insulation (cotton wool, pieces of polystyrene, *etc.*), balance.

Procedure:
1. Find the mass of the copper calorimeter.
2. Approximately half-fill the calorimeter with water and find the mass of the calorimeter and water and hence the mass of the water. Note that there must be enough water in the calorimeter to cover the heating coil.
3. Place the calorimeter in the beaker surrounded by insulation. Place the heating coil in the calorimeter and connect up the circuit as shown in *Fig. 17.3*.
4. Note the temperature of the water. Zero the joulemeter and switch on.
5. Allow the temperature of the water to rise by about 10 °C.

Switch off the current, note the final temperature of the water and the amount of energy supplied, *E*.

6. The s.h.c. of water, c_w, is calculated from

$$E = m_w c_w \Delta\theta + m_c c_c \Delta\theta$$

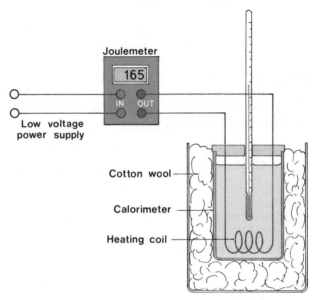

Fig. 17.3

where m_w and m_c are the masses of the water and calorimeter, respectively, $\Delta\theta$ is the increase in temperature of the water and calorimeter and c_c is the s.h.c. of the material of which the calorimeter is made.

RESULTS

Mass of calorimeter	=	kg
Mass of calorimeter + water	=	kg
Mass of water	=	kg
Initial temp. of water and calorimeter	=	°C
Final temp. of water and calorimeter ..	=	°C
Rise in temp. of water and calorimeter	=	°C
Energy supplied	=	J
S.h.c. of water	=	J kg^{-1} K^{-1}.

Note: If a joulemeter is not available the energy supplied can be found by measuring the current flowing, the voltage across the heating coil and the time for which the current flows. The energy supplied is then given by: Energy supplied = Current × Voltage × Time (*see p.234*).

Questions

1. The specific heat capacity of the heating coil was not taken into account in this experiment. Is this likely to be a major source of error? Explain.
2. It is assumed that the temperature of the water and calorimeter are the same. To what extent do you think this assumption is valid? Would it be more or less valid if a beaker were used instead of the calorimeter?

EXPERIMENT 17.2

To Compare the s.h.c.'s of Copper and Water by the Method of Mixtures

Apparatus: Beaker, tripod stand and gauze, test tube, copper calorimeter, thermometer, large beaker with insulation, balance.

Procedure:

1. Find the mass of some copper rivets and put them in the test tube.
2. Fill the smaller beaker with water and place the test tube in it, *Fig. 17.4(a)*.

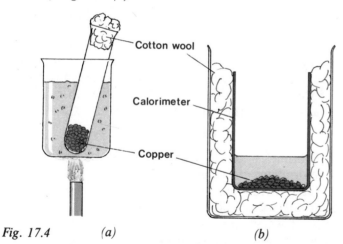

Cotton wool

Calorimeter

Copper

Fig. 17.4   *(a)*   *(b)*

3. Heat the beaker until the water boils. Turn down the gas and allow the water to boil for 10 — 15 minutes. You may now assume that the temperature of the copper is the same as that of the boiling water.
4. Find the mass of the calorimeter. Fill it approximately one quarter full with cold water and find the mass of the calorimeter and water. Hence find the mass of the water. Place the calorimeter in the beaker with the insulation.

5. When the copper is hot measure the temperature of the cold water in the calorimeter.
6. Quickly empty the hot copper into the calorimeter and note the highest temperature reached by the water.
7. The ratio of the s.h.c. of copper, c_1, to the s.h.c. of water, c_2, may then be determined from:

Energy lost by copper = Energy gained by (water + calorimeter)

$$m_1 c_1 \, \Delta\theta_1 = m_2 c_2 \, \Delta\theta_2 + m_3 c_1 \, \Delta\theta_2$$

where m_1, m_2 and m_3 are the masses of the copper rivets, water and calorimeter, respectively, and $\Delta\theta_1$ and $\Delta\theta_2$ are the fall in the temperature of the copper and the rise in the temperature of the water and calorimeter, respectively.

RESULTS

Mass of copper rivets	=	kg
Mass of calorimeter	=	kg
Mass of calorimeter + water	=	kg
Mass of water	=	kg
Temperature of hot copper	=	°C
Temp. of cold water and calorimeter	=	°C
Final temp. of water, copper and calorimeter	=	°C
Fall in temp. of copper	=	°C
Rise in temp. of water and calorimeter	=	°C

Questions

1. Is it valid to assume that the temperature of the boiling water is 100 °C? Explain.
2. Why is it important that the calorimeter be only approximately one quarter full of water? What, do you think, is the minimum amount of water you could use?
3. From your results estimate what would have been the ideal mass of copper to have used.

Heat Capacity

It is sometimes more convenient to talk about the amount of energy required to warm a whole body rather than unit mass (1 kg) of it. This amount of energy is called the heat capacity of a body, *i.e.* the heat capacity of a body is the energy required to change the temperature of the whole body by 1 K. Comparing this definition with the definition of specific heat capacity given earlier we see that the heat capacity of a body is equal to the s.h.c. of the substance from which the body is made, multiplied by its mass.

17.2 Specific Latent Heat

If a piece of ice at, say $-10\ °C$, is heated steadily its temperature rises until it reaches $0\ °C$ and then remains constant until all the ice has melted, *Fig. 17.5*. The temperature then starts to rise again. The energy supplied during the period when the temperature was constant is called **latent heat**. It is used to overcome the forces of attraction which hold the H_2O molecules together in the solid state (ice) so that they become relatively free to move. Similarly, energy is required to change water into water vapour, *Fig. 17.5*.

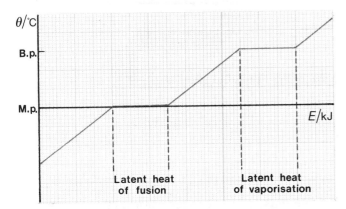

Fig. 17.5 Change in temperature as energy is supplied

> **The specific latent heat (s.l.h.), l, of a substance is the amount of energy which will change the state of 1 kg of the substance.**

The unit of l is the **joule per kilogram**, $J\ kg^{-1}$ and it is a **scalar** quantity. l is called the s.l.h. of **fusion** when the change of state is between liquid and solid. It is called the s.l.h. of **vaporisation** when the change is between liquid and gas.

From the definition we see that

$$s.l.h. = \frac{\text{Energy added or removed}}{\text{mass}}$$

In symbols this becomes

$$l = \frac{\Delta E}{m}$$

S.l.h.'s are, in general, quite large. The s.l.h. of vaporisation of water, for example, is $2.3 \times 10^6\ J\ kg^{-1}$. Since the s.h.c. of water is $4.2 \times 10^3\ J\ kg^{-1}\ K^{-1}$, this means that it takes approximately five times as much energy to convert 1 kg of boiling water to steam as it does to heat 1 kg of water from $0\ °C$ to $100\ °C$. Conversely, when steam condenses to water large amounts of energy are released. This explains why scalds from steam are more painful than scalds from boiling water. Your body's cooling mechanism depends on the large s.l.h. of vaporisation of water. The body secretes sweat from glands just under the skin. On the surface of the skin the sweat evaporates, taking the necessary latent heat from the skin and therefore cooling it.

Example
Calculate the amount of energy required to convert 80 g of water at 100 °C to steam at the same temperature.

$$l = \frac{\Delta E}{m}$$
$$=> \quad \Delta E = l\,m$$
$$= 2.3 \times 10^6 \times 0.08$$
$$= 1.84 \times 10^5\ J \qquad \text{Ans. } 1.8 \times 10^5\ J$$

The Heat Pump

A heat pump is a device which transfers energy from a cold body to a warmer one, thus reducing the temperature of the cold body and raising the temperature of the warmer one. A heat pump thus reverses the normal heating process. The most common application of the heat pump is the refrigerator, where energy is transferred from the inside of the refrigerator (cold) to the room (warm). Heat pumps may also be used to heat buildings by transferring energy from a river or the ground (cold) to the inside of the building (warm).

The principle of the heat pump is illustrated in Fig. 17.6. A special liquid which has a high s.l.h. and a low boiling point is pumped around a closed system. A valve ensures that the pressure on one side of the compressor is much greater than on the other side. When the liquid passes through the valve its pressure is reduced and so it expands and vaporises, absorbing the necessary latent heat from the surroundings (the cold body). When the vapour is compressed it changes to a liquid, releasing its latent heat to the surroundings (the warm body). In this way energy is transferred from the cold body to the warm body.

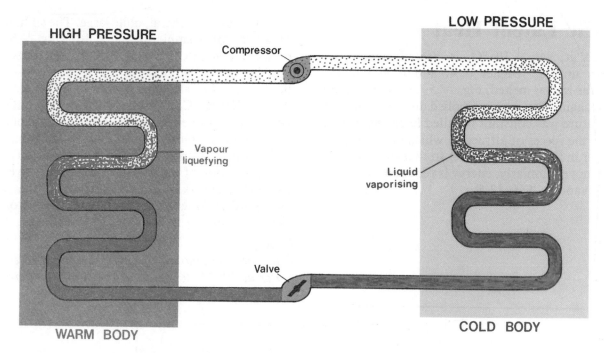

HIGH PRESSURE

LOW PRESSURE

Compressor

Vapour
liquefying

Liquid
vaporising

Valve

WARM BODY

COLD BODY

Fig. 17.6   Heat pump

*Fig. 17.7   Refrigerator showing compressor and condenser
(horizontal pipes)*

EXPERIMENT 17.3

To Measure the s.l.h. of Fusion of Ice

Apparatus: Calorimeter, thermometer, beaker with insulation, balance.

Procedure:
1. Find the mass of the copper calorimeter.
2. Approximately half fill the calorimeter with warm water (at about 25 °C). Find the mass of the calorimeter and water and hence the mass of the water.
3. Place the calorimeter in the beaker surrounded by insulation and note the temperature of the water.
4. Crush some ice into small pieces and dry it with blotting paper or filter paper.
5. Add the ice to the water, a little at a time, stirring with the thermometer until the ice is melted. Continue until the temperature of the water has fallen by about 10 °C. Note the final temperature of the water.
6. Find the mass of the calorimeter and water and hence the mass of ice added.
7. The specific latent heat of fusion of ice is calculated from:

Energy gained by ice = Energy lost by (calorimeter + water).

$$m_1 l + m_1 c_w \Delta\theta_1 = m_2 c_w \Delta\theta_2 + m_3 c_c \Delta\theta_2$$

where m_1, m_2 md m_3 are the masses of the ice, water and calorimeter, respectively, and $\Delta\theta_1$ and $\Delta\theta_2$ are the rise in temperature of the melted ice and the fall in temperature of the water and calorimeter, respectively. c_w is the s.h.c. of water and c_c is the s.h.c. of the material of which the calorimeter is made.

RESULTS

Mass of calorimeter .... =   kg
Mass of calorimeter + water .... =   kg
Initial temp. of water and calorimeter =   °C
Temperature of ice .... =   0°C
Final temp. of water and calorimeter .. =   °C
Rise in temp. of melted ice .... =   °C
Fall in temp. of water and cal. .... =   °C
Mass of cal. + water + melted ice .... =   kg
Mass of ice .... =   kg
Specific latent heat of fusion of ice .... =   J kg^{-1}.

Questions

1. Why should the ice be dry before adding it to the water?
2. Comment on the method used to find the mass of ice added.
3. Do you think it is valid to assume that the temperature of the ice is 0 °C? Explain.

EXPERIMENT 17.4

To Measure the s.l.h. of Vaporisation of Water.

Apparatus: Calorimeter, thermometer, beaker with insulation, round bottomed flask fitted with stopper, insulated glass tube and thermometer *(see Fig. 17.8(a))*, bunsen, retort stand.

Procedure:
1. Find the mass of the copper calorimeter.
2. Approximately half fill the calorimeter with water. Find the mass of the calorimeter and water and hence the mass of the water.
3. Place the calorimeter in the beaker surrounded by insulation and note the temperature of the water.
4. Put some water in the round bottomed flask and heat it until steam is issuing freely from the glass tube. (Take care that the temperature does not rise above 100 °C.)
5. When the steam has been issuing freely for a few minutes place the end of the tube under the surface of the water in the calorimeter and allow the steam to pass into the water until the temperature has risen by about 10 °C. Note the final temperature of the water.
6. Find the mass of the calorimeter and water and hence mass of steam added.

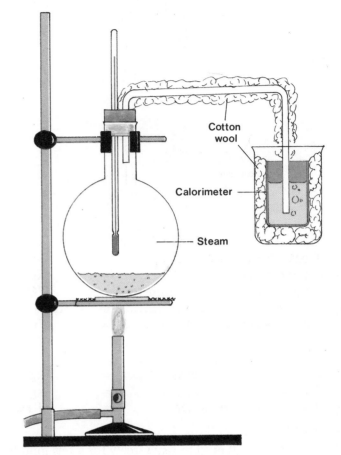

Fig. 17.8(a)

7. The specific latent heat of vaporisation of water is calculated from:
Energy lost by steam = Energy gained by (calorimeter + water).

$$m_1 l + m_1 c_w \Delta'\theta_1 = m_2 c_w \Delta\theta_2 + m_3 c_c \Delta\theta_2$$

where m_1, m_2 and m_3 are the masses of steam, water and calorimeter, respectively, and $\Delta\theta_1$ and $\Delta\theta_2$ are the fall in temperature of the condensed steam and the rise in temperature of the water and calorimeter, respectively. c_w is the s.h.c. of water and c_c is the specific heat capacity of the material of which the calorimeter is made.

Note: The accuracy of the result may be improved by passing the steam through a steam trap, *Fig. 17.8(b)*. This ensures that any water formed by condensation in the tube cannot enter the calorimeter.

181

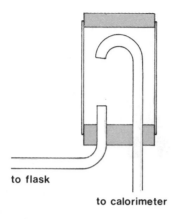

Fig. 17.8(b)

RESULTS

Mass of calorimeter	=	kg
Mass of calorimeter + water	=	kg
Initial temp. of water and calorimeter	=	°C
Temperature of steam	=	100°C
Final temp. of water and calorimeter ..	=	°C
Fall in temp. of condensed steam	=	°C
Rise in temp. of water and cal.	=	°C
Mass of cal. + water + condensed steam	=	kg
Mass of steam	=	kg
S.l.h. of vaporisation of water	=	J kg^{-1}.

Questions

1. Why is the steam allowed to issue freely for a few minutes before the tube is placed in the calorimeter?
2. Suggest two reasons why the temperature of the steam might rise above 100 °C.
3. Would there be any advantage in cooling the water before passing the steam into it? Explain.

17.3 Energy Transfer

There are many ways in which energy may be transferred from one place to another, *e.g.* by a wave motion or by a tanker with a load of oil. When internal energy is transferred from a hot body to a cold body the process is called **heating**. There are three methods of heating and these are called, respectively, **conduction**, **convection** and **radiation**.

Conduction

If you hold the end of a metal rod in a flame the end which you are holding quickly becomes hot. Energy is transferred through the rod. No part of the rod moves; energy is

transferred from one atom to another within the rod. This process is known as conduction.

> **Conduction is the transfer of energy through a substance without any bodily movement of the substance.**

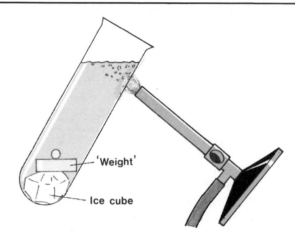

Fig. 17.9 Water is a poor conductor

Substances vary in their ability to transfer energy in this way. As we have just noted metals, generally, are good conductors. Water is a bad conductor, as can be verified by trapping a piece of ice at the bottom of a test tube of water and heating the top, *Fig. 17.9*. The water at the top will boil without the ice melting. The rates of conduction through various solids may be investigated with the apparatus shown in *Fig. 17.10*. Rods

Fig. 17.10 Investigating rates of conduction

of equal length and made from different substances are coated with candle wax and have their ends placed in hot water. The rates at which energy is transferred through the rods may be compared by noting how long it takes the wax to melt on each of them.

Very often we are more interested in preventing the transfer of energy in this way. Materials which are poor conductors are called insulators. Some examples are air and various synthetic materials, *e.g.* polystyrene foam, fibre glass, *etc.* Such materials have a wide range of applications, from handles of kettles and saucepans to heat shields on spacecraft. With the increasing cost of fuel, house insulation has become very important. Windows are insulated by "double glazing" them, *i.e.* using two sheets of glass with a layer of air trapped between them. Walls are insulated by constructing a double wall with the cavity between the walls containing air or polystyrene sheets or foam, *Fig. 17.11.* Attics are insulated by putting down layers of insulating material several cm thick on the floor of the attic, *Fig. 17.12.*

Fig. 17.11 Construction of cavity wall

The insulating properties of a building are indicated by its **U-values**. The U-value of a particular structure is a measure of the rate at which energy is conducted through it for a given temperature difference between the two sides of it. A good insulating structure therefore has a low U-value. The unit of U-value is the watt per metre squared per kelvin, $W\ m^{-2}\ K^{-1}$. In other words, if a structure has a U-value of 1 this means that 1 J per second will pass through each square metre for

each kelvin (degree Celsius) difference in temperature between the two sides of the structure.

Table 17.2 gives typical U-values for different areas of a modern house and shows the effect of added insulation. Note that these are typical values. Actual values depend on the shape, size and location of the house, materials used in its construction and the amount and type of insulation added.

Area		U-value
Roof	No insulation	1.9
	Insulation	0.4
Walls	No insulation	1.7
	Insulation	0.8
Floor	No insulation	0.8
	Insulation	0.4

Table 17.2 U-values for a dwelling house.

Fig. 17.12 Attic insulation

Convection

We saw above that water is a poor conductor. This is true of fluids (*i.e.* liquids and gases) in general. However, energy can be transferred with greater effect through fluids by a process known as **convection.**

> **Convection is the transfer of energy by the circulation of a fluid.**

Convection may be explained as follows. When water, for example, is heated in a flask, *Fig. 17.13*, water at the bottom becomes hot by conduction through the glass. When water is heated it expands and so becomes less dense. The hot water therefore rises towards the surface, its place being taken by colder water which is heated in turn. Thus currents, called **convection currents**, are set up in the water. These currents may be made visible by adding a few crystals of potassium permanganate to the water.

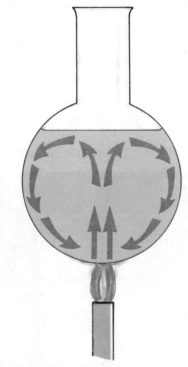

Fig. 17.13 Convection currents

Convection currents in the atmosphere and in the oceans are responsible for most meteorological changes. Clouds are formed when convection currents over the earth's surface carry warm, moist air upwards, where it expands and cools. The Trade Winds are formed when hot air over the equator rises and colder air flows in to take its place. On a hot summer day the air over the land is hotter than the air over the sea. As a result, the hot air over the land rises and the cold air flows in to take its place, thus producing cool sea breezes. At night, the land cools more quickly than the sea and so the situation is reversed, with warm land breezes being formed.

Radiation

The earth is warmed by energy from the sun, yet most of the space between the earth and the sun is virtually empty, *i.e.* it is a vacuum. Since there is neither solid nor fluid between the two bodies neither conduction nor convection can be responsible for the transfer of energy. In this case the process is called **radiation.**

> **Radiation is the transfer of energy between two points by means of electromagnetic waves.**

All bodies absorb and emit (radiate) energy. The rate at which they do so depends on their temperature and on the nature of their surface. At a given temperature a dull black surface will emit and absorb radiation faster than a highly polished surface. If the temperature of a body is the same as its surroundings the rate of emission is the same as the rate of absorption. The average wavelength of the emitted radiation depends on the temperature of the body which is emitting it. The higher the temperature the shorter the wavelength of the emitted radiation. If a piece of metal is heated it first emits infra-red radiation. As it becomes hotter it emits red light. At higher temperatures, the other colours — orange, yellow, green, *etc.* — are also emitted, until the metal becomes "white hot". At even higher temperatures ultra violet radiation may also be emitted. (*See also p.138.*)

The sun radiates energy in all directions but only a very small fraction of it falls on the earth. The amount of energy reaching a particular point on the earth's atmosphere depends on the position of the earth relative to the sun.

> **The solar constant is the amount of energy falling normally on unit area (1 m^2) of the earth's atmosphere per second when the earth is at its mean distance from the sun.**

The value of the solar constant is found experimentally to be 1.35 kW m^{-2}. The proportion of this reaching a particular point on the surface of the earth depends on the weather conditions and on the time of year. In this country the annual average value is approximately 115 W m^{-2}.

Virtually all of the energy "used" on earth comes, or came, from the sun, most of it being accumulated over long periods, *e.g.* in coal and oil. Energy from the sun may also be collected directly, *Fig. 17.14*. In one form of solar heating system,

Fig. 17.14 Solar panels used for domestic heating

radiation from the sun falls on panels containing a network of pipes through which water is pumped. Each time the water passes through the panel it becomes warmer. In this way the radiant energy from the sun is converted into internal energy in the water. (*See also p.176.*)

Solar panels may also be made of photocells which convert the radiant energy directly to electrical energy. Panels such as these are used to power orbiting spacecraft. Solar panels are usually covered with glass to increase the efficiency with which they absorb the incident radiation. Most of the radiation reaching the earth's surface has wavelengths in the visible and near (short wavelength) infra-red regions of the spectrum. Ordinary glass transmits these wavelengths but absorbs longer (and shorter) wavelengths. Solar radiation passes through the glass and is absorbed by the solar panels. The panels re-radiate some of this energy, but at much longer wavelengths since their temperature is relatively low *(see above)*. The glass is opaque to these wavelengths so the energy is effectively trapped. This is known as the "greenhouse effect", as it is the principle on which greenhouses operate.

SUMMARY

The specific heat capacity (s.h.c.) of a substance is the amount of energy required to change the temperature of 1 kg of the substance by 1 K. The heat capacity of a body is the amount of energy required to change the temperature of the body by 1 K.

The specific latent heat (s.l.h.) of a substance is the amount of energy required to change the state of 1 kg of the substance. It is called the s.l.h. of fusion when the change is from solid to liquid and the s.l.h. of vaporisation when the change is from liquid to gas. A heat pump transfers energy from a cold body to warmer one. It does so by first vaporising a liquid within the cold body and then condensing it within the warmer body by increasing the pressure.

Internal energy may be transferred from one body to another by conduction, convection or radiation. Conduction involves no bodily movement of the medium (usually solid); convection involves currents circulating in fluids; radiation is energy transfer by electromagnetic waves.

The U-value of a structure is a measure of its insulating properties — the lower the U-value the better the insulation. Convection currents are responsible for most meteorological changes. The solar constant is the amount of radiant energy from the sun falling normally on unit area of the earth's atmosphere per second when the earth is at its mean distance from the sun.

Questions 17

DATA: s.h.c. of water   = 4180 J kg^{-1} K^{-1}
     s.h.c. of copper   = 390 J kg^{-1} K^{-1}
     s.l.h. of fusion of ice   = 3.3 × 10^5 J kg^{-1}
     s.l.h. of vaporisation of water = 2.3 × 10^6 J kg^{-1}.

Section I

1. What is meant by the internal energy of a body?.......
....
....

2. Define specific heat capacity...................................
....
....

3. The unit of specific heat capacity is ....

4. What is the difference between specific heat capacity and heat capacity?..
....
....

5. Define specific latent heat.......................................

6. The unit of specific latent heat is ....

7. What is the difference between s.l.h. of vaporisation and s.l.h. of fusion?...
....
....

8. What is the function of a heat pump?......................
....

9. In the absence of a refrigerator a carton of milk may be kept cold by wrapping it in a wet cloth and leaving it outside. Explain...
....

10. Leaving the refrigerator door open does not cool the kitchen — in fact it results in the kitchen becoming slightly warmer. Explain...
....

11. The three methods of heating are called ....
....

12. Several thin layers of clothing will keep a person warmer than one thick layer. Explain...............................
....

13. How may the U-value of a structure be reduced?.......
....

14. What is meant by the solar constant?......................
....

15. What is the greenhouse effect?.............................

Section II

1. Calculate the amount of energy required to raise the temperature of 400 g of water from 10 °C to 40 °C.

2. A heating coil, connected to a joulemeter, is placed in a calorimeter of heat capacity 30 J K^{-1} and containing 100 g of a certain liquid at a temperature of 20 °C. When the joulemeter registers 4.8 kJ, the current is switched off and it is found that the temperature of the calorimeter and contents is 40 °C. Calculate the s.h.c. of the liquid. State any assumptions you make and say under what conditions they are justified.

3. When 40 g of warm copper were placed in 25 g of water at 15.2 °C, it was found that the temperature of the water increased to 17.4 °C. Assuming that no energy was lost to, or gained from, the surroundings calculate the original temperature of the copper.

4. A heating coil raises the temperature of 200 g of water by 10 K in five minutes. It also raises the temperature of 420 g of oil by 6 K in the same time. Assuming that the same amount of energy is supplied in both cases calculate the s.h.c. of the oil.

5. Water flows at a rate of 16 cm^3 per second through a pipe which is heated by a 500 W heater. Calculate the change in temperature of the water as it passes through the pipe. (Density of water $= 1.0 \times 10^3$ kg m^{-3}.)

6. Liquid A, of mass $4m$ and temperature 2θ, is mixed with liquid B, of mass $5m$ and temperature 7θ. The final temperature of the mixture is 4θ. Assuming that no energy is lost to the container or the surroundings calculate the ratio of the specific heat capacities of the two liquids.

7. A block of copper is released from rest at a height of 60 m. On striking the ground 20% of the kinetic energy of the block is converted to internal energy in the block. Calculate the rise in temperature of the block.

8. A car of mass 1200 kg travelling at a speed of 25 m s^{-1} is braked to rest. Assuming that 20% of the kinetic energy of the car is converted to internal energy in the steel brake drums, each of mass 1.6 kg, calculate the rise in temperature of the brake drums. (S.h.c. of steel $= 500$ J kg^{-1} K^{-1}.)

9. Calculate the amount of energy required to convert 60 g of ice at 0 °C to water at the same temperature.

10. An aeroboard cup contains 40 cm^3 of water at 25 °C. Ice at 0 °C is added to the water until its temperature falls to 5 °C. Assuming that all the ice has melted and that there is no transfer of energy to or from the surroundings, calculate the mass of ice added. (Density of water $= 1.0 \times 10^3$ kg m^{-3}.)

11. Calculate the total amount of energy required to convert 2.0 kg of ice at -20 °C to steam at 100 °C. (S.h.c. of ice $= 2.1 \times 10^3$ J kg K^{-1}.)

12. A solar panel has a total area of 2.0 m^2. Water is pumped through the panel at a rate of 100 cm^3 s^{-1}. Given that the solar constant is 1.35 kW m^{-2} and that 20% of this is absorbed by the panel, calculate the increase in the temperature of the water as it passes through the panel. Account for the other 80% of the solar energy. (Density of water $= 1.0 \times 10^3$ kg m^{-3})

REVISION EXERCISES C

DATA: s.h.c. of water   = 4180 J kg^{-1} K^{-1}
s.h.c. of copper   = 390 J kg^{-1} K^{-1}
s.h.c. of aluminium   = 910 J kg^{-1} K^{-1}
s.l.h. of fusion of ice   = 3.3 × 10^5 J kg^{-1}
s.l.h. of vaporisation of water = 2.3 × 10^6 J kg^{-1}.

Section I

1. A diffraction grating was used with a spectrometer to determine the wavelength of sodium light. Bright images were seen for the following positions of the telescope: 143° 39′, 150° 42′; 157° 30′; 164° 20′; 171° 9′; 177° 55′; 185° 2′. Assuming that the fourth reading corresponds to the zero order image and given that the grating had 200 lines per mm, calculate the wavelength of the light. Describe the adjustments which must be made to a spectrometer before it may be used.

2. In an experiment to investigate the factors determining the natural frequencies of a stretched string the natural frequencies of a given wire were determined for a series of values of (i) the length, the tension being kept constant at 10.0 N, (ii) the tension, the length being kept constant at 1.00 m. The results obtained are as follows.

l/m	1.00	0.90	0.80	0.70	0.60	0.50	0.40	0.30
f/Hz	212	236	268	303	350	422	531	698

T/N	4.0	6.0	8.0	10.0	12.0	14.0	16.0	18.0
f/Hz	128	158	180	205	224	237	254	273

Draw suitable graphs to illustrate the relationship between the frequency and the length and between the frequency and the tension. From the graphs determine the mass per unit length of the wire, assuming that the frequencies given are the fundamental frequencies.

3. In a resonance tube experiment to determine the speed of sound in air the shortest length of air column which resonated with a series of tuning forks was found. The results obtained were as follows.

f/Hz	256	288	320	341	384	480	512
l/cm	33.3	29.4	26.3	24.7	22.3	17.5	16.4

Calculate, from a suitably drawn graph, the speed of sound in air given that the diameter of the tube was 4.2 cm. What is the most likely source of error in this experiment?

4. In an experiment to verify Boyle's law the following values were obtained for the pressure and volume of a fixed mass of gas.

p/kPa	100	120	140	160	180	200	220	240	260	280
V/cm^3	52	43	37	33	28	26	23	21	20	19

Using the given data plot a suitable graph and explain how this verifies Boyle's law. Give two precautions which should be taken when carrying out this experiment to ensure a more accurate result.

5. The following results were obtained in an experiment to determine the specific heat capacity of a liquid by an electrical method.

    Mass of calorimeter .... = 0.057 kg
    Mass of calorimeter + liquid .... = 0.142 kg
    Initial temp. of liquid and calorimeter = 12 °C
    Final temp. of liquid and calorimeter = 25 °C
    Energy supplied .... = 2.61 kJ

Given that the calorimeter was made of copper (s.h.c. 390 J kg^{-1} K^{-1}) calculate the specific heat capacity of the liquid. Give three precautions which should be taken when carrying out this experiment to ensure a more accurate result.

6. In an experiment to compare the specific heat capacities of copper and water the following results were obtained.

    Mass of copper rivets .... = 0.034 kg
    Mass of calorimeter .... = 0.057 kg
    Mass of calorimeter + water .... = 0.102 kg
    Temperature of hot copper .... = 100 °C
    Temp. of cold water and calorimeter = 15 °C
    Final temp. of water, copper and
    calorimeter .... = 20 °C

Given that the calorimeter was made of copper (s.h.c. 390 $J kg^{-1} K^{-1}$) calculate the ratio of the specific heat capacity of water to that of copper. Apart from steps taken to reduce exchange of energy between the calorimeter and its surroundings give two precautions which should be taken to ensure a more accurate result.

7. Ice was mixed with water in an aluminium calorimeter in an experiment to determine the specific latent heat of fusion of ice. The following measurements were recorded.

Mass of calorimeter	= 0.044 kg
Mass of calorimeter + water	= 0.105 kg
Initial temp. of water and calorimeter =	24 °C
Temperature of ice .... =	0 °C
Final temp. of water and calorimeter =	5 °C
Mass of cal. + water + melted ice	= 0.121 kg

Using the data given calculate the specific latent heat of fusion of ice. What was the least accurate measurement in this experiment? Give three steps which should be taken to improve the accuracy of this experiment.

8. The specific latent heat of vaporisation of water was determined by passing steam into cold water in an aluminium calorimeter. The following results were obtained.

Mass of calorimeter	= 0.044 kg
Mass of calorimeter + water	= 0.122 kg
Initial temp. of water and calorimeter =	8 °C
Temperature of steam .... =	100 °C
Final temp. of water and calorimeter =	23 °C
Mass of cal. + water + condensed steam	= 0.124 kg

From the values given calculate the specific latent heat of vaporisation of water. What was the least accurate value determined in this experiment? How might the accuracy of this particular determination have been improved? What effect would this have had on the overall accuracy of the experiment?

Section II

1. What is meant by the term diffraction effects? Why is it much more difficult to produce diffraction effects with light than with sound?

2. Why is the wave theory of light considered to be superior to the corpuscular theory? Quote experimental evidence in support of your answer.

3. What is monochromatic light? Describe an experiment to determine the wavelength of monochromatic light.

4. A student uses a diffraction grating and a spectrometer to determine the wavelength of one of the lines of the mercury spectrum. If the diffraction grating has 600 lines per mm and she observes that particular line of the spectrum at angles of 10° 15′, 33° 45′, 54° 10′, 74° 30′ and 98° 15′, what is the wavelength of the light? (Assume 54° 10′ is the position of the zero order image.)

*5. Explain how the speed of light may be determined. Why is it important that the speed of light be known accurately.

6. How would you show that sound is a wave motion? In what ways do sound waves differ from light waves?

7. What distinguishes a noise from a musical note? Give three ways in which one note may differ from another and explain how these characteristics are related to the physical properties of the wave.

8. Given that the speed of sound in air is 340 m s^{-1}, what is the wavelength of a note whose frequency is 2.0 kHz?

9. Describe how you would determine the speed of sound in air in the laboratory.

*10. Explain the terms: Sound intensity; Threshold of audibility. What are the advantages of having a logarithmic scale for the measurement of intensity level? If the sound intensity in a particular area doubles, what is the relative increase in intensity in decibels?

11. State Boyle's law and explain how you would verify it experimentally.

12. Explain how a temperature scale is established. Name three types of thermometer and in each case state which physical property is being used to define the temperature scale.

13. In an experiment to calibrate a resistance thermometer, the resistance in a mixture of ice and water was found to be 60 Ω while the resistance in steam over boiling water was found to be 84 Ω. When placed in a beaker of warm

water the resistance was 66 Ω. What was the temperature of the water?

14. Describe an experiment to calibrate a mercury-in-glass thermometer. What are the advantages of using mercury rather than alcohol? What are the disadvantages?

15. The length of the mercury column in a mercury thermometer is 1.5 cm at the ice point and 19.5 cm at the steam point. What is the temperature when the length is 15.0 cm?

*16. Explain what is meant by Avogadro's constant. The relative molecular mass of oxygen is 32. Calculate the number of molecules in one litre of oxygen. (Take density of oxygen = 1.43 kg m^{-3}. Avogadro's constant, N_A = 6.0 × 10^{23} mol^{-1}.)

*17. State the assumptions of the kinetic theory of gases. Explain how these assumptions are verified by experiment. According to the kinetic theory, what is the relationship between the temperature of a gas and the r.m.s. speed of its molecules?

18. Define specific heat capacity and explain how you would determine the s.h.c. of a liquid.

19. When 50 g of a metal at 100 °C are added to 90 g of water at 15 °C, the highest temperature of the mixture is 20 °C. Assuming no energy loss to, or gain from, the container or surroundings, calculate the s.h.c. of the metal.

20. Define specific latent heat. Describe how you would determine experimentally the s.l.h. of fusion of ice.

21. When 10 g of ice at 0 °C are placed in 42 g of water at 25 °C, the final temperature of the water is found to be 5 °C. Assuming that there is no energy gained from, or lost to, the container or surroundings, calculate the s.l.h. of fusion of ice.

22. When 5.0 g of steam at 100 °C are piped into 70 g of water at 12 °C the final temperature of the water is found to be 54 °C. Calculate the s.l.h. of vaporisation of water. What assumptions must be made in doing this calculation?

23. Describe an experiment to determine the s.l.h. of vaporisation of water. If 6.0 g of steam at 100 °C are added to 100 g of water at 15 °C what is the final temperature of the water?

24. When 12 g of ice at 0 °C are added to 80 g of water the final temperature of the water is 8 °C. What was the initial temperature of the water?

18 Electrostatics I

18.1 Electric Charges

Rub a plastic comb or biro on your sleeve and you will find that it has become capable of picking up small pieces of paper. We say that the comb or biro has become **electrically charged**. We can find out more about the properties of charged bodies using the simple apparatus shown in *Fig. 18.1*. A polythene rod is charged by rubbing it with a cloth and is placed in the paper stirrup. If a second polythene rod is charged in the same way and brought near to the first they will be found to repel each other, the suspended one swinging away. However, if a rod of a different material, *e.g.* perspex, is charged and brought near the suspended rod they will be found to attract each other. Obviously, the charge on the polythene rod is different from the charge on the perspex rod. It is clear from these experiments that there are two types of force between charged bodies — attractive forces and repulsive forces. It follows that there must be two types of charge. These two types of charge have been called **positive charge** and **negative charge**. As we have just seen bodies which have the same charge repel each other, while bodies which have different charges attract each other. This important observation is usually summarised as follows.

> **Like charges repel each other; unlike charges attract each other.**

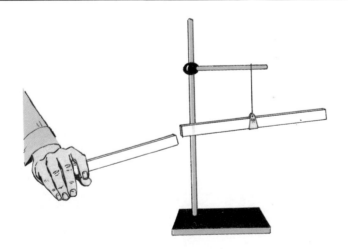

Fig. 18.1 Investigating forces between charges

The forces which charged bodies exert on each other are known as **electrostatic** forces.

If either a positively charged body or a negatively charged body is brought near an uncharged body they will be found to attract each other. As we shall see later an uncharged body contains both positive and negative charge in equal amounts, evenly distributed throughout the body. When, for example, a positively charged rod is brought near such a body the negative charges move to the side nearest the rod, *Fig. 18.2*.

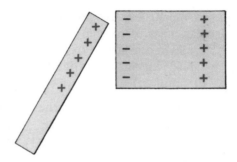

Fig. 18.2

The rod and the negative side of the uncharged body then attract each other. The rod is repelled by the positive side of the uncharged body but because the distance between the positive charges is greater the repulsive force is less than the attractive force (*see p.195*). Attraction between a negatively charged rod and an uncharged body is explained in a similar manner.

The Van de Graaff Generator

The amounts of charge which can be obtained by rubbing plastic rods are relatively small. Because of this, various machines have been designed to produce large amounts of charge reasonably quickly. One of these is the Van de Graaff generator, invented in 1931 by an American physicist, Robert Van de Graaff (1901 — 1967).

In one version of this machine, *Fig. 18.3*, a rubber belt passes over two pulleys. The upper pulley is situated inside a metal dome, while the lower one can be turned, either manually or by a small electric motor. Near the bottom pulley the belt passes close to a sharply pointed metal "comb", B, which is positively charged by a high voltage generator or battery. Due to the point effect (*see p.201*) positive charges are sprayed off the comb onto the belt, which carries them upwards. Inside the dome a second metal comb, C, removes the charge from the belt.

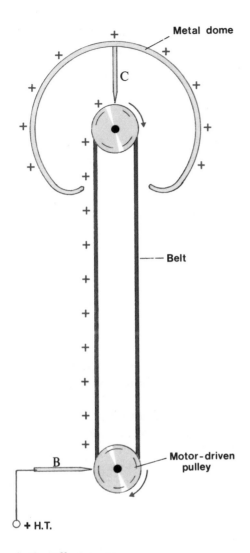

Fig. 18.3 Van de Graaff generator

Fig. 18.4 Lightning: Charge flowing between a highly charged cloud and earth

dutch metal or aluminium) leaves. When charge is placed on the leaves, via the cap, they separate or diverge due to the repulsion between like charges. The leaves are enclosed in a metal case with glass sides which protects the delicate leaves from air currents and also makes the electroscope more sensitive due to the opposite charge induced *(see below)* on the inside of the case.

The charge then flows to the outside of the dome where it builds up *(see p.204)*. With large machines voltages of several million volts may be produced. In many of the smaller machines found in school laboratories the belt is charged by contact with one of the pulleys, otherwise their principle of operation is essentially the same as just described.

The Gold Leaf Electroscope

To learn more about the nature of charges we need a more sensitive instrument than a plastic rod hanging from a thread. Such an instrument is the gold leaf electroscope, *Fig. 18.5*. It consists of a metal cap attached to a pair of thin metal (gold,

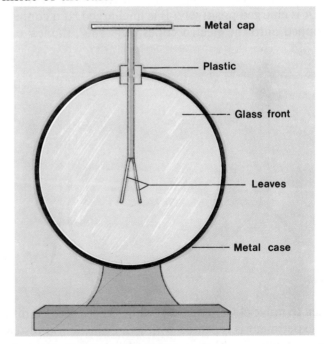

Fig. 18.5 Gold leaf electroscope

Positive and Negative Charge

If a plastic rod is rubbed with a cloth and then placed on the cap of the electroscope the leaves diverge as we noted earlier. Alternatively, if the cloth, rather than the rod, is placed on the electroscope the leaves also diverge. From this we may conclude that, when two bodies are brought into close contact, both become charged. Now, if the rod is rubbed with the cloth and both are placed together on the electroscope nothing happens. Yet, if either one is removed the leaves diverge. (These experiments are best done with the rod and/or cloth inside a tall metal can placed on the cap of the electroscope.) The charge on the rod exactly cancels the charge on the cloth. Thus, if the rod has a positive charge the cloth has an equal negative charge. Since neither body was charged before they were rubbed together it follows that a given amount of charge must have been transferred from one to the other. From these, and similar experiments, we may conclude that all bodies have equal amounts of positive and negative charge. When two bodies are brought into close contact some charge is transferred from one to the other, giving one a net negative charge while the other acquires a net positive charge.

Conductors and Insulators

Consider two electroscopes, A and B (cases not shown), *Fig. 18.6*. A is charged positively; B is uncharged. If a copper wire is dropped onto the electroscopes, *Fig. 18.7*, the leaves of A

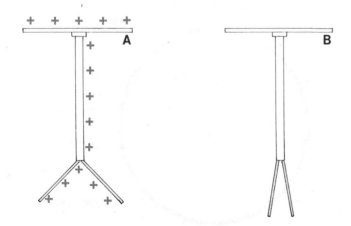

Fig. 18.6

are seen to move closer together, while the leaves of B diverge. If the experiment is repeated with a dry plastic rod instead of the copper wire nothing happens. In the first case, charge

moves from A to B; in the second case it does not. We say that the copper wire is a **conductor** and that the plastic rod is an **insulator.**

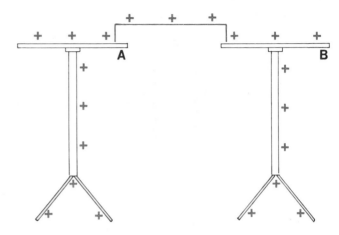

Fig. 18.7

> **Conductors allow charge to flow through them; insulators do not.**

In general, bodies made of metal are conductors; bodies made of glass, rubber, plastic, *etc.*, are insulators. Pure water is an insulator but ordinary tap water conducts charge due to the presence of dissolved impurities.

Another important conclusion regarding the nature of charge may be drawn from this experiment. Charge will flow from one body to another, provided there is a conducting path between them, until it has reached the same "level" in both bodies. (This is very similar to water flowing from a full tank into an empty one until the water reaches the same level in both.) We shall return to this topic in *Chapter 20*.

Earthing

If you charge an electroscope and then touch it with your finger, the leaves immediately collapse. Where does the charge go? The answer is that it flows into the ground; your body is therefore a conductor. This is called earthing the electroscope. Since the earth is so large the addition of the charge from an electroscope has no noticeable effect.

Charging by Induction

At the beginning of this chapter we saw how two bodies, both of which were insulators, could be charged by rubbing them together. This process is known as charging by contact. We shall now investigate a second method of charging a body, in this case a conductor.

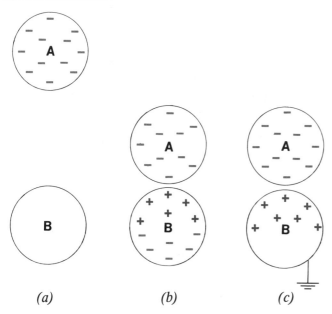

(a)   *(b)*   *(c)*

Fig. 18.8

Consider two metal spheres, A and B, *Fig. 18.8(a)*. A is charged negatively; B is uncharged, *i.e.* it has equal amounts of positive and negative charge. When A is brought near to B, *Fig. 18.8(b)*, some negative charges in B move to the side away from A, leaving the side nearer A positive. If B is now earthed, *Fig. 18.8(c)*, the negative charges will go to ground, leaving B with a net positive charge. B is said to have been **charged by induction**. If A is to be removed the earth connection must first be broken. If the earth connection is not broken the positive charges on B will also go to earth as soon as A is removed. (In practice, as we shall see in the next section, it is usually the negative charges which move while the positive charges are fixed. In this case, negative charges would move up from the ground to cancel the positive charges on B.) Note that no charge was removed from A during the process of charging B by induction, so one charged body could be used to charge any number of other bodies. Obviously, the method would not work if B were made of an insulating material.

18.2 Atomic Theory

Although the properties of charged bodies have been known for over two thousand years, it was not until the latter half of the last century and the beginning of the present one that a theory giving a satisfactory explanation of these properties was developed.

According to this theory, all matter is made up of tiny particles, called **atoms**. Each atom in turn is made up of even smaller particles, of which there are three types — **protons**, **neutrons** and **electrons**. Protons and neutrons form the nucleus of the atom, *Fig. 18.9*. They have approximately the same size and mass and together make up 99.98% of the total mass of the atom. However, they occupy only about 10^{-12}% of the total volume of the atom. The electrons are very small particles which move around at various distances from the nucleus. In any given atom the number of electrons is normally equal to the number of protons.

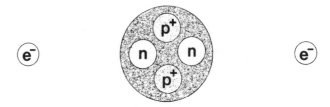

Fig. 18.9

What is of interest to us here is the fact that, according to this theory, the proton has a positive charge while the electron has a negative charge of equal magnitude and the neutron is uncharged. However, when two bodies come into close contact, electrons may be removed from the atoms of one body and become attached to the atoms of the other. Thus, one body becomes negatively charged (the one which gains the electrons) while the other becomes positively charged because it now has more protons than electrons.

We shall now see how the other phenomena, discussed in the previous section, may be explained in terms of this model of the atom.

Conductors and Insulators

In most metals the outer electron(s) in each atom are only weakly attracted to the nucleus and are, therefore, more or less free to move through the metal. If a metal wire is connected from a negatively charged body (A) to an uncharged one (B), the "free" electrons in the wire are repelled by the negative charge on A and move onto B, their place being taken by some of the excess electrons on A. Thus, the negative charge on A is reduced while B gains an negative charge equal to that lost by A. If A is initially positively charged the situation is reversed. Free electrons are attracted into A, reducing the positive charge on A and leaving B with a net positive charge.

In an insulating material all the electrons are strongly attracted to the nuclei; there are no free electrons. Hence, charge cannot be transferred through an insulator. (If sufficient energy is supplied to an insulator some electrons may be freed from their nuclei. The insulator then becomes a conductor.)

Charging by Induction

This process is easily explained in terms of free electrons. Referring back to *Fig. 18.8*, when A is brought near to B the electrons in B are repelled by the excess electrons on A. When B is earthed they are repelled still further, *viz.* to the ground, leaving B with more protons than electrons, *i.e.* with a net positive charge.

18.3 Electric Fields

Consider any positively charged body, A. All charges in the vicinity of A will experience a force — a repulsive force if they also are positive; an attractive force if they are negative. We say that the region around A, or any other charged body, is an **electric field.**

An electric field is a region in which electric charges at rest experience a force.

The direction of the force experienced by a charge in an electric field depends on the sign of the charge itself and on the sign of the charge, or charges, which are the source of the field. It is often useful to draw a diagram of an electric field with lines which show the direction of the force. By convention, these lines go from a positive charge to a negative charge, *i.e.* they give the direction in which a positive charge would move. *Fig. 18.10* shows the field lines for a pair of charged spheres while *Fig. 18.11* shows the lines arising from a pair of charged parallel plates. Where the field lines are closest together the field is strongest and where the lines are furthest apart the field is weakest. Where the lines are equally spaced the field is of uniform strength, *i.e.* over the area in question a charged body would experience the same force. (It is useful to compare electric field lines with contour lines on a map. Where the contour lines are close together the ground is rising steeply; where the contour lines are far apart the ground is almost flat.) Note that field lines cannot stop in "mid-air"; they must begin

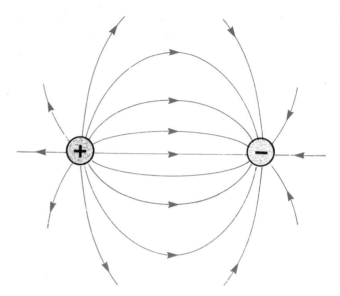

Fig. 18.10 Electric field due to two charged spheres

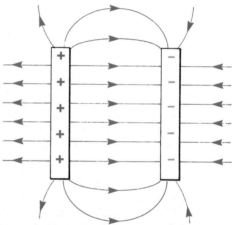

Fig. 18.11 Electric field due to two charged plates

on a positive charge and end on a negative charge. Those whose ends (or beginnings) are not shown in *Figs. 18.10* and *18.11*, end (or begin) on induced charges on surrounding objects, the walls of the room, *etc.*

Electric field lines may be "plotted" using the arrangement shown in *Fig. 18.12*. Two conductors (pieces of aluminium foil) of any required shape are connected to the terminals of a high voltage (c. 2 kV) power supply and are placed in a container of olive oil, on the surface of which is spread some semolina. When the power supply is switched on the pieces of aluminium become charged. The particles of semolina then "line up" along the field lines.

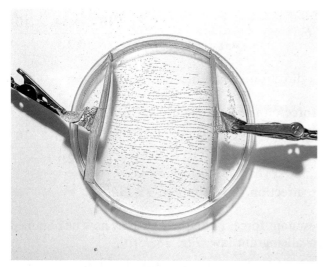

Fig. 18.12 Plotting electric field line

Coulomb's Law

The factors which determine the magnitude of the force between two charged bodies were investigated by Charles Coulomb in 1784. The relationship which he established is now known as **Coulomb's law** and it may be stated as follows.

> **The force between two point charges is proportional to the product of the charges and inversely proportional to the square of the distance between them.**

A point charge is one which may be considered to be concentrated at a point. In symbols, Coulomb's law may be written as

$$F \propto \frac{Q_1 Q_2}{r^2}$$

where Q_1 and Q_2 are the magnitudes of the charges and r is the distance between them. The direction of F is along the line joining the charges. The unit of charge is the **coulomb**, C. *(For a definition of this unit see p.216.)*

Coulomb's law may also be written in the form

$$F = k \frac{Q_1 Q_2}{r^2}$$

where k is a constant.

The value of k is found, experimentally, to depend on the nature of the medium between the charges. For historical reasons k is usually written as $1/4\pi\varepsilon$, where ε is called the **permittivity** of the medium. Thus, we have

$$F = \frac{1}{4\pi\varepsilon} \frac{Q_1 Q_2}{r^2} \qquad \text{(i)}$$

The permittivity may be thought of as a measure of the extent to which the medium "permits" the force between the charges to act through it. From *Eqn. (i)* we see that the larger the value of ε the smaller the force between the charges. For the special case of charges in a vacuum the permittivity is given the symbol ε_0, and is called the **permittivity of free space**. ε_0 has the value 8.9×10^{-12} C^2 N^{-1} m^{-2}. *(For an equivalent, and more commonly used, unit for permittivity see p.210.)* A simple calculation using the given value of ε_0 shows that the value of $1/4\pi\varepsilon_0$ is 8.9×10^9 N m^2 C^{-2}. For practical purposes the permittivity of air is taken to be equal to ε_0. The ratio of the permittivity of a medium to the permittivity of free space is called the relative permittivity, ε_r, of the medium. That is

$$\varepsilon_0 = \frac{\varepsilon}{\varepsilon_r}$$

Coulomb's law, like Newton's law of gravitation (*p.66*), is an example of the **inverse square law**. The force in each case varies inversely with the square of the distance. It should be noted however that, although the two laws are similar in form, the two types of force involved differ greatly in magnitude, the electrostatic force being by far the larger. *(See, for example, Question 2 at the end of this chapter.)* The two types of force also differ in the fact that, while the gravitational force is always an attractive force, the electrostatic force may be either attractive or repulsive.

Example

Calculate the magnitude of the force between two point charges of 4.0 nC each, if the medium between them is air and the distance between them is 5.0 mm.

$$F = \frac{1}{4\pi\varepsilon} \frac{Q_1 Q_2}{r^2}$$

$$= 8.9 \times 10^9 \times \frac{(4 \times 10^{-9})^2}{(5 \times 10^{-3})^2}$$

$$= 8.9 \times 10^9 \times \frac{16 \times 10^{-18}}{25 \times 10^{-6}}$$

$$= \frac{8.9 \times 16}{25} \times 10^{-3}$$

$$= 5.70 \times 10^{-3} \text{ N}$$

$$= 5.70 \text{ mN}$$

Ans. 5.7 mN

Example

Three positive charges, of 2 µC, 3 µC and 4 µC, respectively, are arranged at the vertices of an equilateral triangle of side 20 cm as shown in Fig. 18.13(a). Calculate the magnitude of the resultant force on the charge of 2 µC.
The force on the 2 µC due to the 4 µC is

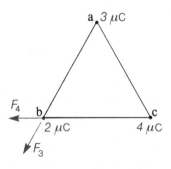

Fig. 18.13(a)

$$F_4 = \frac{1}{4\pi\varepsilon_0} \times \frac{2 \times 10^{-6} \times 4 \times 10^{-6}}{(0.2)^2}$$

in the direction shown in *Fig. 18.13(a)*

The force on the 2 µC due to the 3 µC is

$$F_3 = \frac{1}{4\pi\varepsilon_0} \times \frac{2 \times 10^{-6} \times 3 \times 10^{-6}}{(0.2)^2}$$

in the direction shown in *Fig. 18.13(a)*

The resultant force, F, on the 2 µC may now be obtained using the parallelogram law, *Fig. 18.13(b)*.

$$F^2 = F_4^2 + F_3^2 - 2F_4F_3 \cos 120°$$

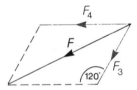

Fig. 18.13(b)

$$= \frac{10^{-12}}{4\pi\varepsilon_0(0.2)^2} \times (8^2 + 6^2 + 2 \times 8 \times 6 \times 0.5)$$

$$F = \frac{10^{-12}}{4\pi\varepsilon_0(0.2)^2} \times \sqrt{148}$$

$$= \frac{8.9 \times 10^{-3}}{0.04} \times \sqrt{148}$$

$$= 2.71 \text{ N}$$

Ans. 2.7 N

Charles Coulomb (1736 — 1806)

Coulomb was born in Angoulême, a town in south-west France. He served for a number of years as a military engineer in the West Indies. He returned to France in 1776 and, on the outbreak of the Revolution, retired to a small estate at Blois, where he devoted himself to scientific research.

In 1777, Coulomb invented the torsion balance, a device for measuring very small forces by the twist they produce in a fine fibre. Using this balance, he investigated the forces between charged spheres and discovered the relationship which is now known as Coulomb's law and for which he is now chiefly remembered. (This relationship was actually discovered by the English scientist, Henry Cavendish, but his results were not published until 1879 and so the credit goes to Coulomb.)

In addition to his electrical experiments, Coulomb also carried out work on magnetism, friction and elasticity. In 1802, just four years before his death, he was appointed an inspector of public education.

SUMMARY

When two bodies come into close contact they sometimes acquire the ability to attract other bodies and are said to be electrically charged. There are two types of charge — positive and negative. Like charges repel each other; unlike charges attract each other. The forces which exist between charges at rest are called electrostatic forces and we say that the region around a charge is an electric field. Bodies which allow charge to flow through them are called conductors; those which do not are called insulators. A conductor may be charged by induction by earthing it and bringing it near to a charged body.

All properties of charged bodies may be explained in terms of the atomic theory which holds that all matter is made up of atoms which in turn consist of a positive nucleus surrounded by negative electrons. Charging a body consists of removing electrons from it (the body is then positively charged) or adding electrons to it (the body is then negatively charged). The magnitude of the force between two point charges, Q_1 and Q_2 which are a distance r apart is given by Coulomb's law

$$F \propto \frac{Q_1 Q_2}{r^2}$$

This law may also be written as

$$F = \frac{1}{4\pi\varepsilon} \frac{Q_1 Q_2}{r^2}$$

where ε is a constant called the permittivity of the medium between the two charges.

Questions 18
DATA: $\varepsilon_0 = 8.9 \times 10^{-12}$ C^2 N^{-1} m^{-2}.

Section I

1. Like charges................. ; unlike charges.................

2. The Van de Graaff generator is a machine for ....

....

3. Name the main parts of a gold leaf electroscope........

....

....

4. If a plastic rod is rubbed with a cloth inside a can on a gold leaf electroscope the leaves do not diverge. Explain.

....

....

5. Explain the terms: conductor; insulator. Give two examples of each..

   ....

   ....

6. What happens when a charged body is earthed?.........

   ....

7. Name two methods of charging a body....................

   ....

8. Which method of charging can be used only with conductors?...

   ....

9. Name the three types of particle found in an atom.....

   ....

10. In the atom, the..

    has a positive charge, the......................................

    has a negative charge and the................is uncharged.

11. What is an electric field?......................................

    ....

12. What is an electric field line?...............................

    ....

13. Give an expression for Coulomb's law....................

    ....

14. What does the symbol ε represent? What is the significance of this quantity?..

    ....

    ....

15. What is meant by saying that a quantity obeys the inverse square law?...

    ....

    ....

Section II

1. Describe how a body may be charged negatively by induction. Why is it not possible to charge an insulator by induction?

2. A charge of 4.5×10^{-9} C is placed on each of two lead spheres. Given that there is air between the spheres and that the distance between their centres is 5.0 cm, calculate the magnitude of the electrostatic force between them. If the mass of each sphere is 200 g, calculate the magnitude of the gravitational force between them. ($G = 6.7 \times 10^{-11}$ N m^2 kg^{-2}.)

3. A negative charge of 2.5 nC and a positive charge of 4.0 nC are placed 50 mm apart in air. Find the resultant force on a positive charge of 2.0 nC placed 20 mm from the negative charge and on the line joining them. At what position would the resultant force be zero?

*4. Three positive charges, each of 0.2 μC, are placed at three corners, A, B and C, of a square of side 40 cm. Calculate the magnitude of the force on a positive charge of 1 μC placed at the fourth corner of the square.

*5. Repeat the calculation in Q.4 for the case where the charge at corner A is negative.

*6. A positive charge of 0.15 μC and a negative charge of 0.25 μC are placed at A and B, respectively, where |AB| is 10 cm. Calculate the magnitude of the resultant force on a positive charge of 1.0 μC placed on the perpendicular bisector of AB at a distance of 12 cm from AB.

7. Two electrons are situated 1.0 pm apart in a vacuum. Calculate the magnitude of their initial acceleration. Explain why it is not necessary to consider the gravitational force between the two electrons in this calculation. (Charge on electron = 1.6×10^{-19} C; mass of electron = 9.1×10^{-31} kg.)

*8. Two pith balls, each of diameter 1 cm and mass 20 mg, are hanging on the ends of two vertical nylon threads, each 1.0 m long, so that their surfaces are in contact. Both are charged simultaneously. They move apart and come to rest when the angle between the threads is 10°. Calculate the magnitude of the charge on each ball, assuming that both carry the same charge. (Acceleration due to gravity, $g = 9.8$ m s^{-2}.)

19 Electrostatics II

19.1 Potential

When water flows from one body to another we say that there is a pressure difference between them; when internal energy flows from one body to another we say that there is a temperature difference between them. Likewise, when charge flows from one body to another we say that there is a **potential difference (p.d.)** between them.

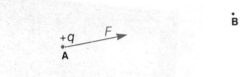

Fig. 19.1

Imagine a small positive charge, q, near to another positive charge, Q, *Fig. 19.1*. The small charge will experience a force, F, in the direction shown (Q will also experience a force of magnitude F in the opposite direction) and, if free to do so, will move in the direction of F. If q moves from A to B a certain amount of work will have been done (*see p.87*). Also, since the charge moves from A to B, there must be a potential difference between the two points. Thus, p.d., V, may defined in terms of the work done on, or by, a charge.

> **The p.d. between two points is the work done in bringing unit charge from one point to the other.**

Since work is a scalar quantity, p.d. is also a **scalar** quantity. It follows from the definition that, if the p.d. between two points is V, the total work done in bringing a charge Q from one point to the other is given by

$$W = VQ$$

Conversely, this is the amount of energy released if a charge Q moves in the direction of the force.

The unit of potential difference is the **volt**, V, so called in honour of the Italian physicist, Alessandro Volta, who invented the first battery. The volt is defined as follows.

> **The p.d. between two points is 1 V if the work done in bringing a charge of 1 C from point to the other is 1 J.**

Example

If the p.d. between two points is 100 V what is the work done in bringing a charge of 2.4 nC from one point to the other?

$$
\begin{aligned}
W &= VQ \\
&= 100 \times 2.4 \times 10^{-9} \\
&= 2.4 \times 10^{-7} \text{ J} \qquad \text{Ans. } 2.4 \times 10^{-7} \text{ J}
\end{aligned}
$$

The Electronvolt

A unit sometimes used to measure very small amounts of energy is the electronvolt, eV.

> **One electronvolt is the energy gained by an electron in being accelerated through a potential difference of 1 V.**

The charge on an electron is found experimentally (*see Chapter 26*) to be 1.6×10^{-19} C. Therefore, the work done (and the energy gained) when an electron is accelerated through 1 V is

$$
\begin{aligned}
W &= VQ \\
&= 1 \times 1.6 \times 10^{-19} \\
&= 1.6 \times 10^{-19} \text{ J}
\end{aligned}
$$

That is, $1 \text{ eV} = 1.6 \times 10^{-19}$ J.

The eV is a convenient unit for measuring very small energies. It is not, however, part of the SI system so energies given in electronvolts must be converted to joules before calculations are done.

Zero Potential

As we have seen, when a charge flows from A to B we say that there is a p.d. between A and B. If the moving charge is positive we say that A is at a higher potential than B, *i.e.* a positive charge will move from a point of higher potential to a point of lower potential (assuming, of course, that there is a conducting path between the two points). Conversely, a negative charge will move from a lower potential to a higher potential. Generally, it is not necessary to know the actual potential of either A or B; only that one is at a higher, or lower, potential than the other. However, it is sometimes convenient to be able to say what the actual potential of a particular point is. To do this we must establish a zero of potential. (This is similar to establishing a temperature scale. The zero of temperature on the Celsius scale is chosen as the freezing point of water. Thus, when we say that the temperature of something is 10 °C we mean that it is 10 degrees higher than the temperature of freezing water.) Since the earth is so large, it is unaffected by the flow of small charges to or from it. For this reason the potential of the earth is taken as being zero.

Measurement of Potential

The gold leaf electroscope in fact measures the p.d. between the leaves and the case, rather than the actual charge on the leaves. That this is so may be verified by connecting the cap to the case with a piece of wire and then charging the electroscope. Since the leaves are connected to the case by a conductor there is no p.d. between them. (If there was, charge would flow from one to the other until both were at the same potential.) It is now found that no matter how much charge is placed on the leaves they do not diverge, *i.e.* the divergence of the leaves is a measure of the p.d. between the leaves and the case rather than of the amount of charge on the leaves. It may be shown that the divergence is approximately proportional to the square of the p.d. between the leaves and the case. It should be noted that this may also be explained in terms of the forces between charges. The leaves do not diverge because the force between the two leaves is equal in magnitude to the forces between the leaves and the case.

The gold leaf electroscope can measure p.d.'s up to several thousand volts and requires only a very small amount of charge for its operation. It is therefore ideal for the measurement of the potentials encountered in electrostatics. It is not, however, suitable for measuring smaller p.d.'s. One device which can be used for measuring such p.d.'s is the cathode ray oscilloscope. In this device a beam of electrons passes between a pair of horizontal parallel plates and onto a screen, *Fig. 19.2*.

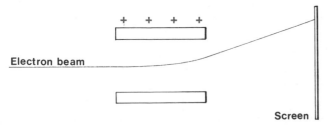

Fig. 19.2 Deflection of electron beam in cathode ray oscilloscope

Where the electrons strike the screen a spot of light is produced (*For a fuller description see p.286.*) The spot moves continuously across the screen at high speed, producing the line shown in the photograph. When a p.d. is applied to the plates the negative electrons are attracted towards the positive

plate and so the line of light on the screen moves up or down, depending on which plate is positive with respect to the other one. The larger the p.d. between the plates the further the line moves, *i.e.* the distance moved by the line is a measure of the p.d. between the plates.

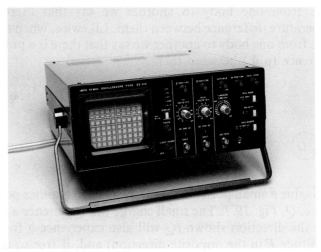

Fig. 19.3 Cathode ray oscilloscope

Electromotive Force (e.m.f.)

A term closely related to potential difference is electromotive force (e.m.f.), E. It is defined as follows.

The e.m.f. in any closed loop is the work done in carrying unit charge around the complete loop.

From this definition we see that, while p.d. refers to two particular points, e.m.f. refers to a complete loop or circuit. Referring to *Fig. 19.4*, the work done in carrying unit charge from A to B is the potential difference between A and B; the work done in carrying unit charge from A, through B and C, and back to A is the electromotive force in the loop ABCA. It should be clear from this that the e.m.f. in any closed circuit is equal to the sum of the p.d.'s around the circuit and that the unit of e.m.f. is the same as that of p.d., *i.e.* the volt.

Work is the process of converting energy from one form to another (*see p.88*), so if there is an e.m.f. around any loop such as ABCA in *Fig. 19.4*, there must be some device in the loop which can convert energy into electrical energy from some other form. Such a device is called a source of e.m.f. and examples are batteries, generators, *etc*. We shall learn more about sources of e.m.f. in subsequent chapters.

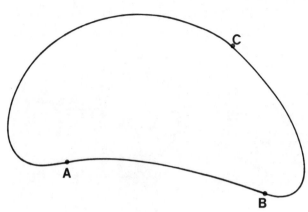

Fig. 19.4

The Point Effect

Fig. 19.5 shows a pear-shaped conductor carrying a positive charge. Although all points on the conductor must be at the same potential, the charge is not evenly distributed over the

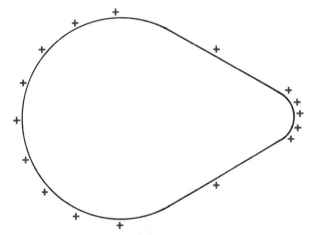

Fig. 19.5 Distribution of charge on a conductor

surface. As shown in the diagram, the charge tends to build up on the sharper end. This may be verified experimentally using a proof plane, *Fig. 19.6*, and an electroscope. The proof plane is charged by placing it on the charged surface. The larger the charge on the surface under the proof plane the larger the charge induced on it. The amount of charge on the proof plane is thus a measure of the density of charge on the surface from which it was charged.

The proof plane is first charged by touching it to one end of the conductor and then discharged on the electroscope (this

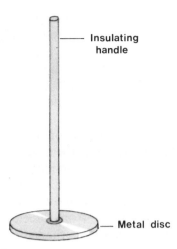

Insulating handle

Metal disc

Fig. 19.6 Proof plane

is best done by touching the proof plane to the inside of a tall can placed on the electroscope). The divergence of the leaves is noted. The process is then repeated with the proof plane being charged from the other end of the conductor. The divergence of the leaves is a measure of the charge on the proof plane and hence of the density of charge on the surface of the conductor. The divergence of the leaves is found to be greater when the proof plane is charged from the sharper end of the conductor, showing that the charge density is greater on that end.

It should be noted that, although the charge is not evenly distributed over the surface of the conductor, the potential is the same at all points on the surface. If this were not so charge would flow from points of higher potential to points of lower potential until all points were at the same potential. This may be verified by connecting various points on the conductor directly to an electroscope. It is found that the divergence of the leaves is always the same, indicating that the potential is the same at all points on the surface. (Remember that the electroscope measures potential. In the first case it was measuring the potential of the proof plane (which was proportional to the charge on it). In the second case the electroscope is measuring the potential of the charged conductor.)

If the concentration of charge on the sharper end of the pear-shaped conductor is high enough the forces in the surrounding electric field may be great enough to ionise the air molecules. When this happens, the positive ions are repelled (assuming the body is positively charged) while the negative ions are

201

attracted to the conductor and discharge it. This phenomenon is known as **point discharge** or the **point effect**. The sharper the point the lower the potential at which it occurs. The effect may be demonstrated by placing a lighted candle in front of a needle attached to a Van de Graaff generator, *Fig. 19.7*. The stream of ions moving away from the point causes the candle flame to be deflected.

Lightning conductors make use of the point effect. When a thunder cloud passes overhead, the negative charge (as much as 40 C) on the bottom of the cloud induces a positive charge on the conductor, *Fig. 19.8*. Point discharge then takes place around the lightning conductor and the ionised air provides an easy path for the discharge from the cloud. The lightning thus strikes the conductor rather than the building and the charge flows to earth through the copper strip.

In the Van de Graaff generator, charge is removed from the belt by the point effect. The "comb" situated inside the dome (*see p.191*) consists of a series of sharp points. When the positively charged belt passes the comb a negative charge is induced on the comb, leaving an equal positive charge on the dome. Point discharge then occurs, effectively transferring the negative charge from the comb to the belt, cancelling the charge on the belt and leaving the dome with a net positive charge.

19.2 Electric Field Intensity

As we have seen, the region around a charge is called an electric field, *i.e.* it is a region in which other charges will experience a force. A useful concept in calculating the force on a given charge in the field is that of the electric field strength or **electric field intensity**, *E*.

> **The electric field intensity at a point is the force per unit positive charge at that point.**

In symbols, this becomes

$$E = \frac{F}{Q}$$

E is a **vector quantity**, its direction being that of the force on a positive charge at the point in question. From the definition, the unit of *E* is the **newton per coulomb**, N C^{-1}. However, an equivalent, and more commonly used, unit is the **volt per metre**, V m^{-1} (*see p.206*)

From Coulomb's law the force between two charges is

$$F = \frac{1}{4\pi\varepsilon} \frac{Q_1 Q_2}{r^2}$$

Fig. 19.7 The point effect - charge streams away from a charged point

Lightning conductor

Copper strip

Fig. 19.8 Lightning conductor

The field intensity is the force on a charge of 1 C. Therefore, the field intensity at a distance r from a point charge Q is

$$E = \frac{1}{4\pi\varepsilon} \frac{Q}{r^2}$$

Example

What is the magnitude of the field intensity at a point 20 cm from a charge of 4.0 nC in air?

$$E = \frac{1}{4\pi\varepsilon} \times \frac{Q}{r^2}$$

$$= 8.9 \times 10^9 \times \frac{4 \times 10^{-9}}{(0.2)^2}$$

$$= 8.9 \times 10^2 \text{ V m}^{-1}$$

Ans. 8.9×10^2 V m^{-1}

Electric Field Flux

Fig. 19.9 shows two large parallel metal plates a small distance apart. One plate carries a positive charge while the other carries

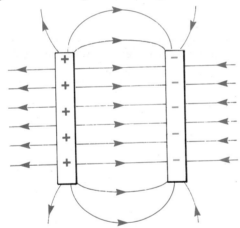

Fig. 19.9

an equal negative charge. If we neglect the edge effects, the field between the plates is uniform, *i.e.* the field intensity is the same at all points between the plates — the field lines are parallel and equally spaced. If we placed a sheet of paper between the plates and parallel to them, we could imagine the field lines passing through the paper as in *Fig. 19.10*. The **electric field flux**, ψ, passing through the paper is then defined as the product of the area of the paper and the field intensity, *i.e.*

$$\psi = EA \qquad (i)$$

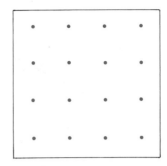

Fig. 19.10

where ψ is the flux passing through an area A perpendicular to the field intensity, assumed uniform. If A is not perpendicular to E then

$$\psi = EA \cos \theta$$

where θ is the angle between E and the normal to the area, *Fig. 19.11*. From *Eqn. (i)* the unit of ψ is N m^2 C^{-1} or V m. Note that electric field flux is sometimes referred to as simply electric flux.

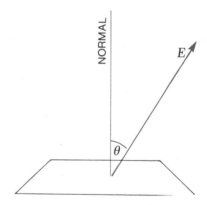

Fig. 19.11

Conservation of Flux

Imagine a point charge Q surrounded by a spherical surface of radius, r, *Fig. 19.12*. The field intensity at any point on this surface is

$$E = \frac{1}{4\pi\varepsilon} \frac{Q}{r^2}$$

The total flux arising from Q passes through the surface normally and is given by

$$\psi = EA$$

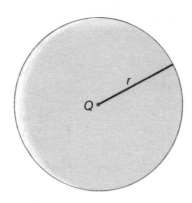

Fig. 19.12

where A is the area of the surface. The area of a spherical surface is $4\pi r^2$, therefore

$$\psi = \frac{1}{4\pi\varepsilon} \ \frac{Q}{r^2} \ 4\pi r^2$$

$$= \frac{Q}{\varepsilon}$$

That is,

$$\psi = \frac{Q}{\varepsilon} \qquad \text{(ii)}$$

This means that the total flux due to a point charge Q is independent of the distance from the charge. In other words, **total flux is conserved**. *Eqn. (ii)* is known as Gauss's Law.

We have derived Gauss's law for a point charge surrounded by a spherical surface. However, it is true for any charge and any closed surface and can be arrived at independently of Coulomb's law. Since we have derived Gauss's law from Coulomb's law it should be clear that it is, in effect, an alternative statement of Coulomb's law. Gauss's law is usually easier to apply than Coulomb's law and we shall now look at some applications of it.

Charged Conductors

Consider an insulated conductor carrying a charge Q, *Fig. 19.13*. Since charges can move through the conductor, they will do so until the resultant force on them is zero, *i.e.* the field intensity at all points in the conductor is zero. If we imagine a surface (called a Gaussian surface) drawn just inside the surface of the conductor, then, since

$$E = 0$$
$$\psi = 0$$

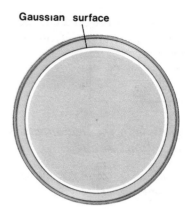

Fig. 19.13

That is, the total flux crossing the Gaussian surface is zero. From Gauss's law, the total charge inside the Gaussian surface is

$$Q = \psi\varepsilon$$

Therefore, $Q = 0$

That is, there is no charge inside the Gaussian surface. In other words,

---
All charge on an insulated conductor resides on its surface.

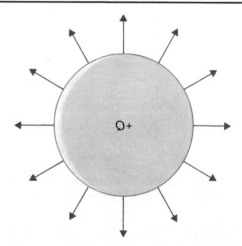

Fig. 19.14

Now consider an insulated positively charged sphere, *Fig. 19.14*. Since all the charge resides on the surface, all field lines must start on the surface. (If the conductor were negatively charged all the field lines would end on the surface.) Since no field line can have a component along the surface all the lines must be normal to the surface, *i.e.* the field outside the surface is the same *as if* all the charge were concentrated at the centre. (**Note:** This applies only to a spherical conductor.)

Field Inside a Hollow Conductor

Consider a hollow charged conductor, *Fig. 19.15*. Since all charge resides on the outer surface and since there can be no field lines in the solid conductor, there can be no field lines inside the conductor.

> **The field is zero at every point inside an empty hollow charged conductor.**

The fact that there can be no electric field inside a hollow conductor (it does not have to be spherical nor does it have

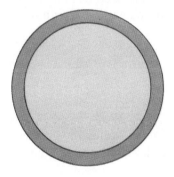

Fig. 19.15

to be solid) has many important applications. For example, instruments which might be affected by stray electric fields are often placed in metal containers. Such containers are called Faraday cages, after Michael Faraday who horrified colleagues by sitting in a wire cage which was then charged to a potential of several thousand volts. The coaxial cable, *Fig. 19.16* (used, for example, to connect a television to an aerial) is another application of a Faraday cage. The central copper wire carrying the signal is surrounded by, but insulated from, a copper wire mesh. This wire mesh prevents external electric fields interfering with the signal in the central wire. Another application of the principle is the suit sometimes worn by electricity board maintenance workers when carrying out repairs to overhead power lines. This suit is made of a special conducting material and completely covers the worker, allowing repair work to be carried out without the necessity of switching off the supply.

Fig. 19.16 Coaxial cable

Field Intensity due to a Charged Metal Plate

Consider a large metal plate carrying a total charge Q. The charge density on the plate is

$$\sigma = \frac{Q}{\text{area of plate}}$$

Imagine a Gaussian surface in the shape of a cylinder as shown in *Fig. 19.17*. Since no lines can pass into the plate and since all lines must be perpendicular to the surface of the plate, all lines due to the charge on the shaded area, A, must pass through the end of the cylinder. The charge on the area, A, is

$$q = A\sigma$$

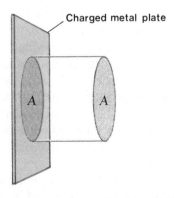

Charged metal plate

Fig. 19.17

The flux passing through the end of the cylinder is

$$\psi = EA$$

From Gauss's law

$$q = \psi\varepsilon$$

$$\Rightarrow \quad A\sigma = \psi\varepsilon$$

$$\Rightarrow \quad A\sigma = EA\varepsilon$$

or,

$$\boxed{E = \frac{\sigma}{\varepsilon}} \qquad \text{(iii)}$$

This is an important result which we shall require later.

Field Intensity and Potential

Consider a sphere of radius r carrying a positive charge Q. We saw earlier that, although all the charge stays on the outside of the sphere, the resulting electric field is the same as if the charge were concentrated at the centre. Thus, the field intensity at the surface is

$$E = \frac{1}{4\pi\varepsilon} \frac{Q}{r^2} \qquad \text{(iv)}$$

It may be shown (using calculus) that the potential at a distance r from a point charge Q is

$$V = \frac{1}{4\pi\varepsilon} \frac{Q}{r}$$

Since the charge on the sphere behaves as if it were at the centre it follows that the potential at the surface of the sphere is given by

$$V = \frac{1}{4\pi\varepsilon} \frac{Q}{r} \qquad \text{(v)}$$

Comparing *Eqns.* (*iv*) and (*v*) we see that

$$\boxed{E = \frac{V}{r}} \qquad \text{(vi)}$$

From this equation it is clear that, as stated earlier, the unit of E may be expressed as volt per metre, V m^{-1}. This equation also helps to explain the point effect. For a sphere at a given potential, the smaller the radius the larger the field intensity. If the field intensity is large enough (about 3 MV m^{-1} for dry air) the air becomes ionised and point discharge takes place as explained earlier. Combining *Eqns.* (*iii*) and (*vi*) gives

$$\frac{\sigma}{\varepsilon} = \frac{V}{r}$$

$$\Rightarrow \quad \sigma \propto \frac{1}{r}$$

That is, the smaller the radius the greater the charge density. This explains the distribution of charge on a pear-shaped conductor (*p.201*).

Michael Faraday (1791 — 1867)

Faraday was born in Surrey, England, one of 10 children. He received only a basic education and at the age of 14 was apprenticed to a bookbinder.

His work with books led Faraday to an interest in their contents, an interest which was encouraged by his employer. He was particularly interested in science and in 1813 he accepted the job of assistant to Humphrey Davy at the Royal Institution, although the salary was less than he earned as a bookbinder. As a scientist he progressed quickly. He became director of the laboratory in 1825 and in 1833 became professor of chemistry at the Royal Institution.

Faraday carried on Davy's work on the passage of electricity through electrolytes and in 1832 established the relationships now known as Faraday's Laws of Electrolysis. He was also interested in electromagnetism and in 1821 he invented a simple type of electric motor.

However, his greatest discovery, and one of the most important discoveries of all time, was that of electromagnetic induction (this is the production of an electric current in a changing magnetic field; it is the principle on which generators in power stations work). (Electromagnetic induction was discovered independently in America by Joseph Henry around the same time. The basic principles had been observed ten years earlier by Ampère but had been ignored by him because they did not fit in with his theories.) In 1831 Faraday invented the first electric generator. In addition to these major discoveries and inventions, Faraday was the first to liquefy gases using pressure; he discovered benzene; he invented the idea of magnetic field lines and made many other discoveries in the areas of electromagnetism and light.

Faraday was a renowned teacher and lecturer. His

demonstration lectures at the Royal Institution were justly famous. He introduced the Christmas Lectures for children, a series of lectures which still continues — these are now usually shown on television. He was a strong advocate of a more important role for science in education. He was a modest man with strong religious beliefs. He placed little value on the numerous honours awarded him other than his membership of the Royal Society, to which he was elected in 1824. He declined the presidency of the Society in 1857 and also the offer of a knighthood. During the Crimean war he refused absolutely a request from the government to develop poison gas for use on the battlefield.

Faraday was possibly the greatest experimental scientist the world has ever known. In 1858 he retired to Hampton Court, outside London, where he died peacefully nine years later.

SUMMARY

The potential difference (p.d.) is the work done in bringing unit charge from one point to another in an electric field. The unit of p.d. is the volt. The potential of the earth is taken to be zero.

Electromotive force (e.m.f.) is the work done in carrying unit charge around a closed loop. The unit of e.m.f. is the volt.

Charge tends to accumulate at points. As a result, the surrounding air may become ionised and charge streams away from the point. This is known as the point effect. Electric field intensity, E, is the force per unit positive charge. It is given by

$$E = \frac{1}{4\pi\varepsilon} \frac{Q}{r^2}$$

Field intensity is a vector quantity and its unit is the volt per metre, $V\ m^{-1}$.

The electric field flux passing through an area A is defined by

$$\psi = EA \cos\theta$$

where θ is the angle between the field intensity and the normal to the area. If the field intensity is at right angles to the area, then $\psi = EA$. The total flux arising from a charge Q is independent of the distance from the charge, *i.e.* flux is conserved, and is given by

$$\psi = \frac{Q}{\varepsilon}$$

This is known as Gauss's law.

Gauss's law leads to the following conclusions.

1. All charge resides on the surface of a charged conductor.
2. The field around a spherical, charged conductor is the same as if the charge were concentrated at the centre.
3. The field is zero inside an empty hollow conductor.
4. The field intensity near the surface of a charged metal plate is given by

$$E = \frac{\sigma}{\varepsilon}$$

Questions 19

DATA: $\varepsilon_0 = 8.9 \times 10^{-12}\ C^2\ N^{-1}\ m^{-2}$
Mass of electron $= 9.1 \times 10^{-31}\ kg$
Charge on electron $= 1.6 \times 10^{-19}\ C$.

Section I

1. Define potential difference..................................

....

2. The unit of potential difference is the ....

3. A charge of $2\ \mu C$ is moved through a p.d. of 40 V. What is the work done?...

4. What is an electronvolt?......................................

....

....

5. The potential of the earth is taken as being zero. Does this mean that the earth is uncharged? Explain.

....

....

6. Name two devices for measuring p.d.......................

7. Define electromotive force.....................................

....

8. The unit of e.m.f. is the ....

9. What is the point effect?......................................

....

10. Name two devices which depend for their operation on the point effect...

....

11. Apart from his actual discoveries in physics, for what is Michael Faraday remembered?..........................

....

*12. Define electric field intensity and give its symbol.

....

*13. The unit of electric field intensity is the................. ....or the..

*14. Define electric field flux.....................................

....

*15. What is meant by conservation of flux?.................

*16. All charge on an insulated conductor resides ....

....

*17. There can be no electric field ....

....

*18. What is the purpose of the outer conductor of a coaxial cable?...

....

*19. Give an expression for the field intensity near the surface of a charged metal plate.......................................

....

*20. Give an expression for the field intensity at the surface of a sphere in terms of the charge and the radius of the sphere..

Section II

1. The work done in bringing a charge of 4.5 μC from A to B is 9.0 mJ. What is the p.d. between A and B?

2. The p.d. between two points is 6 V. How much work is done when a charge of 1000 C is transferred from one point to the other?

3. What is the speed of an electron which has been accelerated through a p.d. of 10 kV?

4. An electron in a TV tube reaches a speed of 3.0×10^7 m s^{-1} before striking the screen. Calculate the p.d. through which the electron was accelerated.

*5. Calculate the magnitude of the field intensity at a point 40 cm from a charge of 2.4 μC in air.

*6. The field intensity at the surface of a charged sphere of radius 15 cm is 0.4 MV m^{-1}. Calculate (i) the charge on the sphere, (ii) the charge density.

*7. Calculate the total flux arising from a charge of 4.4 μC in a medium of relative permittivity 5.0.

*8. Calculate the field intensity at the following points on the straight line passing through a positive charge (A) of 0.24 μC and a negative charge (B) of 0.16 μC, a distance of 20 cm apart: (i) 5 cm, (ii) 15 cm, (iii) 30 cm, from A.

*9. Positive charges of 3 nC, 4 nC and 8 nC are situated at the vertices a, b and c, respectively, of an equilateral triangle of side 10 cm. Calculate the field intensity at the mid-point of bc.

*10. An electron in an X-ray tube is accelerated from rest to a speed of 1.8×10^8 m s^{-1} over a distance of 30 cm. Calculate (i) the acceleration of the electron, (ii) the field intensity of the field in which the electron was accelerated.

*11. An electron is travelling at 2.2×10^6 m s^{-1} when it enters a field of field intensity 4.6 kV m^{-1} in the direction of the velocity of the electron. How far does the electron travel in the field before coming to rest?

*12. If, in the previous question, the field intensity is at right angles to the velocity of the electron, calculate the displacement of the electron from the point where it entered the field when its velocity makes an angle of 45° with its original velocity.

20 Capacitance

20.1 Capacitance

When we speak of the capacity of a container we mean the volume of liquid it will hold without overflowing. Similarly, the capacitance of a conductor is the amount of charge it will hold without "overflowing", *i.e.* discharging. A conductor overflows when its potential reaches a certain value which is determined by the shape and size of the conductor *(cf. Point discharge in previous chapter)*.

The potential of a body is directly related to the amount of charge on it. If a body is given a positive charge its potential rises — work must be done to bring further positive charges to it. The amount of charge which a body will "hold" depends on how quickly its potential rises as charge is added to it. (Think of filling a tank with water. If the tank is narrow the level of the water will rise quickly and the tank will soon overflow.) If the potential rises quickly as charge is added, *i.e.* if a small charge causes a large change in potential, the capacitance is small; if a large charge produces a small change in potential the capacitance is large. In other words, the capacitance depends on the size of the charge compared with the size of the potential produced by the charge. The **capacitance,** *C,* of a body may thus be defined as follows.

> **Capacitance is the ratio of charge to potential.**

In symbols,

$$C = \frac{Q}{V}$$

Capacitance is a scalar quantity. Its unit is the **farad**, F, so called in honour of Michael Faraday.

> **The capacitance of a body is 1 F if the addition of a charge of 1 C raises its potential by 1 V.**

Sine the coulomb is a very large quantity of charge, practical capacitances are usually measured in μF (10^{-6} F), nF (10^{-9} F) or even pF (10^{-12} F) rather than in F. In calculations, of course, all capacitances must be expressed in F.

An isolated conductor has a very small capacitance. If we wish to increase its capacitance we must find some way of holding down its potential as more charge is added to it — like putting water into a tank while applying a force to the surface to keep the water level down.

To understand how this might be done, consider a point, P, in an electric field due to a positive charge, Q_1. P has a certain potential — work would have to be done to bring a positive charge to P from earth. If a negative charge, Q_2, is placed near P its potential decreases — less work is now required to bring a positive charge to P since the positive charge is now attracted to Q_2. Thus, we can see how the potential of a charged body may be reduced by bringing an oppositely charged body near to it. Since the amount of charge on the body remains the same while its potential is decreased, this means that its capacitance has been increased.

The Parallel Plate Capacitor

A capacitor is a device for storing charge. Basically, a parallel plate capacitor consists of a pair of parallel plates which are near together and carry opposite charges, so that the capacitance of the plates is increased as explained above. The factors which determine the capacitance of such a capacitor may be investigated using the arrangement shown in *Fig. 20.1*. One plate is connected to an electroscope and charged, either by induction from a plastic rod or by connecting to a high voltage (c. 2 kV) power supply. The second plate is earthed. If the first plate is charged positively the second plate has a negative charge induced on it, and *vice versa* (*see p.193*). The

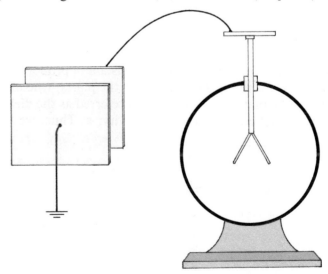

Fig. 20.1   Investigating the factors affecting the capacitance of a parallel plate capacitor

deflection of the leaves of the electroscope is a measure of the potential of the plates (*see p.200*). Keeping the distance between the plates constant, one plate is moved parallel to the other so that the common area, *i.e.* the area of overlap, of the plates is changed. If the common area, A, is increased it is found that the divergence of the leaves decreases, indicating that the potential decreases. Since the charge on the plates remains the same while the potential decreases the capacitance of the plates must increase, *i.e.* as A increases C increases. It may, in fact, be shown that

$$C \propto A$$

By keeping A constant while the distance, d, between the plates is varied it may be shown in a similar way that, as d increases C decreases. Measurements show that, in fact,

$$C \propto \frac{1}{d}$$

Combining these results gives,

$$C \propto \frac{A}{d}$$

or,

$$C = k\frac{A}{d}$$

where k is a constant.
Theory *(see below)* shows that the constant of proportionality, k, depends on the material between the plates, (the material between the plates of a capacitor is referred as the **dielectric**) and is in fact equal to the permittivity, ε. Thus, we have

$$\boxed{C = \frac{\varepsilon A}{d}} \qquad \text{(i)}$$

Rearranging *Eqn. (i)* gives

$$\varepsilon = \frac{Cd}{A}$$

From this equation we can see that ε may be measured in farads per metre, F m^{-1}. This is the unit most commonly used for ε.

Capacitance of a Parallel Plate Capacitor

For a parallel plate capacitor the field between the plates is uniform, provided we can ignore the field at the edges, *Fig. 20.2*. (Edge effects are negligible provided the area of the plates is large compared with the distance between them.)

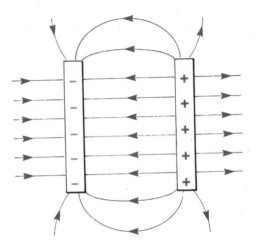

Fig. 20.2

Consequently, from *Eqn. (iii), p.206,* the field intensity between the plates is given by

$$E = \frac{\sigma}{\varepsilon}$$

$$= \frac{Q}{\varepsilon A}$$

where Q is the charge on the plates and A is their area. Since $E =$ force/unit charge, the work done/unit charge in transferring charge from one plate to the other is

$$W = Ed$$

where d is the distance between the plates. Substituting for E gives

$$W = \frac{Qd}{\varepsilon A}$$

But, work/unit charge = potential difference.

$$\therefore \quad V = \frac{Qd}{\varepsilon A}$$

where V is the p.d. between the plates.

Rearranging this equation gives

$$\frac{Q}{V} = \frac{\varepsilon A}{d}$$

But, $\quad C = \dfrac{Q}{V}$

$$\therefore \quad \boxed{C = \frac{\varepsilon A}{d}}$$

Example

Find the capacitance of a parallel plate air capacitor if the common area of the plates is 0.5 m² and the distance between them is 1.0 mm. Find also the charge stored in the capacitor if the p.d. between its plates is 10 V.

$$C = \frac{\varepsilon A}{d}$$

$$= \frac{8.9 \times 10^{-12} \times 0.5}{1 \times 10^{-3}}$$

$$= 4.45 \times 10^{-9} \text{ F}$$

$$= 4.45 \text{ nF}$$

$$C = \frac{Q}{V}$$

$$=> \quad Q = CV$$

$$= 4.45 \times 10^{-9} \times 10$$

$$= 44.5 \times 10^{-9}$$

$$= 44.5 \text{ nC}$$

Ans. 4.45 nF; 44.5 nC

Energy in a Charged Capacitor

If a capacitor is charged by connecting its plates to a battery and the connecting wires are brought near together a spark is seen, and heard, to jump from one to the other — energy in the form of light and sound is released. From this we can conclude that a capacitor stores energy, although the amounts of energy stored are generally small. The amount of energy stored in a given capacitor may be calculated as follows. Suppose the p.d. between the plates of a capacitor is increased from 0 to V by the transfer of a total amount of charge Q. Then, the average p.d. between the plates during the charging process is $V/2$ and the work done in transferring the charge is

$$W = \frac{V}{2} \times Q$$

But, work done = energy stored.
Therefore, the energy stored is given by

$$W = \tfrac{1}{2}VQ$$

But, $\quad Q = CV$

Therefore, substituting for Q we have

$$\boxed{W = \tfrac{1}{2}CV^2} \qquad \text{(ii)}$$

Eqn. (ii) gives the energy stored in the capacitor in terms of the capacitance and the p.d. between the plates. It is sometimes convenient to have this expression in terms of the charge on the capacitor rather than the p.d.

$$C = \frac{Q}{V}$$

$$=> \quad V = \frac{Q}{C}$$

Substituting for V in *Eqn. (ii)* gives

$$W = \tfrac{1}{2}C\frac{Q^2}{C^2}$$

$$=> \quad \boxed{W = \tfrac{1}{2}\frac{Q^2}{C}}$$

Note: The symbol W, rather than E, is used for electrical energy in order to avoid confusion between energy and field intensity.

Example

Calculate the energy stored in a capacitor of 1000 μF when the p.d. between the plates is 20 V.

$$W = \tfrac{1}{2}CV^2$$
$$= 0.5 \times 1000 \times 10^{-6} \times 20^2$$
$$= 0.2 \text{ J}$$

Ans. 0.2 J

20.2 Practical Capacitors

There are many different types of capacitor available. The variable capacitor shown in *Fig. 20.3* consists of two sets of semi-circular plates arranged so that one set may be rotated in and out between the other set, thus varying the common area of the plates and hence the capacitance. The dielectric in such capacitors is usually air. Variable capacitors are used for "tuning" a radio to a particular station. Their maximum capacitance is usually small — less than 1 nF. Variable capacitors may also be made from semiconductor materials *(see p. 308)*

Fig. 20.3 Variable capacitor

Fixed value capacitors, *Fig. 20.4*, are usually named according to the type of dielectric in them. Thus, the mica capacitor consists of two sets of metal plates separated by sheets of mica. Such capacitors have capacitances of the order of 0.1 nF to 1 μF. The plastic foil capacitor consists of two sheets of aluminium foil interleaved with two thin sheets of plastic and rolled into the form of a cylinder which is then enclosed

212

in a plastic case. An electrolytic capacitor is made by depositing a very thin (perhaps 10^{-7} m) layer of metal oxide onto a metal (often aluminium) plate by electrolysis *(see Chapter 23)*.

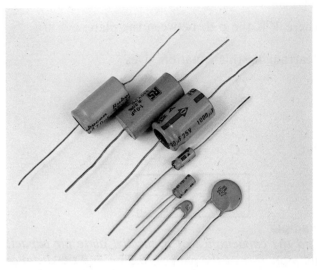

Fig. 20.4 Fixed value capacitors

In this case the dielectric is the oxide layer and, since it is so thin the distance between the plates may be very small and hence the capacitance may be very large, perhaps 100 mF or more. Care must be taken with this type of capacitor when connecting it in a circuit since it is polarised, *i.e.* it has a positive and a negative terminal. Should an electrolytic capacitor be connected the wrong way around in a circuit, the oxide layer may dissolve, thus ruining the capacitor. The symbols for the different types of capacitor are shown in *Fig. 20.5*.

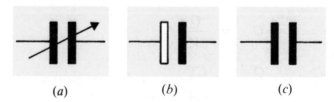

(a)   (b)   (c)

Fig. 20.5 Symbols for capacitors: (a) Variable capacitor; (b) non-electrolytic capacitor; (c) electrolytic capacitor.

Uses of Capacitors

Capacitors are used to separate alternating current (a.c.) from direct current (d.c.) *(p.269)* and to smooth the output from rectifiers *(p.303)*. The time taken for a capacitor to be charged depends on its capacitance and the resistance *(see p.217)* of the circuit. This allows capacitors to be used in circuits which operate a switch after a specified period of time. Variable capacitors are used in the tuning circuits of radio receivers. Adjusting the capacitance matches the natural frequency of the circuit to that of the radiowave to be received, *e.g.* 612 kHz for RTE Radio 2 (*see Resonance p.125*).

SUMMARY

Capacitance is the ratio of charge to potential, *i.e.*

$$C = \frac{Q}{V}$$

The unit of capacitance is the farad, F. The capacitance of a body is increased by bringing an oppositely charged body near to it.

The capacitance of a pair of parallel plates is given by

$$C = \frac{\varepsilon A}{d}$$

where A is the common area of the plates and d is the distance between them.

The energy stored in a capacitor is given by

$$W = \tfrac{1}{2}CV^2 = \tfrac{1}{2}Q^2/C$$

Capacitors are used to separate a.c. from d.c.; to smooth the output of rectifiers and in the tuning circuits of radio receivers.

Questions 20

DATA: $\varepsilon_o = 8.9 \times 10^{-12}$ F m^{-1}.

Section I

1. Define capacitance...
   ....

2. The unit of capacitance is the.............................

   The capacitance of practical capacitors is usually measured

   in..

   or..

3. The capacitance of a charged body may be increased by

   ....

4. Give an expression for the capacitance of a pair of parallel

   plates...

5. How would you demonstrate that a charged capacitor

   stores energy?..

   ....

   ....

*6. Give an expression for the energy stored in a capacitor

in terms of (i) the capacitance and the p.d., (ii) the

capacitance and the charge...................................

....

....

7. Suggest two practical methods of varying the capacitance of a variable capacitor...

   ....

   ....

8. Give two uses of capacitors...................................

   ....

   ....

9. Why are capacitors not normally used to store energy

   instead of, for example, batteries?...........................

   ....

10. Why is it important that the positive terminal of an electrolytic capacitor be connected to the positive terminal

    of the battery or power supply?...............................

    ....

Section II

1. The capacitance of an isolated metal sphere is 3.5 pF. Calculate the charge on it when its potential is 400 V.

2. A capacitor consists of two sheets of aluminium, each 5.0 m long by 3.0 cm wide, separated by an insulating material of permittivity 3.0×10^{-11} F m^{-1} and thickness 25 μm. Calculate the capacitance of the capacitor and the charge on the plates when the p.d. between them is 20 V.

3. A parallel plate capacitor has plates of area 0.10 m^2 and carries a charge of 12 mC. If the relative permittivity of the dielectric is 1200, what is the field intensity between the plates?

4. If the capacitance of the capacitor in Q.3 is 470 μF calculate (i) the distance, (ii) the p.d. between the plates.

*5. Calculate the amount of energy stored in a 4700 μF capacitor when it is charged to a p.d. of 20 V. What is the charge stored in the capacitor?

*6. The energy required to charge a capacitor of capacitance 100 nF is 20 μJ. What is the p.d. between the plates?

*7. Calculate the amount of work required to place a charge of 3.0 mC on a capacitor whose capacitance is 47 μF.

*8. Suppose you had a method of charging a 1000 μF capacitor to a p.d. of 200 V. How many such capacitors would you need to store enough energy to light a 100 W bulb for 5 hours?

21 Electric Circuits

21.1 Electric Current

In *Chapter 19* we saw that a charge will move from a point of higher potential to a point of lower potential, if the two points are connected by a conductor. This flow of charge is called an electric current, *I*.

> **An electric current is a flow of charge.**

Current is a **scalar quantity**. The unit of current is the **ampere**, A. Ampere is often abbreviated to "amp" in everyday speech.

That current is a flow of charge may be demonstrated by connecting the dome of a Van de Graaff generator to earth through a galvanometer (this is a meter which can measure small currents), *Fig. 21.1*. When the Van de Graaff generator is running the meter registers a current; when the generator stops the current stops. If, after the generator is stopped, the dome is connected to an electroscope it will be found that there is no charge on it. From this we can conclude that the current registered by the meter consisted of the charge from the dome flowing to earth. The charge flows because there is a p.d. between the dome of the generator and the earth.

Fig. 21.1 Van de Graaff generator as a source of current

Another way to produce a p.d. between two points is to use a cell or battery *(see Chapter 23)*. *Fig. 21.2* shows a simple circuit consisting of a battery and a switch connected to each other by means of copper wire. The point X is maintained at a higher potential than the point Y by the chemical reactions in the battery. Therefore, when the switch is closed a current will flow from X to Y in the direction shown, *i.e.* from the positive terminal of the battery to the negative terminal. (**Note**: The direction in which a current flows is purely conventional. In fact, a current in a metal consists of a movement of electrons from the negative terminal of the battery to the positive terminal. The conventional direction was adopted by Ampère before the nature of electric current was understood. This apparent contradiction presents no problems as long as we are consistent. If we use the word "current", we are referring to something which goes from positive to negative. If we use the word "electron" we are referring to something which goes from negative to positive.)

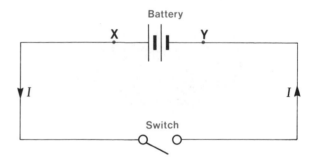

Fig. 21.2

Definition of the Ampere

Ampère discovered that when currents flow through two conductors placed near to each other they either attract or repel each other. This may be demonstrated with the arrangement shown in *Fig. 21.3*. Currents are passed through two long strips of aluminium foil hanging near to each other. If the currents are flowing in opposite directions the strips of foil move apart; if the currents are flowing in the same direction the strips are attracted to each other. This phenomenon is used to define the ampere.

> **The ampere is that constant current which, if maintained in two straight parallel conductors of infinite length, of negligible circular cross-section, and placed 1 m apart in a vacuum, causes each to exert a force of 2×10^{-7} N per metre length of the other.**

Note that the ampere is one of the basic units of the SI system (*see p.6*)

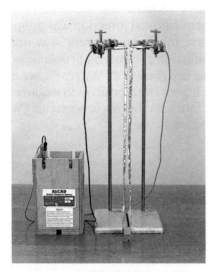

Fig. 21.3 Demonstrating forces between currents

Definition of the Coulomb

In *Chapter 18* we noted that the unit of charge is the coulomb without giving a definition of this unit. Since electric current is a flow of charge the coulomb is defined in terms of the ampere.

> **The coulomb is the quantity of charge transferred when a current of 1 A flows for 1 s.**

From this it follows that, when a current I flows for time t the total charge transferred, Q, is equal to It, *i.e.*

$$Q = It$$

Example

Calculate the charge transferred by a current of 5 A flowing for 10 minutes.

$$
\begin{aligned}
Q &= It \\
&= 5 \times 10 \times 60 \\
&= 3000 \text{ C}
\end{aligned}
$$

Ans. 3000 C

Andre Marie Ampère (1775 — 1836)

Andre Marie Ampère was born at Lyon on January 20, 1775 and died in Marseille on June 10, 1836. He was a child prodigy who mastered all the mathematics then known by the age of 12. He became professor of Physics and Chemistry at Bourg in 1801 and professor of Mathematics in Paris in 1809.

Ampère did much of the pioneering work on the study of current electricity. As well as studying the forces between currents he investigated the magnetic fields produced by currents flowing in coils and solenoids and he was the first person to develop a technique for measuring electric current. He was also the first to explain the distinction between potential difference and current. He also put forward the theory that the magnetism of a magnet was due to tiny currents circulating within the magnet. This theory was mostly rejected at the time but came to be accepted as the atomic structure of matter came to be understood.

Ampère was a classic example of the "absent-minded professor". However, despite his brilliance and success, his private life was very unhappy. His father was guillotined by the revolutionaries in 1793 and in 1804 his wife died after a very few years of marriage. He never really recovered from this latter blow and his outlook on life is probably best summed up by his chosen epitaph — *"Tandem Felix"* (Happy, at last).

21.2 Resistance

The quantity of current flowing through a conductor depends, in a given case, on the potential difference between its ends. The exact relationship between current and potential difference was discovered by Georg Ohm in 1826 and is now known as **Ohm's law.**

> **The current flowing through a conductor is proportional to the p.d. between its ends if the temperature remains constant.**

In symbols, we can write this as

$$V \propto I$$

or,

$$V = RI$$

where R is a constant, provided the temperature is constant. The constant of proportionality, R, is called the **resistance** of the conductor. That is, the resistance of a conductor is given by

$$R = \frac{V}{I}$$

where I is the current flowing through the conductor when the p.d. between its ends is V. The unit of resistance is the **ohm**, Ω, so called in honour of Georg Ohm. The ohm is defined as follows.

> **The resistance of a conductor is 1 Ω if a current of 1 A flows through it when the p.d. between its ends is 1 V.**

Note: Not all conductors obey Ohm's law. Metals, in general, and some liquids do, while other liquids, gases and certain devices do not.

Example
What current flows through a conductor whose resistance is 10 Ω when the p.d. between its ends is 5 V?

$$V = RI$$
$$5 = 10I$$
$$=> \quad I = 0.5 \text{ A}$$

Ans. 0.5 A

EXPERIMENT 21.1

To Investigate the variation of current with p.d. for (i) a length of wire, (ii) a filament lamp.

Apparatus: Length of wire, 12V filament lamp, rheostat, milliammeter, ammeter, voltmeter, power supply.

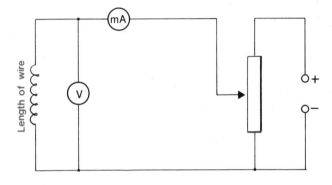

Fig. 21.4

Procedure:
1. Set up the circuit as shown in *Fig. 21.4.*
2. Using the rheostat, set the voltage at 0.2 V. Record the voltage and current.
3. Increase the voltage in steps of 0.2 V, recording the voltage and current at each step.
4. Plot a graph of voltage against current.
5. Replace the wire with the filament lamp and the milliammeter with the ammeter. Repeat Steps 1 to 4. In this case the voltage should be increased in steps of 1 V up to 12 V.

217

RESULTS

Wire		Lamp	
V/V	I/mA	V/V	I/A

Substance	$\rho/10^{-8}\ \Omega$ m
Silver	1.6
Copper	1.7
Gold	2.4
Platinum	11.0
Iron	14.0
Eureka	49
Mercury	94
Nichrome	130

Table 21.1 Resistivities

Questions

1. Does the wire obey Ohm's law? Does the lamp obey Ohm's law? Explain.
2. Why is it preferable to use a small current for the first part of the experiment?
3. Should you use a different voltmeter in the two parts of the experiment? Explain.

Resistivity

It can be shown experimentally that the resistance of a conductor at a given temperature depends on three factors, *viz.* the substance from which it is made, its length and its cross-sectional area.

For a given substance, the resistance is found to be proportional to the length and inversely proportional to the cross-sectional area, *i.e.*

$$R \propto \frac{l}{A}$$

$$\Rightarrow \qquad \boxed{R = \rho\, \frac{l}{A}} \qquad \text{(i)}$$

where ρ is a constant called the **resistivity** of the substance of which the conductor is made. (The reciprocal of the resistivity is called the conductivity of the substance.) Rearranging *Eqn. (i)* gives

$$\rho = \frac{RA}{l}$$

From this we see that the unit of ρ is the **ohm metre**, Ω m. The resistivities of some common materials are given in *Table 21.1*.

Example

What is the resistance of a copper wire of length 50 cm and mean diameter 0.1 mm?

$$R = \frac{\rho l}{A}$$

$$= \frac{1.7 \times 10^{-8} \times 0.5}{\pi(0.5 \times 10^{-4})^2}$$

$$= \frac{85 \times 10^{-8}}{0.785 \times 10^{-8}}$$

$$= 1.08\ \Omega$$

Ans. 1.1 Ω

Georg Simon Ohm (1787 — 1854)

Georg Simon Ohm was born in Erlangen, Bavaria, on March 16, 1787 and died in Munich on July 7, 1854.

Ohm, the son of a master mechanic, started his working life as a highschool teacher. However, his ambition was to secure a university appointment and to that end he undertook research in the new field of current electricity. In 1827 he published the results of his research — the relationship between current and potential difference which is now known as Ohm's law. Although initial reaction to his work was far from favourable — in fact, he was forced to resign from his teaching post — he was eventually awarded the recognition he deserved. In 1849, just five years before his death, he was appointed to a professorship at the University of Munich.

21.3 Resistors

All conductors have a certain resistance, although the resistance of metallic conductors is usually small. The wires, or leads, used to connect together the various components in a circuit are generally assumed to have a negligible resistance. A **resistor**, on the other hand, is a device which is designed to have a certain resistance. Resistors are used to control the current flowing in a circuit or part of a circuit. Radios, televisions, record-players, *etc.*, contain large numbers of resistors. *Fig. 21.5* shows

Fig. 21.5 Resistors

a selection of resistors, while *Fig. 21.6* shows their symbols. Some of these (those which have to carry large currents) are made of wire, but most are made of carbon. Carbon resistors are made in certain *preferred values* only, *e.g.* 220 Ω, 470 Ω, 4.7 kΩ, 2.2 MΩ, *etc.* Therefore, in order to make up a required resistance, a circuit designer may have to use a combination of two or more resistors. There are two methods of combining resistors and we shall now consider each of these in turn.

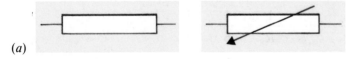

(a)

Fig. 21.6   Symbols for resistors (a) fixed, (b) variable

Resistors in Series

Fig. 21.7 shows three resistors connected in series with each other and with a battery. What is the effective resistance of the three resistors? That is, if we wished to replace the three resistors with a single resistor what would be its resistance?

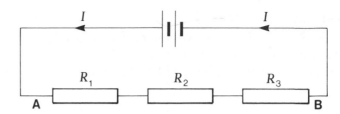

Fig. 21.7 Resistors in series

Let the resistances of the three resistors be R_1, R_2 and R_3, respectively. Let the p.d.'s across the resistors be V_1, V_2 and V_3, respectively. Then, if V is the total p.d. between A and B

$$V = V_1 + V_2 + V_3 \quad (i)$$

The same current, I, must flow through all three resistors. Therefore, from the definition of resistance,

$$V_1 = R_1 I$$
$$V_2 = R_2 I$$
$$V_3 = R_3 I$$
$$V = RI$$

where R is the effective resistance of the three resistors.

Substitution for V, V_1, V_2 and V_3 in *Eqn. (i)* gives

$$RI = R_1 I + R_2 I + R_3 I$$
$$= (R_1 + R_2 + R_3)I$$
$$\Rightarrow \boxed{R = R_1 + R_2 + R_3}$$

That is, the effective resistance of the three resistors is equal to the sum of the resistances of the individual resistors.

If we assume that the resistance of the battery and of the connecting wires in the circuit of *Fig. 21.7* is negligible then there is no work done in carrying charge through them, *i.e.* the p.d. across them is zero. Therefore V, the sum of the p.d.'s across the resistors, is equal to the e.m.f. in the circuit (*see p.200*). Thus, for the complete circuit, we can write

$$E = RI$$

where E is the e.m.f. in the circuit and R is the total resistance of the circuit. Note that this equation is applied to a complete circuit whereas the equation, $V = RI$, is applied to particular parts of a circuit, *e.g.* a resistor.

Resistors in Parallel

Fig. 21.8 shows three resistors connected in parallel with each other and in series with a battery. What is the effective resistance of the three resistors? Let the resistances of the three

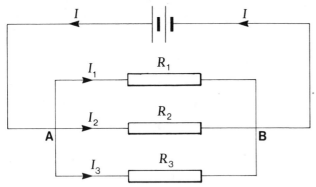

Fig. 21.8 Resistors in parallel

resistors be R_1, R_2 and R_3, respectively. Let the currents flowing through the resistors be I_1, I_2 and I_3, respectively. Then the total current, I, flowing in the circuit is

$$I = I_1 + I_2 + I_3 \quad \text{(ii)}$$

If V is the p.d. between A and B then, from the definition of resistance

$$V = R_1 I_1$$

$$=> \quad I_1 = \frac{V}{R_1}$$

Similarly,

$$I_2 = \frac{V}{R_2}$$

$$I_3 = \frac{V}{R_3}$$

and,

$$I = \frac{V}{R}$$

where R is the effective resistance of the three resistors.

Substitution for I_1, I_2, I_3 and I in *Eqn. (ii)* gives

$$\frac{V}{R} = \frac{V}{R_1} + \frac{V}{R_2} + \frac{V}{R_3}$$

$$= V \left(\frac{1}{R_1} + \frac{1}{R_2} + \frac{1}{R_3} \right)$$

$$=> \quad \boxed{\frac{1}{R} = \frac{1}{R_1} + \frac{1}{R_2} + \frac{1}{R_3}}$$

That is, the reciprocal of the effective resistance is equal to the sum of the reciprocals of the individual resistances.

Example

Fig. 21.9 shows three resistors connected to a 6 V battery. Find (i) the effective resistance of the three resistors, (ii) the total current flowing in the circuit, (iii) the current flowing through the 6 Ω resistor.

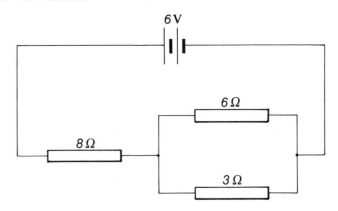

Fig. 21.9

(i) The effective resistance of the two resistors in parallel with each other is given by

$$\frac{1}{R} = \frac{1}{6} + \frac{1}{3}$$

$$= \frac{3}{6}$$

$$=> \quad R = \frac{6}{3}$$

$$= 2 \ \Omega$$

The circuit is therefore equivalent to one having a 2 Ω resistor in series with an 8 Ω resistor. The total effective resistance of the circuit is therefore given by

$$R = 2 + 8$$
$$= 10 \ \Omega$$

(ii) The total p.d. across the resistors is 6 V. Therefore,

$$V = RI$$
$$=> \quad 6 = 10 \ I$$
$$=> \quad I = 0.6 \ A$$

(iii) The effective resistance of the two parallel resistors is 2 Ω. Therefore the p.d. across these resistors is

$$V = RI$$
$$= 2 \times 0.6$$
$$= 1.2 \ V$$

Therefore, the current flowing through the 6 Ω resistor is given by

$$V = RI$$
$$1.2 = 6 \ I$$
$$=> \quad I = \frac{1.2}{6}$$
$$= 0.2 \ A$$

Ans. 10 Ω; 0.6 A; 0.2 A

Variable Resistors

Fig. 21.10 shows two resistors, the resistance of which can be varied continuously. The type found in laboratories, *Fig. 21.10(a)*, are usually called rheostats, while those used in radios, etc., *Fig. 21.10(b)*, are called potentiometers. The principle of each is the same. A length of conductor (wire or carbon, depending on the current to be carried) has three terminals, *Fig. 21.11*. Two of these terminals are fixed, one at either end. The third, C, is moveable and can make contact with the

Fig. 21.10(a) Rheostat   *Fig. 21.10(b) Pontentiometer*

conductor at any point between A and B. Thus, the resistance between C and A or between C and B can have any value between zero and the total resistance of the conductor.

Fig. 21.11

The Potential Divider

A potential divider is a device for obtaining a lower p.d. from a higher one. It consists of two resistors connected in series with a source of e.m.f., *e.g.* a battery, *Fig. 21.12*. If the e.m.f. of the battery (*i.e.* the e.m.f. generated in the circuit by the battery — sometimes called the voltage of the battery, *see p. 245*) is E the current flowing in the circuit, assuming that the resistance of the battery is negligible, is given by

$$E = (R_1 + R_2)I$$
$$=> \quad I = \frac{E}{R_1 + R_2}$$

The p.d. between X and Y is then

$$V = R_2 I$$
$$= \frac{R_2 E}{R_1 + R_2}$$

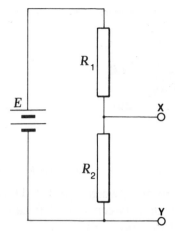

Fig. 21.12 Potential divider — fixed

Thus, the p.d. between X and Y is determined by the values of R_1 and R_2. The minimum value of the p.d. is zero ($R_2 = 0$) and the maximum value is E ($R_1 = 0$). If a continuously variable p.d. is required the two fixed resistors may be replaced with a variable resistor, *Fig. 21.13*. The p.d. between X and Y is now continuously variable between zero and E.

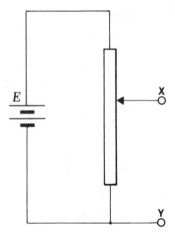

Fig. 21.13 Potential divider — variable

The Potentiometer

The potentiometer is a device for measuring p.d. accurately. It consists of a length of wire, AB, of uniform cross-sectional area and connected to a cell or battery capable of driving a steady current through it, *Fig. 21.14*. This cell· is known as the driver cell. Since the wire is of uniform cross-sectional

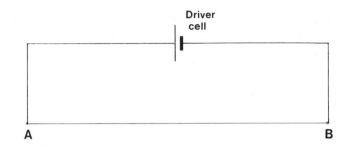

Fig. 21.14

area the resistance between any two points on it is proportional to the distance, d, between them (*see Resistivity, p.218*), *i.e.*

$$R \propto d$$

Since $V = RI$, if the current flowing is constant then the p.d. between any two points is proportional to the resistance between them, *i.e.*

$$V \propto R$$

$$=> \quad V \propto d$$

That is, the p.d. between any two points is proportional to the distance between them. In use, the unknown p.d. is connected between X and Y, *Fig. 21.15*, (note the polarity of the p.d.'s). Y is connected through a galvanometer (sensitive current-measuring meter, *see p.259*) to a moveable contact, C. In the circuit AXYC there are thus two opposing p.d.'s.

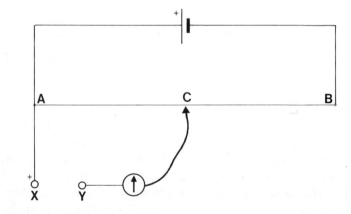

Fig. 21.15 Potentiometer circuit

The p.d. between A and C tries to drive a current in the direction AXYC while the unknown p.d. tries to drive a current in the opposite direction. If no current flows through the galvanometer we can conclude that these p.d.'s are equal. Thus, if the unknown p.d. is V_1 we can say that

$$V_1 \propto |AC|$$

or, $$V_1 = k|AC|$$

where k is a constant. If the unknown p.d. is now replaced with a cell of known e.m.f., E, and a new balance point, D, found, we can say that

$$E = k |AD|$$

Combining these two equations gives

$$\frac{V_1}{E} = \frac{|AC|}{|AD|}$$

Since E is known V_1 may be calculated.

(Note: Some teachers may prefer to take Chapter 23.2 at this point.)

EXPERIMENT 21.2

To Determine the E.m.f. of a Cell using a Potentiometer

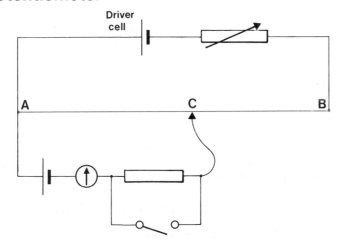

Fig. 21.16

Apparatus: Potentiometer, cell of known e.m.f., cell of unknown e.m.f., low voltage accumulator, centre zero galvanometer, rheostat, resistor of high resistance (~ 1 kΩ), moveable contact.

Procedure:

1. Set up the apparatus as shown in *Fig. 21.16*, with the cell of known e.m.f. connected in the circuit. Make sure that the polarity of the cells is correct. Connect the resistor in series with the galvanometer.
2. Adjust the position of the moveable contact until the galvanometer reads zero. Adjust the rheostat if necessary to ensure that the balance point is near end B of the potentiometer wire.
3. Close the switch to short out the protective resistor and then find the balance point more accurately. Measure the distance $|AC|$, where C is the position of the moveable contact when the galvanometer reads zero.
4. Replace the cell of known e.m.f. with the cell whose e.m.f. is to be found. Find the new balance point, D, by repeating Steps 2 and 3. Measure the distance $|AD|$.
5. The e.m.f. of the cell is calculated from

$$\frac{E_1}{E_2} = \frac{|AC|}{|AD|}$$

where E_1 is the known e.m.f. and E_2 is the unknown one.

RESULTS

Known e.m.f........ =   V
$|AC|$ .... =   cm
$|AD|$ .... =   cm
Unknown e.m.f..... =   V.

Questions

1. What is the purpose of connecting the resistor in series with the galvanometer?
2. Why should the apparatus be arranged so that the balance point is as far as possible from A?
3. Why is it important that the moveable contact not be scraped along the wire?

Kirchhoff's Laws

It is sometimes necessary to be able to calculate the currents flowing in different parts of a circuit made up of a number of loops and branches. To simplify the solution of such problems, the following rules were devised by the German physicist, Gustav Kirchhoff (1824 — 1887). They are now known as **Kirchhoff's laws.**

---

1. **The algebraic sum of the currents at a junction is zero.**

2. **The algebraic sum of the e.m.f.'s in any closed loop is equal to the algebraic sum of the products of current and resistance for each part of the loop.**

The first rule simply states that charge does not accumulate or disappear at a junction, *i.e.* the total current entering a junction is the same as the total current leaving it. This is an example of the fact that charge is conserved. The second rule is merely an extension of Ohm's law. Both rules are best understood by considering how they are applied to a particular case.

Example

Calculate the value of the current flowing in each loop of the circuit shown in Fig. 21.17.

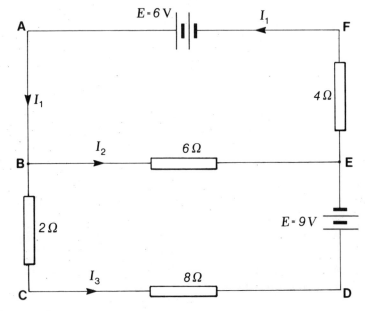

Fig. 21.17

At junction B the first rule tells us that

$$I_1 - I_2 - I_3 = 0 \quad \text{(i)}$$

I_1 is given a positive sign since it flows into the junction; I_2 and I_3 are given negative signs since they flow out of the junction.

Applying the second rule to the loop ACDFA gives

$$6 - 9 = 2I_3 + 8I_3 + 4I_1$$

$$=> \quad -3 = 10I_3 + 4I_1 \quad \text{(ii)}$$

The 9 V is given a negative sign since it is in opposition to the 6 V, *i.e.* it tends to drive a current in the opposite direction to the 6 V. All the quantities on the R.H.S. are given positive signs since in each case the currents are flowing in the same direction as a current driven by the 6 V battery.

Applying the second rule to the loop ABEFA gives

$$6 = 6I_2 + 4I_1 \quad \text{(iii)}$$

From *Eqn. (i)*

$$I_2 = I_1 - I_3$$

Substituting this value for I_2 in *Eqn. (iii)* gives

$$6 = 6I_1 - 6I_3 + 4I_1$$

$$=> \quad 6 = 10I_1 - 6I_3 \quad \text{(iv)}$$

From *Eqn. (ii)*

$$10I_3 = -3 - 4I_1$$

$$=> \quad I_3 = -0.3 - 0.4I_1$$

Substituting this value for I_3 in *Eqn. (iv)* gives

$$6 = 10I_1 + 1.8 + 2.4I_1$$

$$=> \quad 4.2 = 12.4I_1$$

$$=> \quad I_1 = 0.339 \text{ A}$$

Substituting this value for I_1 in *Eqn. (iii)* gives

$$6 = 6I_2 + 1.356$$
$$=> \quad 6I_2 = 4.644$$
$$=> \quad I_2 = 0.774 \text{ A}$$

Substituting for I_1 in *Eqn. (iv)* gives

$$6 = 3.39 - 6I_3$$
$$=> \quad 6I_3 = -2.61$$
$$=> \quad I_3 = -0.435 \text{ A}$$

The negative value for I_3 indicates that the direction in which it is flowing is opposite to that shown in the diagram.

Ans. 0.34 A; 0.77 A; 0.44 A

21.4 Measuring Resistance

The resistance of a resistor may be determined using the circuit shown in *Fig. 21.18*. The p.d. across the resistor is varied by means of the potential divider and measured on the voltmeter, V. The current flowing through the resistor is measured by the ammeter, A, and is recorded for a series of values of the applied p.d. A graph is then plotted of V against I. The resistance of the resistor is then equal to the slope of the graph.

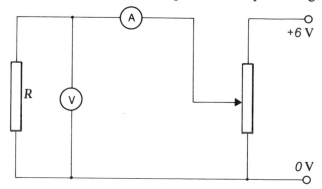

Fig. 21.18

For this method to be accurate the resistance of the voltmeter must be very much greater than the resistance of the resistor. (Since the ammeter measures the total current flowing through both the resistor and the voltmeter the current flowing through the voltmeter must be negligible compared with the current flowing through the resistor. For this to be so the resistance of the voltmeter must be very much greater than the resistance of the resistor.) Also, steps must be taken to ensure that the temperature of the resistor does not change, *e.g.* by using only small currents and/or placing the resistor in a water bath.

The Wheatstone Bridge

A more accurate method of determining resistances uses a Wheatstone bridge circuit, *Fig. 21.19*. The circuit takes its name from the English physicist, Sir Charles Wheatstone (1802 — 1875), who popularised its use. Four resistors are connected in the form of a square, ABCD. The resistance of three of the resistors is known accurately, and the resistance of the other is the one which is to be determined. A battery or cell is connected between A and C and a galvanometer is connected between B and D. The three known resistances are varied until no current flows through the galvanometer.

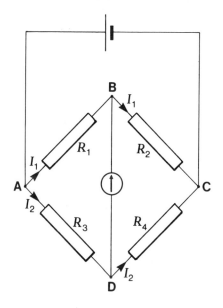

Fig. 21.19 Wheatstone bridge circuit

If no current flows through the galvanometer the same current must flow through R_1 and R_2. Let this current be I_1. Similarly, the current flowing through R_3 and R_4 must be the same. Let this current be I_2. Also since no current flows through the galvanometer, the p.d. between B and D must be zero, *i.e.* B and D are at the same potential. Therefore, the p.d. between A and B must be equal to the p.d. between A and D. If this p.d. is V then

$$V = I_1R_1$$

and

$$V = I_2R_3$$

=>

$$I_1R_1 = I_2R_3 \qquad \text{(i)}$$

Similarly, since the p.d. between B and C must be equal to the p.d. between D and C,

$$I_1R_2 = I_2R_4 \qquad \text{(ii)}$$

Dividing *Eqn. (i)* by *Eqn. (ii)* gives

$$\frac{I_1R_1}{I_1R_2} = \frac{I_2R_3}{I_2R_4}$$

=>

$$\boxed{\frac{R_1}{R_2} = \frac{R_3}{R_4}} \qquad \text{(iii)}$$

If R_2, R_3 and R_4 are known R_1 may be calculated.

The Metre Bridge

A more convenient form of the Wheatstone bridge circuit is the metre bridge, *Fig. 21.20*. In effect, R_3 and R_4 of *Fig. 21.19* have been replaced by a piece of resistance wire, AC, one metre

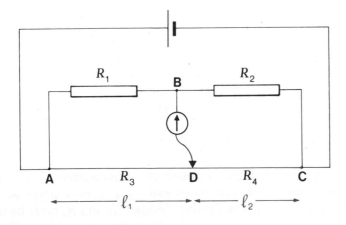

Fig. 21.20 Metre bridge circuit

long and of uniform cross-sectional area. One terminal of the galvanometer is connected to a moveable contact which can be placed at any point, *e.g.* D, on the resistance wire. The resistance of the length $|AD|$ now corresponds to R_3 and the length $|DC|$ now corresponds to R_4. As D is now a moveable contact the values of R_3 and R_4 are now almost infinitely variable.

Suppose that, when the galvanometer reads zero, $|AD| = l_1$ and $|DC| = l_2$. Then, since the wire is of uniform cross-sectional area,

$$R_3 \propto l_1$$

=>

$$R_3 = kl_1$$

where k is a constant. Similarly,

$$R_4 \propto l_2$$

=>

$$R_4 = kl_2$$

Substituting for R_3 and R_4 in *Eqn. (iii)* gives

$$\frac{R_1}{R_2} = \frac{kl_1}{kl_2}$$

=>

$$\boxed{\frac{R_1}{R_2} = \frac{l_1}{l_2}}$$

While the Wheatstone Bridge provides a very accurate method of determining resistances most everyday measurements of resistance are made with an ohmmeter. This device is based on the circuit of *Fig. 21.18*; a known p.d. is applied to the resistor and the resulting current is measured. The ohmmeter circuit is described on *p.261*.

EXPERIMENT 21.3
To Determine the Resistivity of Nichrome
Method 1:

Apparatus: Length of nichrome wire, micrometer, ohmmeter, metre stick.

Procedure:
1. Zero the ohmmeter if necessary. Using the ohmmeter measure the resistance, R, of the wire.

2. Measure the length, *l*, of the wire. Do not measure the parts of the wire which were in contact with the terminals of the ohmmeter.
3. Note the zero error on the micrometer and then use it to measure the diameter of the wire at several points. Find the average diameter, *d*.
4. The resistivity, ρ, is calculated from $\rho = R\pi d^2/4l$

RESULTS

Micrometer Reading/mm

Average reading........... =   mm
Micrometer zero error... =   mm
Diameter of wire......... =   m
Length of wire............ =   m
Resistance of wire........ =   Ω
Resistivity of nichrome.. =   Ω m.

Questions

1. Why is the diameter of the wire measured at several points?

Method 2:

Apparatus: Length of nichrome wire, micrometer, metre bridge, centre zero galvanometer, resistor of large resistance (~ 1 kΩ) to protect the galvanometer, resistors of known resistance (*e.g.* 1 Ω to 10 Ω), battery or low voltage power supply.

Procedure:
1. Set up the circuit as shown in *Fig. 21.20*. Connect the protective resistor in series with the galvanometer. R_1 is the resistance of the length of wire and R_2 is the known resistance.
2. Adjust the position of the moveable contact until the galvanometer reads zero. Close the switch to short out the protective resistor and find the balance point more accurately. Note the position, D, of the moveable contact. Measure the distances |AD| and |DC|.

3. The resistance of the wire is calculated from

$$\frac{R_1}{R_2} = \frac{|AD|}{|DC|}$$

4. Swop the positions of the wire and the resistor. Repeat Steps 2/3 and find an average value for the resistance of the wire.
5. Repeat Steps 1 to 4 using different known resistances. Find an average value, *R*, for the resistance of the wire.
6. Measure the length, *l*, of the wire. Do not measure the parts of the wire which were in contact with the terminals of the metre bridge.
7. Note the zero error on the micrometer and then use it to measure the diameter of the wire at several points. Find the average diameter, *d*.
8. The resistivity, ρ, is calculated from $\rho = R\pi d^2/4l$.

RESULTS

|AC|/cm	|DC|/cm	R/Ω

Micrometer Reading/mm

Average reading........... =   mm
Micrometer zero error... =   mm
Diameter of wire......... =   m
Length of wire............ =   m
Resistance of wire........ =   Ω
Resistivity of nichrome.. =   Ω m.

Questions

1. Why should the metre bridge circuit be arranged so that the balance point is as near the centre of the wire as possible? How might this be achieved?
2. What is the purpose of connecting the resistor in series with the galvanometer?
3. Why is it important that the moveable contact not be scraped along the wire?
4. Why is the diameter of the wire measured at several points?

21.5 Resistance and Temperature

As implied in the statement of Ohm's law *(p.217)* the resistance of a conductor depends on its temperature. For metals, resistance increases with temperature while, for semiconductors *(see Chapter 27),* resistance decreases with temperature.

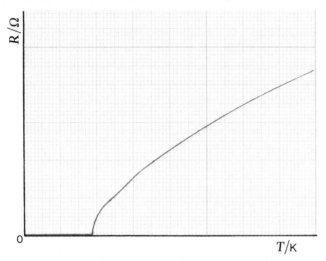

Fig. 21.21 Variation of resistance with temperature for a metal

Fig. 21.21 shows the variation of the resistance of a wire with temperature. Note that at a certain temperature, called the transition temperature and typically of the order of a few kelvins, the resistance falls abruptly to zero. At temperatures below the transition temperature the metal is said to be superconducting. This means that if a circuit, made up from the wire, is kept below the transition temperature a current, once started, will flow indefinitely without a source of e.m.f. in the circuit. (**Note:** Not all metals become superconducting at low temperatures.)

EXPERIMENT 21.4

To Investigate the Variation of the Resistance of a Copper Wire with Temperature

Method 1:

Apparatus: Coil of copper wire on insulating former, test tube containing glycerol, beaker, tripod stand, bunsen burner, thermometer, ohmmeter.

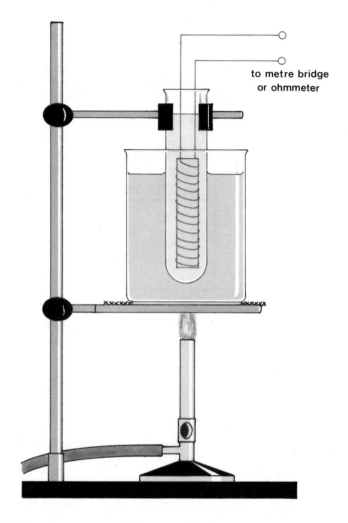

to metre bridge or ohmmeter

Fig. 21.22

Procedure:
1. Place the coil of wire in the test tube containing the glycerol and place the test tube in the beaker full of water, *Fig. 21.22*.
2. Zero the ohmmeter if necessary and connect it to the coil of wire.
3. Note the temperature of the glycerol and the resistance of the wire. Gently heat the beaker and note the resistance and temperature for each 10 K increase in temperature.
4. Plot a graph of resistance against temperature.

RESULTS

R/Ω	$\theta/°C$

Questions

1. Why is the wire immersed in glycerol?
2. Should allowance be made for the leads from the coil to the ohmmeter?

Method 2:

Apparatus: Coil of copper wire on insulating former, test tube containing glycerol, beaker, tripod stand, bunsen burner, thermometer, metre bridge, resistor of known resistance, battery or low voltage power supply, centre zero galvanometer, protective resistor for galvanometer.

Procedure:

1. Set up the apparatus as in the previous method, connecting the coil to the metre bridge instead of the ohmmeter (see Method 2 of previous experiment for setting up the metre bridge).
2. At each 10 K temperature interval note the temperature and find the balance point on the metre bridge. Measure |AD| and |DC| as in the previous experiment. Calculate the resistance of the coil in each case. Plot a graph of resistance against temperature.

RESULTS

Value of known resistance = ___ Ω

|AD|/cm	|DC|/cm	R/Ω	$\theta/°C$

Questions

1. Comment on the relative merits of the two methods of carrying out this experiment.

EXPERIMENT 21.5

To Investigate the Variation of the Resistance of a Thermistor with Temperature

Apparatus: Thermistor, test tube containing glycerol, beaker, tripod stand, bunsen burner, thermometer, ohmmeter.

Procedure:

1. Set up the apparatus as in the previous experiment, connecting the thermistor to the ohmmeter.
2. At temperature intervals of 10 K note the temperature of the glycerol and the resistance of the thermistor.
3. Plot a graph of resistance against temperature.

RESULTS

R/Ω	$\theta/°C$

Note: As in the previous experiment a metre bridge may be used to measure the resistance of the thermistor.

Questions

1. Explain why the results of this experiment should, in principle, be more accurate than those of the previous experiment.
2. What is the major difference between the graphs obtained in these two experiments. Explain the reasons for this difference.

EXPERIMENT 21.6

To Calibrate and use a Thermistor Thermometer

Apparatus: Thermistor, beaker, round-bottomed flask, some ice, retort stand, bunsen burner, ohmmeter.

Procedure:
1. Place some ice and cold water in the beaker. Connect the ohmmeter to the thermistor and place the thermistor in the beaker. After a few minutes note the resistance of the thermistor.
2. Boil some water in the flask. Place the thermistor in the steam above the boiling water and after a few minutes note the resistance of the thermistor.
3. Warm some water in the beaker and place the thermistor in it. Note the resistance of the thermistor. The temperature, θ, of the water is then calculated from

$$\theta = \frac{R_\theta - R_0}{R_{100} - R_0} \times 100$$

where R_0 and R_{100} are the resistances of the thermistor in ice and steam respectively.

4. Measure the temperature of the water in the beaker using a mercury thermometer.

RESULTS

Resistance in ice.......................... = Ω
Resistance in steam...................... = Ω
Resistance in warm water.............. = Ω
Temperature of water (thermistor)... =   °C.
Temperature of water (mercury)...... =   °C.

Questions

1. Comment on the difference or similarity of the temperature reading obtained from the thermistor and that obtained from the mercury thermometer.

SUMMARY

A flow of charge from a point of higher potential to a point of lower potential is called an electric current and is measured in amperes. The ampere is that constant current which, if maintained in two straight parallel conductors of infinite length, of negligible circular cross-section and placed 1 m apart in a vacuum, causes each to exert a force of 2×10^{-7} N per metre length of the other.

The unit of charge, the coulomb, is defined as the amount of charge transferred in 1 s when the current flowing is 1 A. Ohm's law states that, at constant temperature, the current flowing in a conductor is proportional to the p.d. between its ends. (Not all conductors obey Ohm's law.) The ratio of p.d. to current is thus a constant, called the resistance, R, of the conductor, *i.e.*

$$V = RI$$

For a complete circuit this becomes

$$E = RI$$

where E is the e.m.f. in the circuit and R is the total resistance of the circuit.

The resistance of a conductor is proportional to its length and inversely proportional to its cross-sectional area; the constant of proportionality is called the resistivity, ρ, of the material of the conductor, *i.e.*

$$R = \rho \frac{l}{A}$$

The effective resistance, R, of a number of resistors in series is given by

$$R = R_1 + R_2 + R_3$$

For resistors in parallel the effective resistance is given by

$$\frac{1}{R} = \frac{1}{R_1} + \frac{1}{R_2} + \frac{1}{R_3}$$

The resistance of a conductor varies with temperature, increasing for metals and decreasing for semiconductors. At temperatures of a few kelvins some metals become superconducting, *i.e.* their resistance becomes zero. The resistance of a metal or a thermistor may be used as a thermometric property. Resistance may be measured using a Wheatstone bridge, a metre bridge or an ohmmeter. P.d. and e.m.f. may be measured accurately using a potentiometer. Kirchhoff's laws state that, in any circuit loop (i) the algebraic sum of the currents at a junction is zero, (ii) the algebraic sum of the e.m.f.'s is equal to the algebraic sum of the products of p.d. and current.

Questions 21

Section I

1. What is an electric current?................................

....

2. Define the unit of current, the ampere...................

....

....

3. Define the unit of charge, the coulomb.................

....

....

4. State Ohm's law...

....

....

5. The ratio of the p.d. across a conductor to the current flowing through it is known as...................

....

6. Define the ohm..

....

....

7. The distinction between p.d. and current was first explained by ...., a................. scientist who worked at the beginning of the .... century.

8. What is the difference between resistance and resistivity? Give the unit of resistivity...................

....

....

9. If a substance has a low resistivity it has a high

....

10. Give an expression for the effective resistance of two resistors connected in parallel with each other.

....

11. What is a potential divider used for?....................

....

....

*12. What is a potentiometer used for?......................

....

....

*13. State Kirchhoff's laws....................................

....

....

....

14. A meter used to measure resistance is called an

....

*15. In what way does a metre bridge differ from a basic Wheatstone bridge?...

....

16. Sketch a graph to show how the resistance of a wire varies with its temperature in the range 0°C to 100°C.

Section II

1. Two wires, each of length 25 cm, are placed parallel to each other and 1.0 m apart in a vacuum. If each wire carries a steady current of 2.0 A what is the magnitude of the force exerted by one on the other?

2. A sensitive meter connected between the dome of a Van de Graaff generator and earth shows a steady current of 1.2 μA. How much charge is carried to the dome by the belt in one minute?

3. A capacitor of capacitance 470 μF is charged to a voltage of 12 V and then totally discharged through a small resistance. This process is repeated 50 times per second. What is the average current flowing in the resistor?

4. *Fig. I* shows two resistors connected in series with a 9 V battery. From the values given in the diagram calculate (i) the current, I, (ii) the p.d. between A and B.

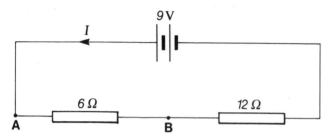

Fig. I

5. Using the values given in the diagram of *Fig. II* calculate (i) the total current, I, flowing in the circuit, (ii) the current flowing in each resistor.

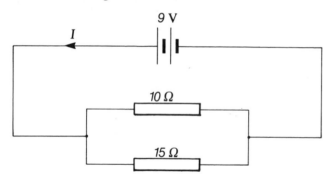

Fig. II

6. How may four 2 Ω resistors be connected to give an effective resistance of (i) 2 Ω, (ii) 0.8 Ω?

7. Using the values given in *Fig. III* calculate (i) the total current, I, flowing in the circuit, (ii) the current flowing in the 9 Ω resistor, (iii) the p.d. between A and B.

8. Calculate the diameter of a copper wire 2.5 m long if its resistance is 0.1 Ω. (Resistivity of copper = 1.7 $\times$ 10^{-8} Ω m.)

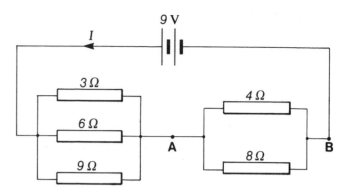

Fig. III

9. What is the length of the copper wire in a galvanometer coil if its diameter is 0.1 mm and the resistance of the coil is 100 Ω?

10. A potential divider consists of a fixed resistor of resistance 10 Ω and a variable resistor whose resistance may be varied from 0 to 40 Ω, *Fig. IV*. If the voltage of the battery is 9 V what are the maximum and minimum p.d.'s between X and Y? What is the advantage of having the fixed resistor in the circuit?

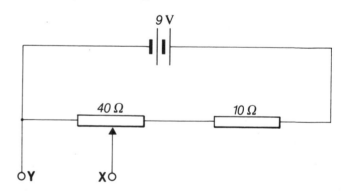

Fig. IV

*11. In a potentiometer experiment to determine the e.m.f. of a cell the balance point for a cell of e.m.f. 1.5 V was 80 cm from the end of the wire. For the second cell the balance point was 69 cm from the end. What was the e.m.f. of the second cell? If the p.d. across the full wire (1 m long) is 90% of the e.m.f. of the driver cell what is the e.m.f. of the driver cell?

*12. Calculate the current flowing in each branch of the circuit shown in *Fig. V*.

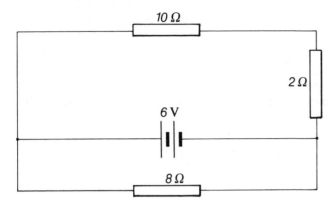

Fig. V

*13. Calculate the current flowing in each branch of the circuit shown in *Fig. VI*.

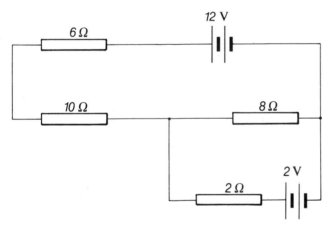

Fig. VI

*14. In the circuit shown in *Fig. VII*, $I_1/I_2 = 3/4$. Calculate the e.m.f. of the battery, B, and the current flowing through the 10 Ω resistor.

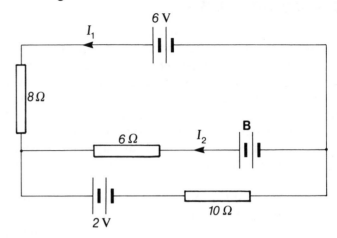

Fig. VII

*15. The metre bridge in *Fig. VIII* is balanced when the moveable contact is at the 40 cm mark on the scale. What is the value of X?

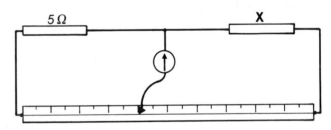

Fig. VIII

22 Electrical Energy

22.1 Joule's Law

When a current flows through a wire the wire becomes hot. Many domestic appliances are based on this principle — electric kettles, irons, heaters, *etc.* In general, when a current flows through a resistor, some electrical energy is converted into internal energy in the resistor. This energy conversion results from collisions between the moving electrons which make up the current, and the metal ions. These collisions cause the ions to vibrate faster, *i.e.* their average kinetic energy is increased and so the temperature of the metal is increased *(cf. Chapter 16)*.

The p.d. between two points is the work done in carrying unit charge from one point to the other *(p.199)*. Therefore, if the p.d. across the resistor is V the electrical energy converted in time t is

$$W = VQ$$

where Q is the charge which passed through the resistor in the time t. Then, since

$$Q = It$$

$$W = VIt \qquad \text{(i)}$$

From the definition of resistance *(p.217)*

$$V = RI$$

where R is the resistance of the resistor. Substituting for V in *Eqn. (i)* gives

$$W = RI^2t \qquad \text{(ii)}$$

Rather than the total energy converted in a given time, it is frequently more convenient to talk of the rate at which energy is converted, *i.e.* the amount of energy converted in one second or, the power.

From *Eqns. (i)* and *(ii)* the power is given by

$$\boxed{P = VI} \qquad \text{(iii)}$$

and

$$\boxed{P = RI^2} \qquad \text{(iv)}$$

Note: *Eqn. (iii)*, since it is derived from the definition of p.d., can be applied in all cases involving the conversion of electrical energy to other forms. *Eqn. (iv)* can be applied only to the conversion of electrical energy in a resistor.

If the resistance of the resistor is constant *Eqn. (iv)* is a statement of **Joule's Law.**

> **The rate at which electrical energy is converted to internal energy in a resistor is proportional to the square of the current flowing through the resistor.**

In symbols, Joule's law may be stated as

$$\boxed{P \propto I^2}$$

Example
Calculate the current flowing through a 1 kW heater when it is connected to the mains supply (220 V). Find also the resistance of the element of the heater.

$$P = VI$$

$$1000 = 220\,I$$

$$\Rightarrow \quad I = \frac{1000}{220}$$

$$= 4.55 \text{ A}$$

$$V = RI$$

$$220 = R \times 4.55$$

$$\Rightarrow \quad R = \frac{220}{4.55}$$

$$= 48.4\ \Omega \qquad \text{Ans. 4.55 A; 48.4 }\Omega$$

EXPERIMENT 22.1
To Verify Joule's Law
Method 1:
Apparatus: Lagged beaker or calorimeter, coil of wire (manganin or constantan), battery or low voltage power supply, rheostat, ammeter, thermometer.

Procedure:
1. Put some cold water in the beaker and find the mass of the beaker and water. Note the temperature of the water.
2. Set up the circuit as shown in *Fig. 22.1* and place the coil in the water.

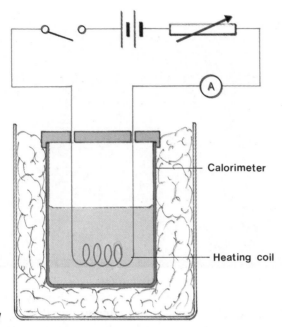

Fig. 22.1

3. Note the time, switch on the current and allow it to flow until the temperature has risen by about 2°C. Note the current, the temperature and the time for which the current flowed. Make sure the current stays constant throughout; adjust the rheostat if necessary.

4. Refill the beaker with cold water. The mass of water must be the same as before. Repeat Step 2, using a different current but allowing it to flow for the same length of time.
5. Repeat Step 3 four or five times. In each case calculate the rise in temperature, $\Delta\theta$, of the water. Plot a graph of $\Delta\theta$ against I^2, where I is the current flowing in each case.

RESULTS
Mass of beaker and water .... =   kg
Time for which current flowed .... =   s.

$\theta_1/°C$	$\theta_2/°C$	$\Delta\theta/°C$	I/A	I^2/A^2

Questions
1. Why is it not necessary to know the mass of the beaker or calorimeter?
2. Why must the same mass of water be used each time?
3. Why is it important that the current be kept constant?

James Joule (1818 — 1889)

Joule was born in Lancashire, England, the son of a wealthy brewer. He suffered from poor health and as a result devoted most of his time to study. His father encouraged him in this, even to the extent of supplying him with a laboratory.

Joule was mainly interested in work and changes in energy, particularly internal energy. (At that time it was not realised that heating involves the transfer of energy just as work does.) While still in his teens, he published papers on the rises in temperature produced by electric motors and by 1840 he had worked out the relationship between the current in a resistor and the resulting rise in temperature (a measure of the internal energy produced). This relationship is now known as Joule's law. He also measured the rise in temperature of water when it was churned with paddles, when it was passed through small holes and when it fell from a height. He concluded, in a paper published in 1847, that the rise in temperature

of a body is proportional to the work done on it. He also arrived at the basis of the Principle of Conservation of Energy, although he did not formulate it as such.

Joule had difficulty having his ideas accepted by the scientific establishment, probably because he had no formal academic qualifications (he took over the running of his father's brewery while still a teenager and remained a brewer for the rest of his life). However, in 1847 he managed to attract the interest of William Thompson (Lord Kelvin) and from then his reputation was made. Much of Joule's success was due to dedicated and painstaking experimental work. He made thermometers which could be read to 0.005 °F (0.003 °C) and while on honeymoon he made a special thermometer for measuring the temperature at the top and bottom of waterfalls he and his wife were to visit. He was elected to the Royal Society in 1850 and died at Sale, in Cheshire, in 1889.

The Kilowatt-hour

The unit used by the E.S.B. in calculating the amount of energy "used" (*i.e.* the amount of electrical energy converted to other forms) by a customer is not the joule, since the joule is a relatively small unit, but the **kilowatt-hour**, kW h.

> **1 kW h is the amount of electrical energy converted to other forms in one hour when the rate of conversion, *i.e.* the power, is 1 kW.**

From this definition we see that to calculate the number of kW h of electrical energy "used" we must multiply the number of kW by the number of hours for which the energy has been used at that rate.

Example

Calculate the total cost of using a 100 W bulb, a 2 kW heater and a 60 W television for 5 hours if 1 Unit (1 kW h) costs 8 p.

$$
\begin{aligned}
\text{Total power} &= 100 + 2000 + 60 \\
&= 2160 \text{ W} \\
&= 2.16 \text{ kW}
\end{aligned}
$$

$$
\begin{aligned}
\text{Energy converted in 5 h} &= 2.16 \times 5 \\
&= 10.8 \text{ kW h}
\end{aligned}
$$

$$
10.8 \text{ kW h @ 8 p} = 86.4 \text{ p} \qquad \text{Ans. 86 p}
$$

Note that the kW h is not part of the SI system. In calculations the standard unit of energy, *i.e.* the joule, must be used.

$$
\begin{aligned}
1 \text{ kW} &= 1000 \text{ W} \\
&= 1000 \text{ J s}^{-1} \\
1 \text{ h} &= 3600 \text{ s}
\end{aligned}
$$

Therefore,
$$
\begin{aligned}
1 \text{ kW h} &= 1000 \text{ J s}^{-1} \times 3600 \text{ s} \\
&= 3\ 600\ 000 \text{ J} \\
&= 3.6 \text{ MJ}.
\end{aligned}
$$

Transmission of Electrical Energy

Electrical energy is transmitted from one place to another via metal cables. Since these cables have a resistance a certain amount of energy will always be converted into internal energy in them. It is important that the amount of energy converted

Fig. 22.2 High tension transmission lines

should be small — for two reasons. Firstly, a rise in temperature could cause a fire. Secondly, the energy thus converted is effectively lost, making what remains more expensive.

In overhead cables the problem is solved by increasing the p.d. to several thousand volts. An example best illustrates the effect of this.

Suppose it is required to transmit electrical energy from one place to another through cables which have a total resistance of 10 Ω at a rate of 1 MW, *i.e.* 1 million joules per second. If the energy is transmitted at 5000 V the current flowing may be calculated from

$$
\begin{aligned}
P &= VI \\
\Rightarrow \quad 1 \times 10^6 &= 5000\ I \\
\Rightarrow \quad I &= 200 \text{ A}
\end{aligned}
$$

That is, the current flowing through the cables is 200 A. The energy converted to internal energy per second in the cables is given by

$$
\begin{aligned}
P &= RI^2 \\
&= 10 \times 200^2 \\
&= 4 \times 10^5 \text{ W} \\
&= 0.4 \text{ MW}
\end{aligned}
$$

In other words, 40% of the energy supplied is wasted in heating the cables. On the other hand, if the same energy were transmitted at 250 kV a similar calculation shows that the current flowing would be 4 A and the energy converted into internal energy per second in the cables would be 160 W — a mere 0.016% of the total supplied.

In homes and factories it is not possible to use electricity at such high p.d.'s because of the difficulty in providing adequate insulation. Because the p.d. is lower, usually 220 V, the currents flowing are larger. The resistance of the cables used must therefore be small enough to ensure that there will be no significant rise in the temperature of the cables due to the current.

22.2 Domestic Wiring

There is scarcely a home in the country which does not use electricity for something — lighting, heating, radio, television, *etc*. It is a clean and convenient form of energy; it is also potentially dangerous. Therefore, everyone should learn to use it properly and safely and so prevent needless injury and loss of life.

Electric Shock

All electric appliances have, when connected to the mains, some parts which are at a potential different from that of the earth (zero). Therefore, if such parts are connected to earth by a conductor a current will flow.

The human body is an electrical conductor. Its response to the passage of a current depends on the size of the current and on the path taken by the current. Thus, a current passing from, say, the finger to the elbow, while possibly painful, is unlikely to be dangerous. On the other hand, a current of around 50 mA passing through the chest area will make breating very difficult and may cause death. A current of around 100 mA is usually fatal, the most common cause of death being a form of heart failure due to the body's own electrical signals being "swamped" by the current.

Fuses

A fuse is a thin piece of wire which melts when a current greater than a certain value passes through it. Fuses are connected in series in the "live" wire so that, if a fault develops anywhere in the circuit, causing the current to increase, they melt, switching off the current before damage or injury is caused. Fuses are manufactured with certain current ratings, *e.g.* 3 A, 13 A, *etc*. This means that the fuse will melt if the current in

the circuit in which they are connected goes above these values. The correct fuse to use in a particular case is determined by the power likely to be used by the appliance, or in the case of a distribution box (*p. 238*), appliances, connected to it. **Note:** A fuse does **not** protect from electric shock. As noted above a current of only 50 mA may be lethal and, in any case, the fuse would not melt fast enough.

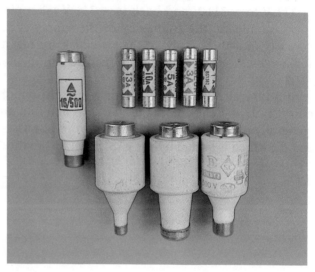

Fig. 22.3 Fuses

Example
What fuse should be used in a plug which is to be connected to a 500 W drill?

$$P = VI$$
$$=> \quad 500 = 220\,I$$
$$=> \quad I = 2.3\text{ A}$$

The nearest, higher rating to this is 3 A.

Ans. 3 A

A fuse is a safety device, designed to prevent not only damage to an appliance, but also injury or death. Consequently, **under no circumstances** should you replace a fuse with a piece of ordinary wire or with aluminium foil. To do so could result in another part of the circuit overheating and starting a fire. This is very likely to happen where the wires are concealed under floorboards, in the attic, *etc.*, and so the fire is unlikely to be discovered until it is too late to put it out.

House Circuits

Electrical energy is supplied to all homes and many other buildings in this country at 220 V a.c. (a.c. = alternating current — *see Chapter 25*) via two cables. The potential of one of these, the **neutral**, is always at or near earth potential, *i.e.* zero. The potential of the other, the **live** or **phase**, varies from +311 V to —311 V relative to the neutral. The variation from +311 V to —311 V and back to +311 V takes place 50 times per second, *i.e.* the frequency of the supply is 50 Hz. When we say that the supply voltage is 220 V we mean that the effective p.d. (*see p. 269*) between the live and the neutral is 220 V.

Part of a typical wiring diagram for a modern building is shown in *Fig. 22.4*. (The colours used in this diagram are not those generally used in practice.) The live wire, L, passes first to a fuse, F, which is sealed by the E.S.B. on installation and then, with the neutral, to a meter, M, which records (in kW h) the total amount of energy supplied. From the meter the two wires go to a distribution box, B, which contains a set of fuses or circuit breakers. (Circuit breakers are electromechanical devices which switch off the current should it exceed some pre-set value, *i.e.* they fulfil the same function as fuses.) From B electricity is delivered to the various appliances. An appliance which requires a large current, *e.g.* a cooker, is connected directly to the distribution box. Smaller appliances, *e.g.* televisions, heaters, *etc.*, are connected by plugs to sockets, S, in a ring main circuit. This circuit, like the one to the cooker, includes a third wire. This is called the earth wire and its purpose is explained in the next section. The third type of circuit is the one connecting the lights to the distribution box. This circuit is arranged so that a number of lights are connected in parallel as shown in the diagram. Note that the

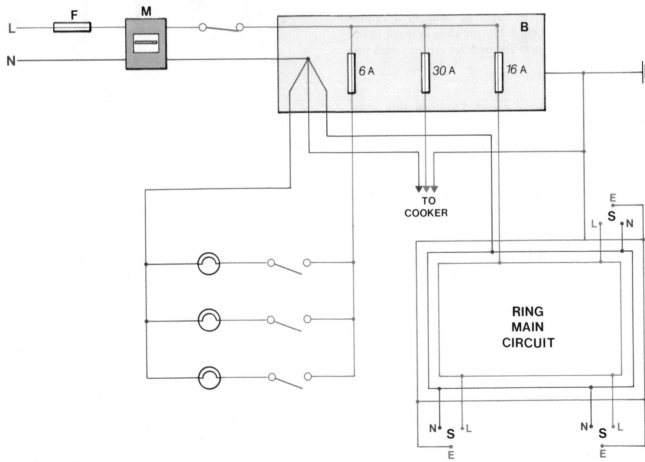

Fig. 22.4 Domestic circuits

switch must **always** be on the live wire. Not all the lights or all the sockets in a house would be on one circuit — there would normally be a number of each type of circuit.

A particular type of lighting circuit is required on stairs, corridors, *etc.* In these situations it must be possible to switch the light on or off at either of two positions. This requires two two-way switches, *Fig. 22.5*. The two switches are joined by two wires, AA and BB, as shown in the diagram. For the light to be "on" both switches must be in the same position, *i.e.* both in position A or both in position B. Thus, in the diagram, the light may be switched on either by moving switch 1 to position B or by moving switch 2 to position A.

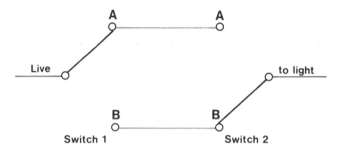

Fig. 22.5 Two-way switch circuit

Earthing

In addition to the two wires carrying the current many circuits have a third wire. This is the earth wire and, as its name suggests, it is connected directly to the earth. This is a safety device and all appliances which have a metal body should have an earth wire connected to the metal body. In the event of a fault developing such that the body of the appliance comes in contact with the live wire, the current will flow harmlessly to earth through the earth wire rather than through the body of a person touching it. Usually, when such a fault occurs, the current flowing to earth is large enough to blow the fuse.

However, even if this is not the case, it is still safe to touch the appliance since the current will take the path of lowest resistance, *i.e.* the earth wire.

Plugs

A plug is a device for connecting a portable appliance, *e.g.* a television , a vacuum cleaner, *etc.*, into a ring main circuit. Plugs have three terminals — one for the live wire, one for the neutral and one for the earth. It is essential that each terminal be connected to the correct wire. To help ensure that this is done the wires attached to all appliances have a standard colour code. The live wire is brown, the neutral is blue and the earth has green and yellow stripes. The wires should be connected as shown in *Fig. 22.6*. Looking at the plug from the back with the cover removed, the earth is connected to the central terminal, the neutral to the left hand terminal and the live to the right hand terminal — the one attached to the fuse.

The cable grip at the point where the cable enters the plug is designed to keep the cable firmly in place and so prevent the connections from becoming loose as a result, for example, of someone pulling on the cable. The cable grip should be tightened firmly but not over-tightened as this could damage the insulation and possibly cause a short-circuit.

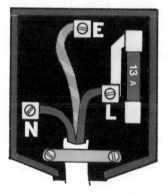

Fig. 22.6 Three pin plug

SUMMARY

When a current flows through a resistor electrical energy is converted to internal energy in the resistor at a rate given by
$$P = RI^2 = VI$$
Joule's law states that the rate at which energy is converted in a resistor is proportional to the square of the current, *i.e.*
$$P \propto I^2$$

The unit of energy used by the E.S.B. is the kW h — the amount of energy supplied in one hour when the rate of supply is 1 kW.

Electrical energy is transmitted at high voltages to minimise energy loss in the cables carrying the current.

A fuse is a thin piece of wire, connected in series in a circuit, which melts if the current in the circuit rises above a certain value. All circuits should have a fuse of the correct current rating. All metal appliances should be earthed so that, even in the event of a fault, the body of the appliance is always at the same potential as the earth, *i.e.* zero.

In domestic circuits a number of lights are connected in parallel in a circuit; sockets are connected in a ring main circuit and larger appliances, *e.g.* cookers, are on individual circuits.

Questions 22

Section I

1. State Joule's law..

....

2. In the equation $W = RI^2t$, what does each of the symbols represent?...

....

3. James Joule is chiefly remembered for his work on ...

....

4. What is a kilowatt-hour?.......................................

....

5. Why is electrical energy transmitted at high voltages?..

....

6. What is the purpose of a fuse?..............................

....

7. In a three core cable the live wire is coloured............

the neutral is................and the earth is................

8. What is the advantage of a ring circuit?.................

....

9. Why should the switch always be on the live wire in a circuit?...

....

10. Why would it be dangerous, when wiring a plug, to connect the blue wire to the live and the brown wire to the neutral?

....

Section II

1. Calculate the amount of energy "used" by a 2 kW electric heater in 3 hours.

2. Calculate the amount of energy converted to internal energy when a current of 2.5 A flows for 10 minutes through a wire whose resistance is 20 Ω.

3. Calculate the current flowing in each of the following appliances when it is connected to the mains: 2.8 kW kettle; 100 W bulb; 1250 W toaster.

4. What is the resistance of the element of each of the following mains appliances: 60 W bulb; 2 kW heater; 250 W lamp.

5. A bulb is marked 12V 45W. What do these markings mean? What is the resistance of the filament of the bulb? What current normally flows through it?

6. In an X-ray tube operating at 50 kV the current flowing is 20 mA. How much energy is transferred across the tube in 200 ms?

7. Calculate the total cost of using a 3.5 kW immersion heater for 3 hours; a 1250 W toaster for 10 minutes; a 150 W bulb for 10 hours and a 2 kW heater for 5 hours if one "unit" costs 8p.

8. What fuse should be used in a plug connected to each of the following mains appliances: 2.5 kW kettle; 100 W standard lamp; 450 W hair drier?

9. A coil of wire is placed in an aeroboard vessel containing 100 g of water. When the coil is connected to a power supply the p.d. across it is 12 V and the current flowing through it is 1.5 A. If the current is kept constant for 10 minutes, and assuming that there is no energy lost to the surroundings, calculate the rise in temperature of the water. (S.h.c. of water $= 4.2 \times 10^3$ J kg^{-1} K^{-1}.)

10. A 5 kW immersion heater takes 75 minutes to raise the temperature of 50 kg of water from 10 °C to 70 °C. Calculate the amount of energy lost to the surroundings. Given that a lagging jacket would reduce this loss by 80% calculate the money saved by fitting such a jacket if one unit costs 8p. (S.h.c. of water $= 4.2 \times 10^3$ J kg^{-1} K^{-1}).

23 Chemical Effect of Current

23.1 Electrolysis

When an electric current is passed through water with a little acid in it, as in the arrangement shown in *Fig. 23.1*, bubbles of gas are seen to appear at the plates. Obviously, chemical reactions are taking place. This process is known as **electrolysis.**

> **When conduction in a liquid is accompanied by chemical reactions the process is known as electrolysis.**

Liquids in which electrolysis takes place are known as **electrolytes**. Examples of electrolytes are solutions of acids, bases and salts in water. All salts in the molten state also act as electrolytes. The carbon plates, *Fig. 23.1*, through which the current enters and leaves the electrolyte are called **electrodes**; the one connected to the positive terminal of the battery is called the **anode** and the other is called the **cathode**. The whole arrangement of container, electrolyte and electrodes is called a **voltameter**.

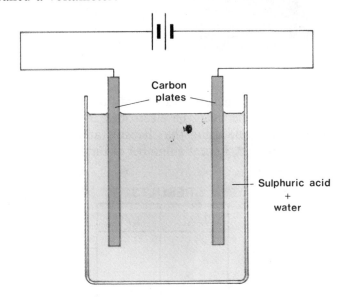

Fig. 23.1 Voltameter

Ions

When an atom or molecule (two or more atoms joined together) gains or loses an electron, and hence becomes negatively or positively charged respectively, it is called an **ion.**

When sodium and chlorine combine to form sodium chloride (common salt) each sodium atom loses an electron while each chlorine atom gains an electron. Solid sodium chloride thus consists of positive sodium ions and negative chloride ions joined together in a regular structure. When sodium chloride is dissolved in water or is melted, the sodium and chloride ions become free to move independently of each other.

Conduction in Electrolytes

In a voltameter, the current in the electrolyte consists of a movement of ions. The negative ions move towards the positive plate (the anode) and the positive ions move towards the negative plate (the cathode). At the anode, the negative ions give up their extra electrons and become neutral atoms. The electrons freed in this way then move around the circuit, through the wires and the battery, to the cathode. At the cathode the positive ions pick up the electrons and so also become neutral atoms. This process, for a voltameter in which the electrolyte is molten sodium chloride (NaCl), is illustrated in *Fig. 23.2*.

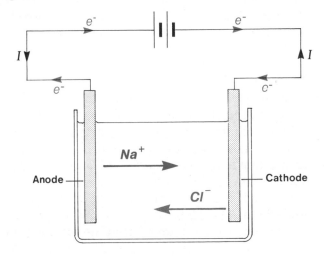

Fig. 23.2 Conduction in an electrolyte

The relationship between the current flowing through a voltameter and the p.d. between the electrodes may be investigated using the arrangement shown in *Fig. 23.3*. There are two cases to consider. If the electrodes are active, *i.e.* they

241

take part in the chemical reactions, a graph similar to that shown in *Fig. 23.3(a)* is obtained. This graph is a straight line through the origin, showing that the current is proportional to the p.d. between the electrodes. In other words, a voltameter with active electrodes obeys Ohm's law. An example of such a voltameter is one consisting of copper electrodes in a copper sulphate solution. When a current is passed through such a voltameter copper from the anode goes into solution while copper metal is deposited on the cathode.

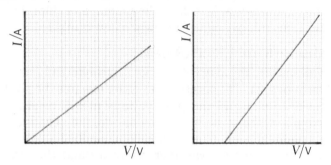

Fig. 23.3 *Variation of current with p.d. for an electrolyte (a) active electrodes, (b) inert electrodes*

If inert electrodes, *i.e.* electrodes which do not take part in the chemical reactions, are used a graph similar to that shown in *Fig. 23.3(b)* is obtained. In this case no current flows unless the p.d. is above a certain value. The graph does not pass through the origin — Ohm's law is not obeyed. An example of such a voltameter is one having carbon electrodes in dilute hydrochloric acid. In this case hydrogen is released at the cathode and oxygen at the anode; neither of the electrodes takes part in a chemical reaction. The hydrogen on the cathode tends to go back into solution, giving up its electrons to the cathode. In other words, this effect is tending to drive electrons in the opposite direction to the battery *(cf. Fig. 23.2)*. The hydrogen in fact forms part of a simple cell *(see p. 245)* which generates a p.d. which tends to drive a current in the opposite direction to the applied p.d. Thus, if the applied p.d. is not greater than the p.d. generated in the voltameter itself no current can flow. This is illustrated by the graph of *Fig. 23.3(b)*. The point at which the line cuts the p.d. axis represents the p.d. generated by the voltameter.

EXPERIMENT 23.1
To Investigate the Relationship between P.d. and Current for an Electrolyte

Apparatus: Beaker, two copper electrodes, two carbon electrodes, copper sulphate solution, dilute hydrochloric acid solution, battery or low voltage power supply, voltmeter, ammeter, rheostat.

Procedure:
1. Fill the beaker with the copper sulphate solution and set up the circuit, using the copper electrodes, as shown in *Fig. 23.4*.

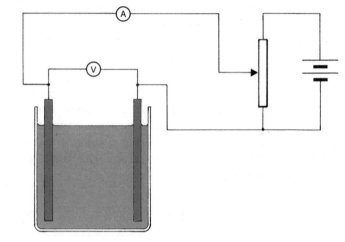

Fig. 23.4

2. Switch on the current and record the current flowing, I, and the p.d. between the electrodes, V.
3. Adjust the p.d. using the rheostat and record the new current and p.d. Repeat a number of times and plot a graph of I against V.

RESULTS

I/A	V/V

4. Replace the copper electrodes with the carbon electrodes and replace the copper sulphate with the hydrochloric acid.
5. Repeat the procedure given above and again plot a graph of I against V.

RESULTS

I/A	V/V

Questions

1. Comment on the difference between the two graphs.
2. In both cases the currents should be kept small. Why do you think this is so?

Faraday's First Law of Electrolysis

When a current is passed through an electrolyte one, or more, elements are "liberated" from the liquid. For example, in the voltameter in *Fig. 23.2* the element sodium is released at the cathode and the element chlorine is released at the anode. Michael Faraday suspected that a relationship existed between the mass of an element liberated and the charge passed through the voltameter. From experiments with a number of different voltameters, he deduced the following relationship which is now known as **Faraday's First Law of Electrolysis.**

> **The mass of an element liberated during electrolysis is proportional to the quantity of charge passed through the electrolyte.**

In symbols,

$$m \propto Q$$

$$=> \quad m = zQ$$

where z is a constant.

But, $$Q = It \text{ (p. 216)}$$

Therefore, *Eqn. (i)* becomes

$$\boxed{m = zIt}$$

The constant z is called the **electrochemical equivalent (e.c.e.)** of the element liberated.

Example

Calculate the e.c.e. of copper given that a current of 2.0 A liberates 1.2 g of copper in 30 minutes.

$$m = zIt$$

$$=> \quad z = \frac{m}{It}$$

$$= \frac{1.2 \times 10^{-3}}{2 \times 30 \times 60}$$

$$= \frac{1.2 \times 10^{-3}}{3.6 \times 10^{3}}$$

$$= 3.33 \times 10^{-7} \text{ kg C}^{-1}$$

Ans. 3.3×10^{-7} kg C^{-1}

EXPERIMENT 23.2

To Measure the Electrochemical Equivalent of Copper

Apparatus: Beaker, two copper electrodes, copper sulphate solution*, battery or low voltage power supply, ammeter, rheostat, stopclock.

Procedure:
1. Clean the cathode thoroughly. Fill the beaker with the copper sulphate solution and set up the circuit as shown in *Fig. 23.5*.
2. Switch on the current and adjust the rheostat to give a small current.
   (About 0.02 A per cm^2 of cathode area is suitable.)
3. Switch off the current. Remove the cathode, dry it and find its mass.

243

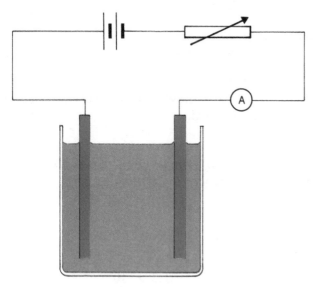

Fig. 23.5

4. Replace the cathode in the beaker, switch on the current and allow it to flow for 15 — 20 minutes. Make sure that the current remains constant throughout, adjusting the rheostat if necessary. Switch off and note the time, t. Record the current, I.

5. Remove the cathode. Wash it carefully with methylated spirits and dry it by warming it gently. Find its mass and hence the mass of copper, m, deposited on it. Calculate the electrochemical equivalent, z, of copper from $m = zIt$. If time permits repeat and find an average value for the e.c.e.

RESULTS

Mass of cathode .... =   kg
Current .... =   A
Time .... =   s
Mass of cathode + copper .... =   kg
Mass of copper deposited .... =   kg
E.c.e. of copper .... =   kg C^{-1}.

Questions

1. What would be the likely consequence of using a larger current in this experiment?
2. Why is it necessary to let the current flow for 15 — 20 minutes?

3. Why is it important to keep the current constant? Why are variations in the current likely?

* The copper sulphate solution may be made up by adding approximately 15 g of copper sulphate to 100 cm^3 of warm water. To this should be added 2.5 cm^3 of conc. sulphuric acid.

Electroplating

One important industrial application of electrolysis is electroplating. We noted earlier that when a current is passed through a copper voltameter copper is deposited on the cathode. In electroplating a cheap metal, *e.g.* iron or steel, is coated in this way with a more expensive one, *e.g.* chromium or silver. This is done to prevent the cheaper metal rusting and/or to improve its appearance. Examples of this are the chromium plating on bicycle handlebars and car bumpers and the silver plating on cutlery.

We noted in *Chapter 20* that the dielectric in electrolytic capacitors is produced by electrolysis. In this process the electrodes are aluminium and the electrolyte is aluminium borate solution. When a current is passed through the voltameter a layer of aluminium oxide is formed on the anode. Aluminium oxide is an insulator and so acts as a dielectric between the plates of the capacitor.

Fig. 23.6 Electroplating: Chrome plating electric kettles

23.2 Cells

A cell is a device in which energy is stored as chemical energy which can be released, when required, as electrical energy. It may also be thought of as a device which generates a p.d. between two points — the terminals of the cell — by means of a chemical reaction.

There are two main types of cell, *viz.* those which cannot be recharged and those which can. Cells which cannot be recharged are called **primary cells.** Cells which can be recharged are called **secondary cells** or accumulators. A number of cells connected together, usually in series, is called a battery.

Alessandro Volta (1745 — 1827)

Volta was one of nine children of an Italian nobleman. As a baby he was so slow to develop and his parents feared he was retarded. However, as he grew older he quickly improved and at the age of fourteen he decided to be a physicist.

Volta was interested in the new phenomenon of electricity. He gained international recognition for his discoveries involving static electricity and, in 1779, received a professorship at the University of Pavia in northern Italy. However, his greatest achievement was his invention, in 1794, of the electric battery. The principle had been discovered by his friend and compatriot, Luigi Galvani, but it was Volta who constructed the first batteries. These included his famous "Voltaic Pile", which consisted of a pile of copper and zinc plates separated by sheets of cardboard soaked in salt solution.

Volta's invention brought him fame and honour throughout Europe. Today, the SI unit of electromotive force and potential difference, the volt, is named in his honour.

Primary Cells

The first primary cell was invented by Alessandro Volta in 1794 at Pavia University in Italy. It consisted of a copper plate and a zinc plate separated by a pad soaked in brine. In fact, Volta constructed a battery of such cells and found that when he touched the two end plates he got an electric shock. A number of primary cells were developed during the 19th century , each an improvement on its predecessor. However, the basic construction is the same in each case, *viz.* two plates of different types of metal, or carbon, immersed in an electrolyte.

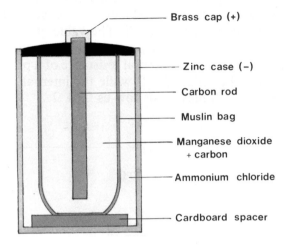

Fig. 23.7 Dry cell

The structure of one type of dry cell commonly found in "transistor" radios, torches, *etc.*, is shown in *Fig. 23.7*. The container, which is made of zinc, is also the negative electrode. The positive electrode is a carbon rod and the electrolyte is ammonium chloride, made into a paste with zinc chloride, water, flour and gum. In order to prevent the build-up of hydrogen gas around the positive electrode, the electrode is surrounded by manganese dioxide which oxidises the hydrogen as it is formed. If the hydrogen is allowed to build up a new "cell" is formed, the carbon electrode being effectively replaced by a hydrogen electrode. The e.m.f. of this "cell" tends to drive a current in the opposite direction to the original current. Thus the net e.m.f., and hence the available current, are greatly reduced. When this happens the cell is said to be **polarised.** (*Cf. conduction in electrolytes with inert electrodes, p. 242.*) The manganese dioxide is mixed with carbon to reduce its resistance.

The e.m.f. of this type of cell is 1.5 V. This means that, when the cell is connected in a circuit such as the one shown in *Fig. 23.8*, the work done in carrying unit charge around the whole circuit, *i.e.* through the resistor and the cell itself, is 1.5 J. (*Cf. definition of e.m.f., p. 200.*) A 9 V battery thus consists of six cells in series, a 6 V battery consists of four cells, *etc.*

Internal Resistance of a Cell

When a cell drives a current through a circuit, that current has to flow through the cell itself. Like any other part of the circuit the cell has a certain resistance. This resistance is called the internal resistance of the cell. The value of the internal resistance depends on the shape and size of the cell and on how it has been used. The internal resistance of a primary cell is typically of the order of 1 Ω while the internal resistance of a secondary cell may be as low as 0.01 Ω.

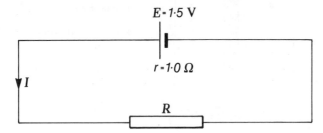

Fig. 23.8

The effect of the internal resistance is to limit the maximum current which may be obtained from the cell. Consider the circuit of *Fig. 23.8*. A cell of e.m.f. 1.5 V, and internal resistance 1 Ω, is connected in series with a resistor of resistance R. The total resistance of the circuit is $R + 1$ Ω. The current flowing in the circuit is given by

$$E = RI$$
$$1.5 = (R + 1)I$$
$$=> \quad I = \frac{1.5}{R + 1}$$

From this we can see that the maximum current obtainable from the cell is 1.5 A, the current which flows when $R = 0$.

If the value of R is, say 1 Ω, then the current flowing is 0.75 A. The p.d. across the resistor, *i.e.* between the terminals of the cell, is then

$$V = RI$$
$$= 0.75 \times 1$$
$$= 0.75 \text{ V}$$

The p.d. between the terminals is less than the e.m.f. of the cell.

In general, if E is the e.m.f. of the cell and r is the internal resistance, then

$$E = (R + r)I$$
$$=> \quad I = \frac{E}{R + r}$$

The p.d. across the resistor, *i.e.* between the terminals of the cell is

$$V = RI$$
$$= R \times \frac{E}{R + r}$$
$$= \left(\frac{R}{R + r}\right) E \quad \text{(i)}$$

If $R \gg r$ then $(R + r) \approx R$ and $V \approx E$. That is, if the resistance of the resistor is very much greater than the internal resistance of the cell the p.d. between the terminals of the cell is approximately equal to the e.m.f. of the cell. This implies that, if a voltmeter is used to measure the e.m.f. of a cell, *Fig. 23.9*, its resistance must be very much greater than the internal resistance of the cell.

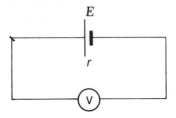

Fig. 23.9

From *Eqn. (i)*, E is exactly equal to V when R equals infinity. If R equals infinity then the current flowing through the cell is

$$I = \frac{E}{\infty}$$
$$= 0$$

That is,

> **The e.m.f. of a cell is equal to the p.d. between its terminals when no current is being drawn from it.**

In *Experiment 21.2 (p. 223)* we saw how a potentiometer could be used to determine the e.m.f. of a cell. The potentiometer is, in principle, more accurate than a voltmeter since it draws no current from the cell whose e.m.f. is being determined.

EXPERIMENT 23.3
To Measure the Internal Resistance of a Cell

Method 1:

Apparatus: Cell, resistors of known resistance, high resistance voltmeter.

Procedure:
1. Connect the voltmeter directly to the cell and record the voltage. This is equal to the e.m.f., E, of the cell.
2. Connect one of the resistors in parallel with the voltmeter, *Fig. 23.10*, and record the new voltage, V. Calculate the internal resistance, r, from

$$V = \frac{RE}{R + r}$$

where R is the resistance of the resistor.

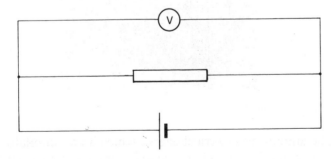

Fig. 23.10

3. Repeat a number of times using different resistors. Calculate an average value for the internal resistance of the cell.

RESULTS

E.m.f. of cell =   V.

V/V	R/Ω	r/Ω

Internal resistance of cell = Ω.

Questions
1. Explain the reasons for using a voltmeter which has a high resistance.
2. What happens when a resistor of relatively low resistance is connected in parallel with the voltmeter? What can you deduce from this regarding the nature of the internal resistance of a cell?

Method 2:
Apparatus: Cell, resistors of known resistance, potentiometer, centre zero galvanometer, protective resistor ($\sim$ 1 kΩ) for galvanometer, accumulator or low voltage power supply.

Procedure:
1. Set up the apparatus as shown in *Fig. 23.11*. Connect the cell to the potentiometer, making sure that the polarity is correct. Connect the protective resistor in series with the galvanometer.

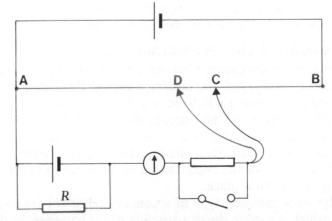

Fig. 23.11

2. Adjust the position of the moveable contact until the galvanometer reads zero. Close the switch to short out the protective resistor and find the balance point more accurately. Measure the distance |AC|, where C is the position of the contact when the galvanometer reads zero.

3. Connect one of the resistors in parallel with the cell and find the new balance point, D. Measure the distance |AD|. Calculate the internal resistance, r, from

$$\frac{R + r}{R} = \frac{|AC|}{|AD|}$$

4. Repeat a number of times using different resistors. Calculate an average value for the internal resistance of the cell.

RESULTS

|AC|/cm	R/Ω	|AD|/cm	r/Ω

Internal resistance of cell =   Ω.

Questions

1. What happens when a resistor of relatively low resistance is connected in parallel with the cell? What can you deduce from this regarding the nature of the internal resistance of a cell?

2. Why should the potentiometer circuit be arranged so that the balance point is always a reasonable distance from A? How might this be achieved in practice?

Secondary Cells (Accumulators)

When a secondary cell has given up some of its chemical energy, it may be "recharged" by passing a current through it in the opposite direction to that in which it produced current, *Fig. 23.12*. Obviously, the source of the recharging current must have a higher e.m.f. than that of the cell being recharged. This process converts electrical energy into chemical energy and, in effect, reverses the chemical reactions which released the energy in the first place.

The most common type of secondary cell is the lead-acid cell, which is used in the construction of car batteries. In this type of cell, the positive electrode is lead dioxide and the negative electrode is lead. The electrolyte is sulphuric acid. As the cell delivers current the acid becomes more dilute and hence less dense. The state of charge of the battery may therefore be determine by measuring the density of the acid using a hydrometer (*see p.83*).

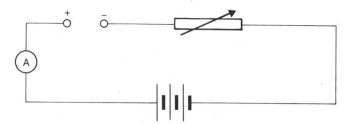

Fig. 23.12 Recharging an accumulator

A lead-acid cell produces an e.m.f. of 2 V. A car battery usually consists of six such cells in series. Its internal resistance is very low, being of the order of 0.05 Ω. It is therefore capable of supplying a large current, such as is required by the starter motor of a car.

Fig. 23.13 Car battery

The amount of electrical energy which an accumulator can store is usually given as being so many ampere-hours, typically around 50 A h for an average size car. This means that it can supply 5 A for 10 h; 10 A for 5 h; *etc.* These figures are only approximate as a battery usually lasts longer at lower currents, *e.g.* a battery rated as 50 A h might give 1 A for 52 h. (**Note:** The A h is a unit of charge. 1 A h = 3600 C.)

When a lead-acid accumulator is being recharged, oxygen

and hydrogen are produced. It is therefore important that there should be no sparks or naked lights in its vicinity. Never use a lighted match when checking the level of acid in a car battery.

In addition to the lead-acid cell, there is a number of other types of secondary cell available. For example, the rechargeable batteries in calculators, *etc.*, often consist of "nicad" cells. In this type of cell the electrodes are nickel oxide and cadmium and the electrolyte is potassium hydroxide. The e.m.f. of this type of cell is 1.2 V.

SUMMARY

When the flow of current through a liquid is accompanied by chemical reactions the process is known as electrolysis and the liquid is called an electrolyte. Conduction in electrolytes is by means of positive and negative ions which move towards the cathode and anode, respectively.

Faraday's first law of electrolysis states that the mass of an element liberated in electrolysis is proportional to the quantity of charge which liberated it. In symbols,

$$m = zQ$$

or,

$$m = zIt$$

where z is a constant called the electochemical equivalent of the element liberated.

A cell generates a p.d. between two points by means of chemical reactions. The principal components of a simple cell are two electrodes of different types of metal (or one metal and one carbon) and an electrolyte. The e.m.f. of a cell is equal to the p.d. between its terminals when no current is being drawn from it. A primary cell cannot be recharged when it has yielded its store of energy; a secondary cell, or accumulator, can.

The internal resistance of a cell is determined by measuring the p.d. between its terminals when a resistor of known resistance, R, is connected across its terminals. This may be done using a high impedance (resistance) voltmeter or a potentiometer. In either case the internal resistance, r is calculated from

$$\frac{E}{V} = \frac{R + r}{R}$$

where E is the e.m.f. of the cell and V is the p.d. between its terminals when the resistor is connected across them.

Questions 23

DATA: e.c.e. of copper $= 3.3 \times 10^{-7}$ kg C^{-1}.

Section I

1. The principal parts of a voltameter are the ....
   .... the................. and the.................
2. What are ions?...
   ....
3. The charge carriers in metals are...........................
   The charge carriers in electrolytes are.....................
   ....
4. Do electrolytes obey Ohm's law. Explain..................
   ....
   ....

5. State Faraday's first law of electrolysis...................
   ....
   ....
6. What is meant by the electrochemical equivalent of an element?..
   ....
   ....
7. Give two industrial applications of electrolysis..........
   ....
8. Alessandro Volta is chiefly remembered for his invention, towards the end of the.......... century, of the..........
   ....

9. What is the essential difference between a primary cell and a secondary cell?...

....

*10. A secondary cell is capable of giving a larger current than a primary cell of the same e.m.f. because of its........

....

11. How many cells are there in a 22.5 V zinc-carbon battery?

....

12. What is meant by saying that the capacity of a car battery is 45 A h?...

....

Section II

1. Calculate the mass of copper liberated in a copper voltameter when a current of 2.5 A flows for 20 minutes.

2. Calculate the e.c.e. of silver if a current of 2.0 A flowing for 30 minutes liberates 4.0 g of the metal.

3. What is the time taken for a current of 3.5 A to deposit 1.4 g of copper in a copper voltameter?

4. In a copper voltameter the cathode consists of a rectangular piece of copper 3.0 cm wide by 4.0 cm long and of mass 4.50 g. A current of 0.15 A is passed through the voltameter for 20 minutes. Calculate (i) the mass of the cathode at the end of the experiment, (ii) the thickness of the layer of copper deposited, assuming it to have been deposited uniformly on both sides of the cathode. (Density of copper $= 8.9 \times 10^3$ kg m^{-3}.)

5. In a copper sulphate solution the copper ion has a double positive charge, *i.e.* it has lost two electrons. How many copper ions are required to transport a charge of 12 C? If the process takes 2 minutes what is the current flowing in the circuit? (Charge on electron $= 1.6 \times 10^{-19}$ C.)

6. In a copper voltameter 0.54 g of copper is deposited in 15 minutes. Calculate (i) the current flowing, (ii) the total charge transferred, (iii) the number of copper atoms deposited. *(See previous question.)*

7. A current of 8.0 A, passing through a silver voltameter for 15 minutes, is found to deposit 8.1 g of silver on the cathode. What is the specific charge (*i.e.* the charge per unit mass) of the silver ion?

*8. Calculate the maximum current obtainable from (i) a 9 V torch battery of internal resistance 6 Ω, (ii) a 12 V car battery of internal resistance 0.3 Ω. What is the essential difference between the two types of battery?

*9. A cheap voltmeter has a resistance of 500 Ω. It is to be used to measure the e.m.f. of a battery of internal resistance 10 Ω. If the e.m.f. of the battery is 9.0 V calculate (i) the reading on the voltmeter, (ii) the error in the reading as a percentage of the true value.

*10. What is the maximum current obtainable from a 2 V lead-acid cell whose internal resistance is 0.05 Ω?

*11. It is required to drive a current through a resistor using two "dry" cells, each of e.m.f. 1.5 V and internal resistance 1.0 Ω. Draw circuit diagrams showing the two possible ways of arranging the circuit. If the resistance of the resistor is 4 Ω, what is the current flowing in each case? For what value of the resistance would the current be the same for both arrangements?

*12. Two similar cells, each of internal resistance r, are to be used to drive a current through a resistance R. Show that a parallel arrangement of the cells will give a larger current than a series arrangement if $R < r$.

24 Current and Magnetism

24.1 Current and Force

We saw in *Chapter 21* that currents (moving charges) exert forces on each other. Since the conductors in these cases can easily be shown to have no net charge, the forces cannot be electrostatic forces and must therefore represent a new type of force.

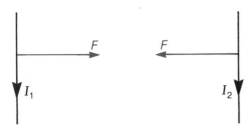

Fig. 24.1 Forces between currents

The forces which are due to moving charges are called **magnetic forces** and the region around a moving charge is therefore a **magnetic field**. Thus, we may consider that the force on I_1, *Fig. 24.1*, is due to the fact that it is in the magnetic field of I_2, and *vice versa*. It can be shown experimentally that the force on a current in a magnetic field is proportional to the current, to the length of the conductor in the field and to the sine of the angle between the field and the current, *i.e.*

$$F \propto Il \sin \theta$$
$$=> \quad F = kIl \sin \theta$$

where k is a constant.
The **magnetic flux density, B,** of the field is defined to be equal to k. That is,

$$F = BIl \sin \theta$$

or
$$B = \frac{F}{Il \sin \theta}$$

If the current is flowing at right angles to the field, then $\theta = 90°$, $\sin \theta = 1$, and the equation for the magnetic flux density becomes

$$B = \frac{F}{Il} \quad \text{(i)}$$

That is,

> The magnetic flux density, B, of a magnetic field is the force per unit current per unit length of conductor at right angles to the field.

The magnetic flux density is thus the force that would be experienced by a current of 1 A flowing at right angles to the field in a conductor 1 m long. B is therefore a measure of the strength of the field at a particular point. It is a **vector** quantity. (Compare this definition with the definition of the field intensity of an electric field, *p. 202*.) The unit of magnetic flux density is the **tesla**, T. From *Eqn. (i)* the tesla is defined as

> The magnetic flux density is 1 T if a current of 1 A flowing at right angles to the field experiences a force of 1 N per metre length of conductor.

Rearranging *Eqn. (i)* gives

$$F = IlB \quad \text{(ii)}$$

for the force on a current I flowing in a conductor of length l in a magnetic field of magnetic flux density B. This is an important equation which we shall require later.

24.2. Magnetic Fields

Certain objects have a magnetic field associated with them, *i.e.* they can exert forces on moving charges. Such objects can also attract pieces of iron and, when placed near to each other, they may attract each other or repel each other, depending on which parts of them are nearest together. Bodies which have these properties are called **magnets**. Magnets are made of alloys containing one or more of the elements iron, nickel and cobalt, and are available in different shapes, *Fig. 24.2*.

If two bar magnets are placed end to end they will repel each other. If one of them is now reversed they will attract each other. Opposite ends of a magnet have different effects and, to distinguish between them, one is called the **north pole** of the magnet and the other is called the **south pole** and the bar may thus be referred to as a **magnetic dipole**.

If a bar magnet is hung from a retort stand, one end of it will always point in the same direction, *viz.* towards the earth's North Pole. The end of the magnet which points towards the North Pole is therefore called the north-seeking pole or, simply,

Fig. 24.2 Magnets

the north pole, of the magnet. Experiment shows that two north poles or two south poles always repel each other, while a north and a south pole always attract each other, *i.e.*

Like poles repel each other; unlike poles attract.

Magnetic poles occur in pairs, some magnets having 4, 6, or more poles. The position of the poles depends on the shape and structure of the magnet. For example, the disc-shaped magnet shown in *Fig. 24.2* has both poles on one side at opposite ends of a diameter.

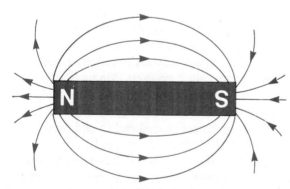

Fig. 24.3 Magnetic field lines due to a bar magnet

The region around a magnet, in which magnetic forces are experienced, is called a **magnetic field**. A magnet field may be represented by magnetic field lines just as an electric field is represented by electric field lines. (Remember that, in both cases, the lines are purely imaginary. They are simply an aid

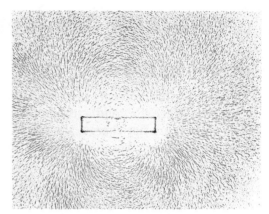

Fig. 24.4 Plotting field lines using iron filings

to understanding electric and magnetic effects.) It is agreed that the direction of the magnetic field at any point is the direction in which a north pole would move. Thus, magnetic field lines run from the north pole of a magnet to the south pole, *Fig. 24.3*. The shape of the magnetic field lines for a given magnet may be demonstrated by placing a page over the magnet and sprinkling it with iron filings, *Fig. 24.4*. Alternatively, the lines may be demonstrated using plotting compasses, *Fig. 24.5*. As in the case of electric field lines the field is strongest where the lines are closest together. In the case of magnets this means that the field is strongest near the poles, *Fig. 24.3*. However, whereas electric field lines start on a positive charge and end on a negative charge, magnetic field lines are closed loops.

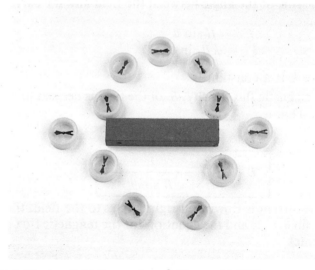

Fig. 24.5 Plotting field lines using plotting compasses

Although they seem to seem to start at a north pole and end at a south pole they in fact continue within the magnet.

We noted at the beginning of this chapter that magnetic effects were due to moving charges, *i.e.* electric currents. In the case of magnets the magnetic effects arise from movements of the electrons within the atoms of the magnet. The effects are due to the spin of the electrons and, to a lesser extent, to the rotation of the electrons around the nuclei of the atoms. In other words, the movements of the electrons constitute current loops which act as tiny magnetic dipoles. (The idea that magnetism in a magnet was due to tiny currents circulating within the magnet was first proposed by Ampère in 1820.) In a magnetic substance the atoms form groups in which all the "atomic magnets" are pointing in the same direction. These groups of atoms are called **magnetic domains.** Each domain then, is a tiny magnet. In an unmagnetised piece of iron, for example, these domains point in all directions so that, overall, they cancel each other out and the piece of iron shows no magnetic effects. When the piece of iron is magnetised all of the domains are lined up in the same direction. The magnetic effects of the domains now add up to form one large magnet and so the piece of iron is now a magnet.

Magnetic Field Due to a Current

The magnetic effect of a current was discovered by the Danish physicist, Hans Christian Oersted (1777 — 1851) in 1819. He noticed that a compass needle placed near a current-carrying conductor turned so that it was at right angles to the conductor.

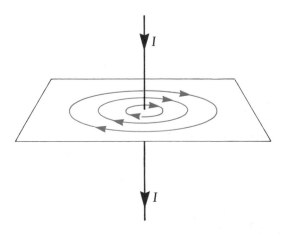

Fig. 24.6 Magnetic field lines due to a current in a straight wire

The shape of the magnetic field lines due to a current depends on the shape of the conductor in which the current is flowing. When the current is flowing in a straight conductor, *Fig. 24.6*, the field lines are circles concentric with the conductor. The direction of the field lines may be remembered by the right hand rule. If the right hand is held as in *Fig. 24.7*, the fingers give the direction of the field lines when the current is flowing in the direction of the thumb.

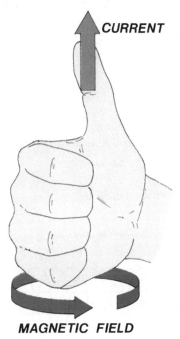

Fig. 24.7 Right Hand Rule

The magnetic field lines due to a current in a coil are shown in *Fig. 24.8*. The right hand rule may again be used to remember the direction of the lines. The magnetic field due to a current in a solenoid is shown in *Fig. 24.9*. Note the similarity between the field due to a solenoid and that due to a bar magnet, *Fig. 24.3*. Both the coil and the solenoid behave as magnetic dipoles, *i.e.* they have two distinct poles.

The magnetic field lines around a given conductor may be demonstrated, as in the case of magnets, using iron filings or plotting compasses. The arrangement used for a solenoid is shown in *Fig. 24.10*, a suitable source of current being a lead-acid or nicad cell. A similar arrangement may be used for a straight conductor and for a coil.

A solenoid with an iron core is called an **electromagnet.** When there is a current flowing in the solenoid the

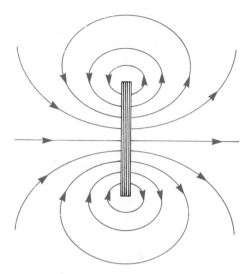

Fig. 24.8 Magnetic field lines due to a current in a coil

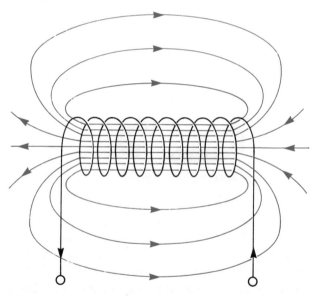

Fig. 24.9 Magnetic field lines due to a current in a solenoid

electromagnet acts as a magnet; when the current stops the magnetic field, for practical purposes, disappears. (The iron core remains weakly magnetic for some time after the current is switched off.) Large electromagnets, attached to cranes, are used in scrapyards for moving heavy loads. Smaller electromagnets are used in electric bells and in electromagnetic relays *(see next section)*.

A solenoid may be used to make a magnet from, for example, a piece of steel. The steel is placed inside the solenoid

254

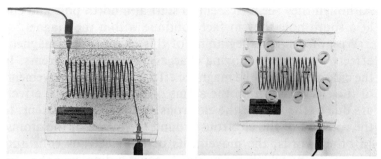

Fig. 24.10 Demonstrating the magnetic field of a solenoid

and a direct current passed through the solenoid. The steel becomes magnetised and, unlike iron, retains its magnetism. A piece of steel may also be magnetised by rubbing it repeatedly in one direction with another magnet.

The Electromagnetic Relay

This is a device by means of which a current in one circuit may be used to switch on or switch off a current in another circuit, *Fig. 24.11(a)*. It consists of an electromagnet and an armature, A, which is free to rotate on a fixed pivot, P. When the switch is closed a current flows through the solenoid and the lower end of the armature is attracted to the electromagnet, thus closing the contacts, C. This completes the secondary circuit, S, allowing a current to flow through the load. The symbol for a relay is shown in *Fig. 24.11(b)*.

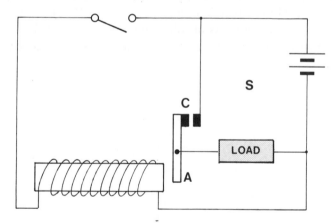

Fig. 24.11   (a) Electromagnetic relay circuit

The electromagnetic relay has various applications. In a car the starter motor is operated through a relay, the current in the solenoid being controlled by the ignition switch. Relays are also often included in other circuits in a car, *e.g.* the rear window heater and the heater fan. In telephone exchanges,

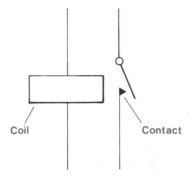

Fig. 24.11 (b) *Symbol for electromagnetic relay*

although they are being replaced by electronic switches, relays are used for "connecting" a subscriber to the number he/she has dialled.

The Earth's Magnetic Field

We saw earlier (*p. 251*) that a freely suspended bar magnet always points in a particular direction. Since this happens at all points on, and near, the earth's surface it means that the earth is surrounded by a magnetic field.

The shape and direction of the earth's magnetic field lines is the same as *if* there were a bar magnet through the centre of the earth, *Fig. 24.12*. (There is not, of course, a bar magnet through the earth. In fact, the core of the earth is molten and, while the source of the earth's magnetic field is not completely understood, it is thought to be due to electric currents flowing in the core.) The earth's magnetic poles do not coincide with the geographic poles; in fact, the magnetic poles move (slowly). At present, the earth's south pole is located in northern Canada at a latitude of approximately 79° N and a longitude of approximately 69° W. (Although this is a south pole since the north pole of a magnet is attracted towards it, it is often referred to as the Magnetic North Pole.) The angle between the geographic N-S and the magnetic N-S at a particular point is called the **magnetic declination** or **magnetic variation**. In Ireland its value is about 10° at present (1987). By about the year 2080 it should have decreased to zero.

A bar magnet, freely suspended at its centre of gravity, does not remain horizontal but lies at an angle to the surface of the earth, the north pole being lower than the south pole. (In most of the southern hemisphere the situation is reversed, the south pole being lower than the north pole.) Reference to *Fig. 24.12* shows why this should be so. The earth's magnetic field lines are not parallel to the surface. Rather, they cut the surface at various angles. The angle between the field line and the surface is called the **angle of dip** and its value varies from 90° at the magnetic poles to 0° at the magnetic equator. In Ireland, its value is about 70°. The angle of dip may be measured with a dip circle, *Fig. 24.13*. This consists of a compass needle mounted on a horizontal axis and free to move around a vertical scale.

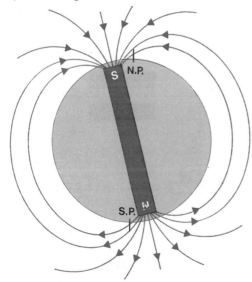

Fig. 24.12 *Earth's magnetic field*

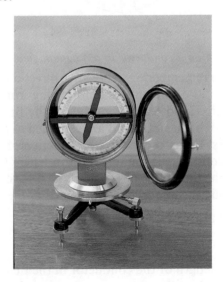

Fig. 24.13 Dip circle

24.3 Force on a Current in a Magnetic Field

The force on a current in a magnetic field may be demonstrated as follows. A long strip of aluminium foil is placed between the poles of a magnet, *Fig. 24.14*. It is found that, when the ends of the foil are connected to a battery, the foil jumps away from the magnet, *i.e.* at right angles to field. From this we conclude that, when a current flows through a magnetic field, it experiences a force. Since forces always occur in pairs, it should be clear that the magnet also experiences a force which is equal and opposite to the force experienced by the current (Newton's third law).

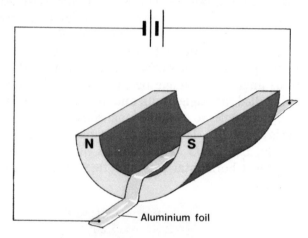

Fig. 24.14 Force on a conductor in a magnetic field

As an aid to remembering the relative directions of the force, the field and the current, the following rule, known as Fleming's Left-Hand Rule, is useful. If the first two fingers and the thumb of the left hand are extended as shown in *Fig. 24.15*, the first finger represents the direction of the magnetic field, the second finger the direction in which the current is flowing and the thumb the direction of the force on the current. The magnitude of the force, as explained at the beginning of this chapter, is given by

$$F \quad = \quad IlB$$

where l is the length of the conductor in the field and B is the magnetic flux density.

The factors affecting the force on a current may be investigated using the circuit shown in *Fig. 24.16*. The conductor, C, is fixed and the magnet is placed on the pan

of a top pan balance. When a current flows through the conductor the magnet experiences a downward force, giving a reading on the balance. The reading on the balance is therefore a measure of the force between the current and the magnet. The reading may then be taken for different values of the current, for different lengths of conductor in the magnetic field and for different magnets.

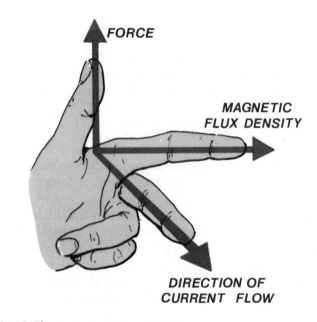

Fig. 24.15 Fleming's Left Hand Rule

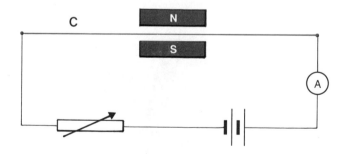

Fig. 24.16

Force on a Moving Charge

An electric current consists of a flow of charge, so the equation $F = IlB$ may also be expressed in terms of the charge and the velocity of the charge. For simplicity, we shall confine ourselves to the case of charges moving at right angles to the magnetic field.

Consider a section of conductor of length l containing N free electrons, each carrying a charge e in a magnetic field of uniform flux density, B, *Fig. 24.17*. Let the average velocity of the electrons in the direction shown be v. (This velocity is known as the drift velocity of the electrons.) B passes perpendicularly into the page, while v is in the plane of the page. Let the number of free electrons per unit length be n, *i.e.*

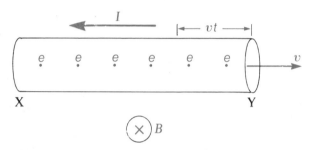

Fig. 24.17

Let Q be the total charge which passes the end Y in time t. Then,

$$Q = It$$

where I is the current flowing in the conductor. Then,

$$F = IlB$$

$$\Rightarrow \quad F = \frac{QlB}{t} \qquad \text{(iii)}$$

Since v is the speed of the electrons, an electron will travel a distance vt in time t. Therefore, all the electrons within a distance of vt of the end Y will pass through Y in a time t. The total number of electrons in a length vt is nvt. Since the charge on each electron is e the total charge passing through Y in time t, *i.e.* Q, is

$$Q = envt$$

Substituting for Q in *Eqn. (iii)* gives

$$F = \frac{envtlB}{t}$$

$$= envlB$$

But, $$N = nl$$

where N is the total number of electrons in the length l. Therefore,

$$F = NevB$$

Therefore, the force on each electron in the conductor is

$$F = evB$$

We have derived this relationship in terms of the charge on an electron. However, similar reasoning could be applied to any moving quantity of charge, q. Thus, in general, the force on a moving charge is

$$\boxed{F = qvB}$$

Since the direction of the force is always at right angles to the direction of the current, it follows that F is always perpendicular to v. Applying Fleming's Left-Hand Rule to the situation shown in *Fig. 24.17* shows that the force on the electrons is down the page, at right angles to both B and v. It should be noted that the force is on the moving charges not on the conductor (the charges in turn exert an equal force on the conductor). A beam of electrons, for example, passing through a vacuum would still experience a force given by $F = evB$.

Example
Calculate the force on a charge of 1.6×10^{-19} C travelling at a speed of 2×10^7 m s^{-1} at right angles to a magnetic field of uniform flux density 2.5 T. What shape of path will the charge follow?

$$\begin{aligned} F &= qvB \\ &= 1.6 \times 10^{-19} \times 2 \times 10^7 \times 2.5 \\ &= 8.0 \times 10^{-12} \text{ N} \end{aligned}$$

Ans. 8×10^{-12} N. Since F is always perpendicular to v, the charge will follow a circular path.

The Moving Coil Loudspeaker
The moving coil loudspeaker, *Fig. 24.18*, has a small coil of wire positioned between the poles of a powerful permanent magnet and attached to a moveable diaphragm. When a current flows through the coil it experiences a force which causes it, and consequently the diaphragm, to move. If the current is oscillating, *i.e.* flowing first in one direction and then in the

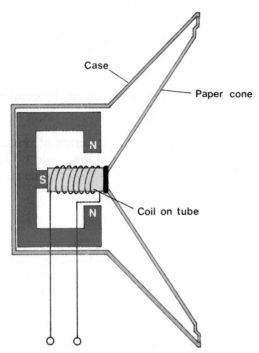

Fig. 24.18 Moving coil loudspeaker

Fig. 24.19 Magnet and coil of loudspeaker

opposite direction, at an audio frequency the coil and diaphragm will vibrate with the same frequency. As a result a sound wave is produced in the air in front of the loudspeaker (*cf. p. 143*).

24.4 The D.C. Motor

The electric motor is perhaps the most common electrical device we have — washing machines, tape recorders, refrigerators, clocks, record players, *etc.*, all depend to a greater or lesser extent on electric motors. While there are many types of electric motor, the basic principle is the same in all — a current-carrying conductor in a magnetic field experiences a force. We shall consider here only the simple d.c. motor.

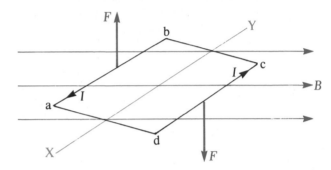

Fig. 24.20 Couple on a coil

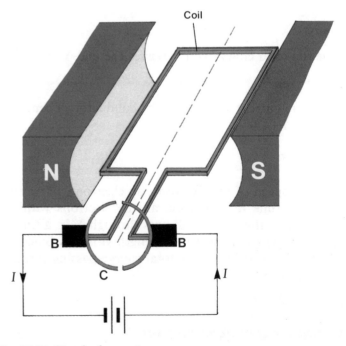

Fig. 24.21 Simple d.c. motor

Consider firstly a rectangular coil of wire with a current, I, flowing in it. The coil is placed in a uniform magnetic field of flux density, *B*, and is free to rotate about an axis, XY, *Fig. 24.20*. Applying Fleming's Left-Hand Rule we see that the side ab will experience an "upward" force while cd will experience a "downward" force of equal magnitude. Such a pair of forces constitutes a couple (*p.77*), the effect of which is to turn the coil about the axis, XY.

In a simple d.c. motor, *Fig. 24.21*, a rectangular coil of wire is placed between the cylindrical pole pieces of a permanent magnet (in larger motors an electromagnet is used). The coil is wound on an iron core called the armature. The effect of the iron core with the cylindrical pole pieces is to produce a radial magnetic field, *Fig. 24.22*. This ensures that the torque of the couple on the coil is constant.

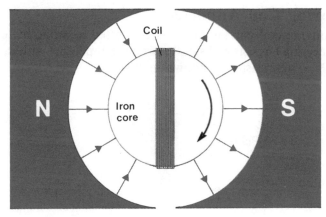

Fig. 24.22 Radial magnetic field

When the coil rotates past the gap between the poles the direction of the current must be reversed if the torque is to remain in the same direction. This is achieved by means of the split ring commutator, C, *Fig. 24.21*. Current enters and leaves the coil *via* the two carbon brushes, B, which are held in place against the commutator by means of springs.

The torque produced by a motor depends on the current flowing, *I*, the number of turns, *N*, of wire in the coil, the flux density of the magnetic field, *B*, and the area, *A*, of the coil. It can be shown *(see Appendix B)* that the torque is given by

$$T = BIAN$$

| *(a)* | *(b)* |

Fig. 24.23(a) Coils and magnet (electro) of a motor. (b) Brush and commutator. (Note number of segments on commutator — two for each coil on the armature.)

24.5 Meters

In this section we shall discuss the structure of meters which measure current (ammeters), meters which measure p.d. (voltmeters) and meters which measure resistance (ohmmeters). Basically, all three are very similar.

Moving Coil Galvanometer

As we saw in the last section a current-carrying coil in a magnetic field experiences a torque. In a moving coil galvanometer a coil of wire hangs from a special suspension wire between the cylindrical pole pieces of a permanent magnet, *Fig. 24.24*. When a current flows through the coil it turns, twisting the suspension wire. The wire tends to untwist, *i.e.* it produces a torque in the opposite direction. The magnitude of this restoring torque, T_r, is proportional to the angle, θ, through which the wire has been twisted, *i.e.*

$$T_r \propto \theta$$

The torque, T_I, due to the current is, from $T = BIAN$, proportional to the current, *i.e.*

$$T_I \propto I$$

When a current flows through the coil, the coil will turn until the torque due to the suspension is equal in magnitude to the torque due to the current. Therefore, when the coil comes to rest

$$T_r = T_I$$

$$\Rightarrow \quad \theta \propto I$$

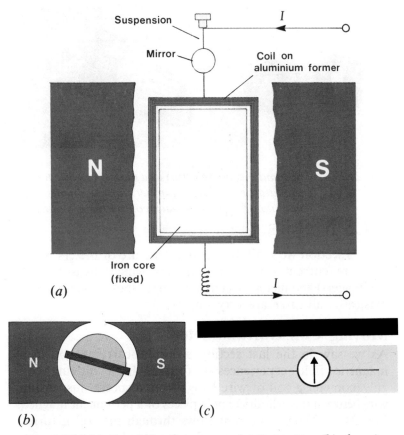

(a)

(b)

(c)

Fig. 24.24 *Moving coil galvanometer (a) front view, (b) plan view,*
(c) symbol

That is, the angle through which the coil turns is proportional
to the current flowing. Hence, by measuring the angle we can
determine the current. The deflection of the coil is detected
by reflecting a narrow beam of light from the mirror onto a
graduated scale, *Fig. 24.25*.

The galvanometer described above is very sensitive, *i.e.* it
is capable of giving a large deflection for a small current and
so can be used to measure very small currents. It is also very
delicate. In a more robust, though less sensitive, version the
coil is mounted in jewelled bearings and the restoring torque
is provided by a pair of spiral springs, one above and one below
the coil. The movement of the coil is detected by means of
a pointer which is attached to the coil and moves over a
graduated scale. Meters of this type typically have a full scale
deflection (f.s.d.) of 1 mA, *i.e.* the maximum current which
they can measure is 1 mA.

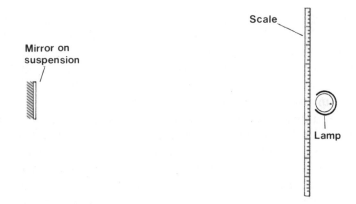

Fig. 24.25

Moving Coil Ammeter

A moving coil galvanometer is converted to an ammeter by
connecting a resistor of small resistance in parallel with it, *Fig.
24.26*. This resistor is known as a **shunt**.

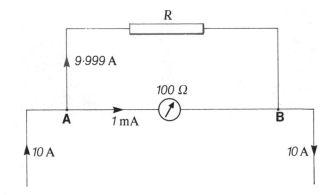

Fig. 24.26 *Using a galvanometer as an ammeter*

Suppose we have a moving coil galvanometer of full scale
deflection 1 mA and resistance 100 Ω and we wish to convert
it to an ammeter of f.s.d. 10 A. Let the shunt required have
a resistance R. Since only 1 mA may flow through the
galvanometer, the current through the shunt must be 9.999 A.
Let the p.d. between A and B be V. Then,

$$V = RI$$
$$\Rightarrow \quad V = R \times 9.999$$
$$\text{and} \quad V = 100 \times 0.001$$

Combining these two equations gives

$$R \times 9.999 = 100 \times 0.001$$

$$=> \quad R \quad = \quad \frac{0.1}{9.999}$$

$$= \quad 0.010001 \ \Omega$$

Thus, when a resistor of resistance 0.010001 Ω is connected in parallel with the galvanometer, a reading of 1 mA on the scale indicates a total current of 10 A flowing in the circuit. Similarly, a reading of 0.1 mA indicates a current of 1 A, and so on.

From this example it should be clear that it is possible to convert a galvanometer to an ammeter of any required sensitivity by the use of a suitable choice of shunt.

Moving Coil Voltmeter

A galvanometer may also be used as a voltmeter by connecting a resistor of large resistance in series with it, *Fig. 24.27*. The resistor in this case is called a **multiplier**.

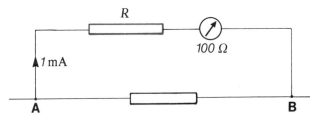

Fig. 24.27 Using a galvanometer as a voltmeter

Suppose we wish to convert a galvanometer of f.s.d. 1 mA and resistance 100 Ω into a voltmeter of f.s.d. 10 V. Let the resistance of the multiplier required be R. The total resistance of the galvanometer and the multiplier is then $R + 100 \ \Omega$. Then, since the current through the galvanometer must be 1 mA when the p.d. between A and B is 10 V, we have

$$V \quad = \quad RI$$

$$10 \quad = \quad (R + 100) \times 0.001$$

$$= \quad 0.001R + 0.1$$

$$=> \ 0.001R \quad = \quad 9.9$$

$$=> \quad R \quad = \quad \frac{9.9}{0.001}$$

$$= \quad 9900 \ \Omega$$

$$= \quad 9.9 \ \text{k}\Omega$$

Thus, when a resistor of resistance 9.9 kΩ is connected in series with the galvanometer, a reading of 1 mA on the scale indicates that there is a p.d. of 10 V between A and B, a reading of 0.5 mA indicates a p.d. of 5 V, and so on.

From this example we see that it is possible to convert a galvanometer to a voltmeter of any required sensitivity by the use of a suitable multiplier.

The Ohmmeter

The ohmmeter is a meter used for measuring resistance directly. It consists of a galvanometer connected in series with a variable resistance and a cell or battery, *Fig. 24.28*. The resistance to be measured is connected between X and Y. The scale on the galvanometer is adapted to read ohms directly. Note that, since a large resistance between X and Y means a small current flowing through the galvanometer and *vice versa*, the ohmmeter scale reads from right to left. In other words, full scale deflection corresponds to zero resistance, while zero deflection indicates an infinite (very large) resistance. Also, the ohmmeter scale is not linear, *i.e.* the deflection of the needle is not proportional to the resistance and so the divisions on the scale are not of equal width. The reason for this is the presence in the circuit of the variable resistor. The resistance of this resistor is chosen to ensure that, when there is no resistance between X and Y, the current flowing through the galvanometer is equal to its f.s.d.

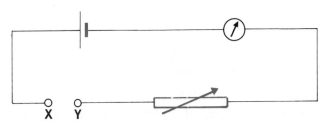

Fig. 24.28 Using a galvanometer as an ohmmeter

In use, the terminals X and Y are first short-circuited, *i.e.* they are connected together by a wire of negligible resistance, and the variable resistor is adjusted until the meter reads zero. The terminals are then connected directly to the device whose resistance is to be determined.

The Multimeter

A multimeter, *Fig. 24.29*, is a device which may be used as an ammeter, voltmeter or ohmmeter. One type in common use consists of the basic moving coil galvanometer plus all the necessary shunts and multipliers and a battery, or batteries, in a single unit. The various resistors are connected as required by moving a single multi-position switch. The meter has a separate scale for each position of the switch.

A more modern type of multimeter is based on the digital voltmeter. This is an electronic instrument which gives a digital readout. When used as a voltmeter it has a much higher resistance than a voltmeter based on a moving coil galvanometer and is therefore more accurate. As an ohmmeter it can measure small resistances accurately. The fact that the measurement is given as a digital readout means that there is no possibility of reading the wrong scale, as can happen with a conventional multimeter.

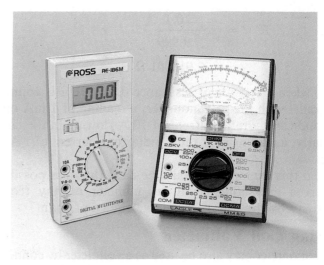

Fig. 24.29 Multimeters

SUMMARY

Magnetic forces are forces between charges in motion, *i.e.* between electric currents. Magnetic flux density is the force, per metre length of conductor, on a current of 1 A flowing at right angles to the magnetic field, *i.e.*

$$B = \frac{F}{Il \sin \theta}$$

The force on a current-carrying conductor, at right angles to a magnetic field is therefore given by

$$F = IlB$$

The force on a charge q moving with velocity v at right angles to a magnetic field of flux density B, is given by

$$F = qvB$$

The magnetic field of a permanent magnet is caused by the movements of electrons in the atoms of the magnet. When magnets are placed near each other like poles repel each other and unlike poles attract each other.

The shape of the magnetic field lines due to a current depends on the shape of the conductor in which the current is flowing. The field of a solenoid is similar to that of a bar magnet. A solenoid with an iron core is called an electromagnet. A relay is a device, based on an electromagnet, for using a current in one circuit to switch on, or off, a current in another circuit.

The earth is surrounded by a magnetic field. The magnetic declination (or magnetic variation) is the angle between the magnetic N-S and the geographic N-S. The angle of dip is the angle between the earth's magnetic field and the surface of the earth.

A current-carrying coil in a magnetic field experiences a torque. This is the principle of the simple d.c. motor and of the moving coil galvanometer.

A galvanometer may be converted to an ammeter by means of a shunt (a resistor of small resistance connected in parallel with the galvanometer). A galvanometer may be converted to a voltmeter by means of a multiplier (a resistor of large resistance connected in series with the galvanometer). A galvanometer may also, with a battery and variable resistor in series, be used as an ohmmeter.

Questions 24
Section I

1. Magnetic forces are those caused by ....

2. Define the term magnetic flux density.....................

....

3. The unit of magnetic flux density is the ....

....

4. .... poles repel each other; ....poles attract.

5. Show in a sketch the shape and direction of the field lines around a current flowing in a straight wire.

6. What is the principle of the electromagnetic relay?.....
....

7. The angle of dip is the angle.................................
..............................; the magnetic declination is
....

8. In the equation $F = IlB$, what does each symbol represent?
....

*9. Write down an expression for the force on a moving charge in a magnetic field.................................
....

10. What is the principle of the moving coil loudspeaker?.
....

11. Name the principal parts of a simple d.c. motor........
....

12. How is a radial field produced in a moving coil galvanometer?...
....

13. The resistor used to convert a galvanometer to an ammeter is called a .... and it is connected in .... with the galvanometer.

14. The resistor used to convert a galvanometer to a voltmeter is called a....................................and it is connected in....................................... with the galvanometer.

15. If a galvanometer is to be used as an ohmmeter it must have a .... and a ....
connected in series with it.

Section II

1. A straight horizontal wire of length 50 cm and carrying a current of 5.0 A is placed at right angles to a horizontal magnetic field of flux density 0.80 T. Calculate the force on the wire.

*2. A straight wire of length 20 cm and carrying a current of 7.5 A makes an angle of 30° with a magnetic field. If the force on the wire is 0.25 N, what is the flux density of the field?

*3. A straight wire of length 25 cm and carrying a current of 6.5 A experiences a force of 0.45 N in a field of flux density 0.50 T. What is the angle between the field and the current?

*4. Calculate the magnitude of the force on a charge of 3.2×10^{-19} C moving at right angles to a magnetic field of flux density 0.24 T with a speed of 8.2 Mm s^{-1}.

*5. A proton in an accelerator enters a magnetic field of flux density 0.50 T with a velocity of 4.6×10^7 m s^{-1} at right angles to the field. Calculate the radius of the path which the proton follows in the field. (Charge on proton $= 1.6 \times 10^{-19}$ C; mass of proton $= 1.7 \times 10^{-27}$ kg.)

*6. A helium ion enters a magnetic field of flux density 0.22 T with a velocity of 1.6×10^7 m s^{-1} at right angles to the field. As it leaves the field its velocity makes an angle of 45° with the velocity it had on entering the field. Calculate the distance travelled by the ion in the field. (Charge on helium ion $= 3.2 \times 10^{-19}$ C; mass of helium ion $= 6.7 \times 10^{-27}$ kg.)

7. A moving coil galvanometer has an internal resistance of 100 Ω and an f.s.d. of 1 mA. Calculate the resistance of the shunt required to convert it into an ammeter of f.s.d. (i) 100 mA, (ii) 5 A.

8. Calculate the resistance of the multiplier required to convert a moving coil galvanometer of f.s.d. 2 mA and internal resistance 75 Ω into a voltmeter of f.s.d. (i) 5 V, (ii) 100 V.

9. The total resistance of a circuit is 25 Ω and there is a current of 0.40 A flowing in it. If an ammeter of resistance 2.0 Ω is inserted in the circuit to measure this current what reading will it give? Express the error in the reading as a percentage of the true value.

10. A galvanometer of f.s.d. 10 mA is to be used as an ohmmeter. If the e.m.f. of the battery is 1.5 V what should be the resistance of the variable resistor?

25 Electromagnetic Induction

25.1 The Laws of Electromagnetic Induction

Fig. 25.1 shows a coil of wire connected to a sensitive galvanometer. When the magnet is moved into, or near, the coil the galvanometer shows a deflection, indicating that a current is flowing in the coil. When the magnet stops, the current stops. If the coil, rather than the magnet, is moved a current again flows in the coil. Such currents are called **induced** currents and the e.m.f.'s which drive them are called **induced** e.m.f.'s.

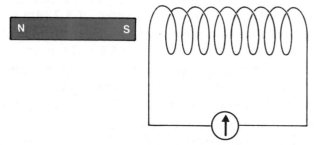

Fig. 25.1

If an electromagnet, rather than a permanent one, is used other effects, in addition to the ones already referred to above may be investigated, *Fig. 25.2*. When the current to the electromagnet is switched on the galvanometer shows a momentary deflection; likewise when the current is switched off. Note that there is no deflection while a steady current is flowing. The deflection occurs only at the instant when the current is switched on or switched off. If the current to the electromagnet is varied, using the rheostat, a deflection is noticed while the change is taking place.

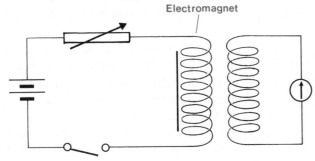

Fig. 25.2

All of the above experiments indicate that, when a conductor is situated in a magnetic field which is changing in any way, a current flows through it (if it is part of a complete circuit). A simple extension of these experiments shows that the size of the current induced in this way is directly related to the rate at which the magnetic field is changing.

These phenomena were discovered in 1831 by Michael Faraday. To explain them he introduced the idea of **magnetic flux**, Φ. Imagine a magnetic field of uniform flux density, B, perpendicular to the page, *Fig. 25.3*. Then, the magnetic flux in any area, A, is defined as the product of B and A, *i.e.*

$$\boxed{\Phi = B \times A}$$

Magnetic flux is a **scalar** quantity and its unit is the **weber**, Wb. (In general, if the flux density makes an angle θ with the area the magnetic flux is given by $\Phi = BA \sin \theta$.)

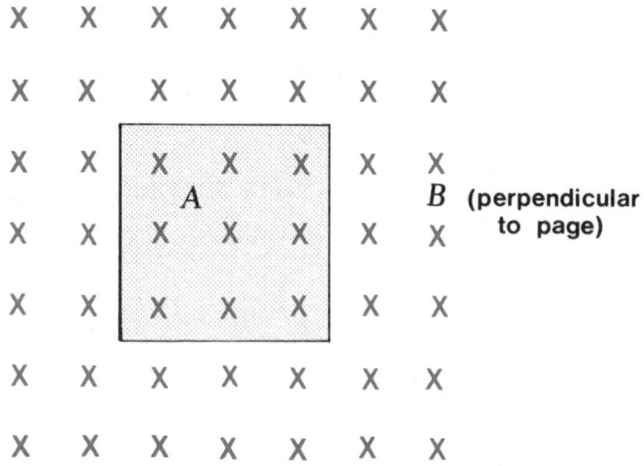

$$
\begin{array}{ccccccc}
\times & \times & \times & \times & \times & \times & \times \\
\times & \times & \times & \times & \times & \times & \times \\
\times & \times & \times & \times & \times & \times & \times \\
\times & \times & \times & \times & \times & \times & \times \\
\times & \times & \times & \times & \times & \times & \times \\
\times & \times & \times & \times & \times & \times & \times \\
\times & \times & \times & \times & \times & \times & \times \\
\end{array}
$$

A B (perpendicular to page)

Fig. 25.3

When a coil is placed in a magnetic field the magnetic flux passing through the coil is said to **thread** or **link** the coil. The flux threading a coil may be thought of as the total number of field lines passing through the area bounded by the coil. Using these ideas, Faraday summarised his discoveries as follows.

When there is a change in the magnetic flux threading any closed loop an e.m.f. is induced in the loop.

The e.m.f. induced in any closed loop is proportional to the rate at which the flux threading it is changing.

These statements are known as **Faraday's Laws of Electro-magnetic Induction** and they apply to all the cases of changing magnetic fields referred to above.

In all of these cases we are dealing with loops of wire, *i.e.* of conducting material and, in practice, this is generally the case. However, it should be noted that Faraday's laws apply whether or not the loop in question is made of a conducting material. If the loop is a conductor then a current will flow. The size of the induced current depends on the induced e.m.f. and on the resistance of the loop.

The statement of the law which gives the direction in which an induced current will flow is attributed to a Russian-born physicist, Heinrich Lenz (1804 — 1865). This law, now known as **Lenz's law,** states

> **The direction in which an induced current flows is such that its effects oppose the change which caused it.**

This law is illustrated in the diagrams of *Fig. 25.4*. In *(a)* the induced e.m.f. is caused by the magnet moving towards the coil. The induced current in the coil will flow in a direction to produce a magnetic field such that the magnet will be repelled. In *(b)* the induced e.m.f. is caused by the magnet moving away from the coil. In this case, the induced current in the coil will produce a magnetic field in which the magnet will be attracted towards the coil.

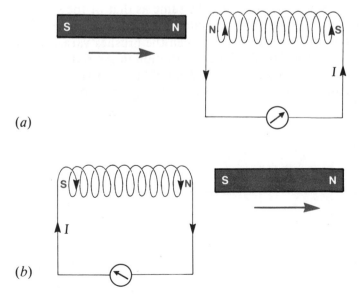

(a)

(b)

Fig. 25.4

Lenz's law may be demonstrated experimentally in a number of ways. A simple way to do it is to suspend an aluminium ring from a long thread and move a strong magnet towards it, *Fig. 25.5*. As the magnet approaches the ring it swings away, showing that the induced current produces a magnetic field to oppose the approaching magnet. Conversely, if we start with the magnet in the ring and try to move the magnet away the ring will tend to follow, again showing that the effect of the induced current is to oppose the change which created it.

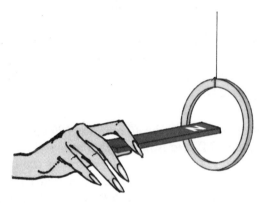

Fig. 25.5

Another method is to set up an aluminium pendulum between the poles of an electromagnet, *Fig. 25.6*. When no current flows through the coils of the electromagnet the pendulum swings freely. When the current is switched on the pendulum quickly comes to rest. Again, the effects of the induced currents in the pendulum oppose the changes which create them.

Fig. 25.6 Demonstrating Lenz's law

Lenz's law is, in fact, another way of stating the Principle of Conservation of Energy. To see this more clearly, consider what would happen if the law were not true. A magnet moving towards a coil would be attracted as a result of the induced current. This would cause the magnet to move faster, so producing a larger current, *etc.* In other words, energy would be created.

Faraday's laws of Electromagnetic Induction are summarised in the equation

$$E = \frac{d\phi}{dt}$$

where E is the induced e.m.f. Note that the constant of proportionality is equal to one. This results from the definition of B *(Chapter 24)*.

Example

A rectangular coil, 20 cm by 40 cm, has 100 turns of wire of total resistance 10 Ω and is placed in a magnetic field of flux density 0.6 T at right angles to the plane of the coil. If the coil is withdrawn from the field in 0.2 s calculate (i) the average induced e.m.f., (ii) the average current, in the coil.

The flux threading each loop of the coil is

$$\phi = BA$$
$$= 0.6 \times 0.2 \times 0.4$$
$$= 4.8 \times 10^{-2} \text{ Wb}$$

The e.m.f. induced in each loop is, therefore,

$$E_1 = \frac{d\phi}{dt}$$
$$= \frac{4.8 \times 10^{-2}}{0.2}$$
$$= 0.24 \text{ V}$$

The total e.m.f. induced in the coil is

$$E = E_1 \times \text{(no. of loops)}$$
$$= 0.24 \times 100$$
$$= 24 \text{ V}$$

The current in the coil is given by

$$E = RI$$
$$=> \qquad 24 = 10I$$
$$=> \qquad I = 2.4 \text{ A}$$

Ans. 24 V; 2.4 A

Note: For a coil containing N turns the product of N and the flux linking each turn is sometimes referred to as the **flux linkage**, *i.e.*

$$\text{Flux linkage} = N\phi$$

We can therefore say that the total e.m.f. induced in a coil is equal to the rate of change of the flux linkage.

25.2 Generators

An electrical generator is a device which converts kinetic energy to electrical energy. A generator is therefore the reverse of an electric motor. Indeed, a motor will usually work as a generator. If the terminals of a small motor are connected to a meter and the motor is turned by hand, a current will flow through the meter. The faster the motor is turned the larger the current flowing, *i.e.* the greater the kinetic energy supplied, the greater the electrical energy produced. The structure of a simple generator is essentially the same as that of the simple d.c. motor. If a split-ring commutator is used, as in the motor (*Fig. 24.21, p.258*), the p.d. between the brushes varies with time as shown in *Fig. 25.7*. (The variation in the p.d. can be reduced by using two coils, wound at right angles to each other on the same armature. In this way, when the p.d. due to one

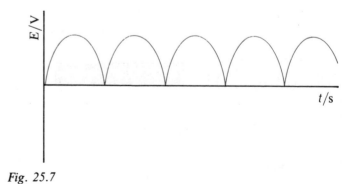

Fig. 25.7

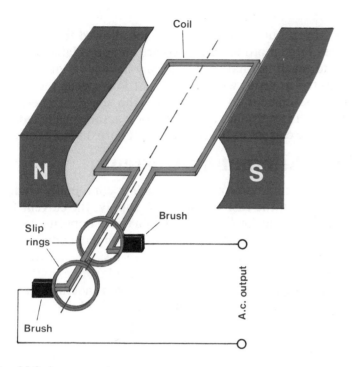

Coil

N   S

Slip
rings

Brush

Brush

A.c. output

Fig. 25.8 A.c. generator

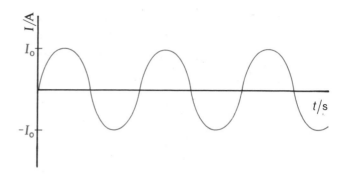

Fig. 25.10 Variation of an alternating current with time

The Alternator

We saw earlier in this chapter that bringing a magnet near to a coil caused a current to flow in the coil. From Lenz's Law it is clear that, if bringing the north pole of a magnet near the coil causes a current to flow in one direction, then bringing the south pole to the coil will cause a current to flow in the opposite direction. If this process is repeated continuously we have a current which flows first one way and then the other. Such a current is an alternating current (a.c.) and the arrangement of moving magnet and coil is a rather crude type of alternator.

The effect of moving a magnet repeatedly past a coil may be further investigated by attaching the magnet to a small motor and connecting the coil to an oscilloscope. The trace on the screen, *Fig. 25.11*, shows that the induced e.m.f. is a sinusoidally alternating one. The height of the trace is a

is a maximum, the p.d. due to the other is a minimum and *vice versa,* and so the total p.d. is approximately constant.) If the split-ring is replaced by two slip-rings, *Fig. 25.8*, the p.d. varies as shown in *Fig. 25.9*. Such a p.d. is said to vary sinusoidally and the resulting current is referred to as a (sinusoidally) **alternating current, a.c.,** *Fig. 25.10*. In practice, generators such as these are only suitable where large currents are not required. Large currents cause excessive sparking at the brushes and shorten the life of the generator. We shall now consider the most important type of generator, the alternator.

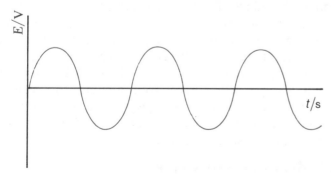

Fig. 25.9   Variation of an alternating voltage with time

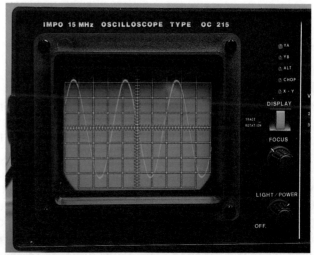

Fig. 25.11   Using an oscilloscope to display an alternating voltage

measure of the size of the induced e.m.f. so the effect of using different magnets and different coils can be investigated. It is found that the size of the induced e.m.f. depends on the strength of the magnet (the magnetic flux density, *B*) and the number of turns of wire on the coil. (Our "design" of alternator is not, of course, a practical one since we are using electrical energy to produce electrical energy!)

It should now be clear that an alternator consists of a magnet which can be rotated beside a coil. In a small alternator the magnet is a permanent one while in larger alternators, *e.g.* in a car, the magnet is an electromagnet, current being fed to it through carbon brushes and slip rings, *Fig. 25.12*.

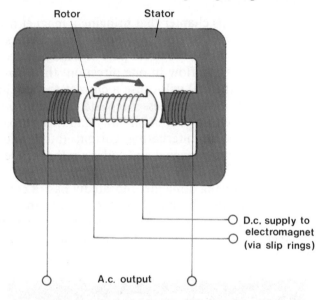

Fig. 25.12 Alternator

Back E.M.F. in Motors

We saw earlier that a motor can be used as a generator, *i.e.* when the coil is turned an e.m.f. is induced in it. Thus when a motor is being used, as a motor, an e.m.f. is induced in the coil. Acccording to Lenz's law, the induced e.m.f. will be such as to oppose the motion, *i.e.* it will try to drive a current in the opposite direction to the applied e.m.f. This induced e.m.f. is referred to as a back e.m.f. and it reduces, very considerably, the total e.m.f. in the coil of the motor. This means that the current flowing in the coil when the motor is running is much less than when it is stopped. The coil is designed to carry the normal "running" current and so will quickly become hot and

may "burn out" if the motor is not allowed to run freely.

Very large motors, which require an appreciable length of time to reach normal speed, usually have a variable resistor connected in series with the coil. At low speeds the resistor limits the current to the coil. As the speed of the motor increases, the resistance of the resistor is automatically reduced, so that the current to the coil remains approximately constant.

25.3 Alternating Current

An alternating current is one which is continually changing direction. In the case of the mains supply, the current flows in one direction for 1/100 s and then in the opposite direction for 1/100 s. In other words, it goes through a complete cycle in 1/50 s or it completes 50 cycles in 1 s, *i.e.* the frequency of the supply is 50 Hz. It should be noted from *Fig. 25.10* that the change from one direction to the other is not instantaneous. Rather, the size of the current increases gradually from zero to a maximum value, I_0, in one direction, decreases to zero and then increases to a maximum value, I_0, in the opposite direction — the size of current varies *continuously*.

Measuring A.C.

The moving coil galvanometer cannot be used for measuring alternating currents except in the rare cases where the frequency is less than 1 or 2 Hz. Since an a.c. changes direction continually, in most cases 100 times per second, the torque of the couple on the coil also changes direction 100 times per second. The inertia of the coil prevents such rapid changes in its motion, so it simply remains at rest.

As we have seen, an oscilloscope may be used to measure alternating voltages, the height of the trace on the screen being directly related to size of the voltage. It may also be used to determine the values of alternating currents by measuring the p.d. produced across a resistor of known resistance. (An oscilloscope may also be used to measure d.c. voltages. Its particular advantage as a voltmeter, both a.c. and d.c., is that it has a very high internal resistance — about 1 MΩ. This means that it draws very little current from a circuit to which it is connected.) The more common method used is to convert the a.c. to a d.c. using a rectifier (*see p.302*) and then to measure the resulting d.c. using a d.c. meter.

Root-Mean-Square Values

Earlier in this chapter we noted that the current from a simple generator or alternator varies sinusoidally with time. It can be shown that the value of the current, I, at any instant t, is given by

$$I = I_0 \sin 2\pi ft$$

where I_0 is the maximum value, or **peak value**, of the current and f is the frequency. Since the actual value of the current changes continuously with time it is necessary to establish what the effective value is. It is an experimental observation that an alternating current flowing in a resistor causes a rise in temperature just as a direct current does. This fact is used to define the effective value of an a.c.

The effective value of an a.c. is that steady current which, when flowing through a purely ohmic resistance, converts electrical energy to internal energy at the same rate as the a.c.

In other words, an a.c. is 1 A if it heats a resistor at the same rate as a d.c. of 1 A would.

At any given instant the rate at which an a.c. converts energy in a resistor is

$$P = RI^2$$
$$= R(I_0 \sin 2\pi ft)^2$$

Since the value of P varies continuously with time, we must find the average value of P over a full cycle. It may be shown that the average value of $\sin^2 x$ is 1/2, where the average is taken over all values of x between 0 and 2π. Therefore, the average value of P is

$$P_{av} = \frac{RI_0^2}{2}$$

Since a d.c. converts energy at a rate given by

$$P = RI^2$$

for the average power produced by the a.c. to be the same as power produced by the d.c. we have

$$I^2 = \frac{I_0^2}{2}$$

$$=> \qquad I = \frac{I_0}{\sqrt{2}}$$

That is, the effective value of the a.c. is its peak value divided by the square root of 2. The effective value of an a.c. is called the **root-mean-square (r.m.s.)** current, $I_{r.m.s.}$ Thus,

$$I_{r.m.s.} = \frac{I_0}{\sqrt{2}}$$

By a similar argument it may be shown that the r.m.s. value of an alternating voltage is given by

$$V_{r.m.s.} = \frac{V_0}{\sqrt{2}}$$

where V_0 is the peak, or maximum, value of the voltage.

In discussing alternating currents and voltages it is normally the r.m.s. values which are being referred to. For example, the mains voltage is quoted as being 220 V. This is the r.m.s. value and so the peak value is $220\sqrt{2}$, or 311 V.

A.C. and Capacitors

If the circuit of *Fig. 25.13* is set up the lamp will not light. From this we conclude that a direct current will not flow through a capacitor. This is not surprising when we remember that a capacitor consists of a pair of metal plates with an insulating material (the dielectric) between them. However, if the cell in the circuit is replaced with an a.c. source the lamp lights — a capacitor ''conducts'' alternating current. In fact what happens is that for the first half of the cycle a positive charge builds up on one plate while an equal negative charge builds up on the other. In effect, charge is being transferred from one plate to the other through the circuit, *i.e.* a current

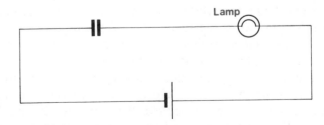

Fig. 25.13

flows. For the second half of the cycle the e.m.f. in the circuit drives a current in the opposite direction. Thus, the plate which had been charged positively loses its positive charge and becomes negatively charged, while the second plate loses its negative charge and becomes positively charged. Again, as in the first half of the cycle, charge is transferred from one plate to the other through the circuit. So, while no charge passes through the capacitor a current flows to and fro in the rest of the circuit just as if the capacitor were a conductor.

If capacitors of different capacitance are used in the circuit it will be noticed that the larger the capacitance the brighter the light. In other words, the larger the capacitance the lower the effective resistance of the capacitor to a.c. Thus, a capacitor may be used, not only to block d.c. in a circuit while allowing a.c. to pass, but also to control the size of an a.c. in the same way as a resistor may be used to control a d.c. (or an a.c.). It may also be noted that the effective resistance of a capacitor to a.c. depends on the frequency of the a.c. - the higher the frequency the lower the effective resistance. This allows capacitors to be used in frequency filters, *i.e.* circuits which allow alternating currents of only certain frequencies to pass.

25.4 Inductance

We saw at the beginning of this chapter that a change in the current in one coil induced an e.m.f. in a second coil nearby. This phenomenon is known as **mutual induction**. Furthermore, a coil carrying a changing current is surrounded by a changing magnetic field and so there is an e.m.f. induced in the coil itself. This phenomenon is known as **self-induction**. In both cases the size of the induced e.m.f. depends on the sizes of the coils and the number of turns of wire on them. The size of the induced e.m.f. is also greatly increased by having the coils wound on an iron core.

Self-induction, including the effect of an iron core, can be demonstrated using the circuit of *Fig. 25.14*. The neon lamp flashes each time the switch is opened or closed; it does not remain lighting after the switch has been closed or opened. This is explained as follows. The lamp requires a p.d. of approximately 100 V to light. The e.m.f. of the battery is only a few volts so the lamp does not light while the switch is closed. However, while the switch is closed a current flows through the coil, producing a magnetic field. When the switch is opened the magnetic field collapses, inducing an e.m.f. in the coil. The size of this e.m.f. depends on the self-inductance of the coil

and is high enough to light the lamp. At the instant when the switch is closed an e.m.f. is also induced in the coil. However, it takes the current some time to reach its maximum value and so the induced e.m.f. is smaller when the switch is closed than when it is opened. Using the circuit shown in *Fig. 25.14* the effect of removing the iron core and of using coils with different numbers of turns of wire may also be investigated.

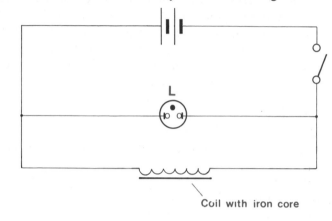

Coil with iron core

Fig. 25.14

A.C. and Inductors

If the circuit of *Fig. 25.13* is set up with a solenoid with an iron core instead of the capacitor, it will be found that the lamp lights brightly when the supply is d.c. but only dimly, if at all, when the supply is a.c. This shows that a coil has a high effective resistance to a.c. but a relatively low resistance to d.c. The higher effective resistance to a.c. arises from the e.m.f. induced in the coil. According to Lenz's law the induced e.m.f. will always oppose the applied e.m.f. and so the net e.m.f. in the circuit is reduced. As a result the current flowing in the circuit is smaller. Hence the idea that the coil has a higher effective resistance to a.c. Removing the iron core from the solenoid causes the lamp to brighten. This is as expected since removing the core reduces the self-inductance of the coil and so reduces the induced e.m.f.

The Induction Coil

An induction coil is a device for producing a high voltage from a low-voltage d.c. source. The induction coil was invented by Nicholas Callan at Maynooth College in 1836. Essentially, it consists of two coils of insulated wire wound, one on the top of the other, around a soft iron core, *Fig. 25.15*. The primary coil consists of a few turns of thick wire and is connected through a switch to a battery or low voltage power supply.

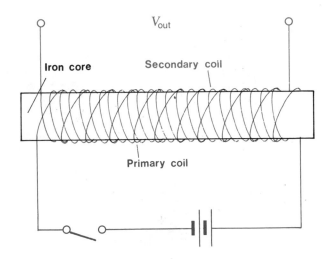

Fig. 25.15 Induction coil

The secondary coil consists of many thousands of turns of thin wire and is wound on top of the primary. When the current in the primary is switched on or switched off the magnetic field due to the current changes. This changing flux threads the secondary in which, as a result, an e.m.f. is induced. (An e.m.f. is also induced in the primary since the changing flux also threads it.) Since there is a very large number of turns on the secondary the induced e.m.f. is very large. We noted earlier that when the current in a coil is switched on it builds up relatively slowly and so the e.m.f. induced in the secondary when the primary current is switched on is much smaller than that induced when the primary current is switched off. Indeed, for practical purposes the current induced in the secondary flows in one direction only. *Fig. 25.16* shows an induction coil connected to a make-and-break mechanism similar to that in an electric bell and in an electromagnetic relay (*p.254*). This allows the coil to operate continuously when the switch is closed. The e.m.f. induced in the primary when the contacts open causes a spark to jump from one to the other. To reduce this sparking and so prolong the life of the contacts a capacitor is connected in parallel with the contacts. The current induced in the primary then charges the capacitor rather than causing a spark at the contacts. The relationship between the current in the primary and the e.m.f. induced in the secondary is illustrated in *Fig. 25.17*.

The most common use of the induction coil is in the petrol-fuelled engine of a car. In this type of engine a mixture of petrol

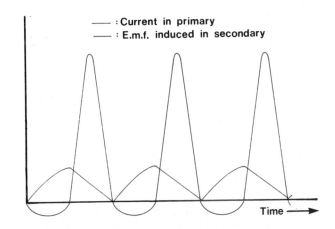

Fig. 25.17 Relationship between current in the primary and the e.m.f. induced in the secondary of an induction coil

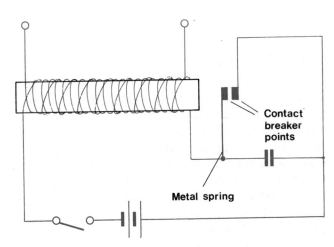

Fig. 25.16 Induction coil with make-and-break mechanism

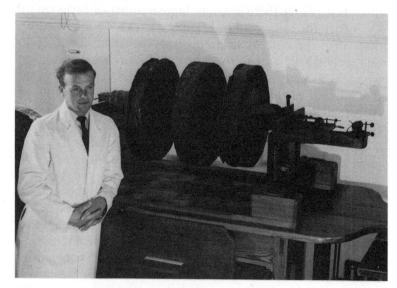

Fig. 25.18 Callan's 'Great' induction coil which gave 15" sparks in 1857 (Courtesy of Maynooth College)

vapour and air is ignited in the cylinder by a spark caused by an electric current crossing between the electrodes of a "spark plug", Fig. 25.19. The gap between the electrodes is typically of the order of 0.6 mm and, in order to make a current cross

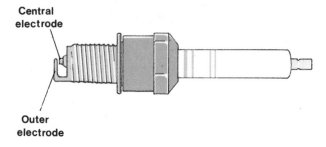

Fig. 25.19 Spark plug

such a gap under the conditions which exist in a cylinder, a p.d. of several thousand volts is required. This high p.d. is obtained from an induction coil whose primary is connected to the low-voltage (normally 12 V) system of the car. The p.d. developed by the secondary is typically of the order of 15 000 V. The current in the primary of the induction coil is switched on and off by a cam driven by the engine, Fig. 25.20, and timed so that a current is delivered to each cylinder at precisely the right time to impart the maximum impulse to the piston.

272

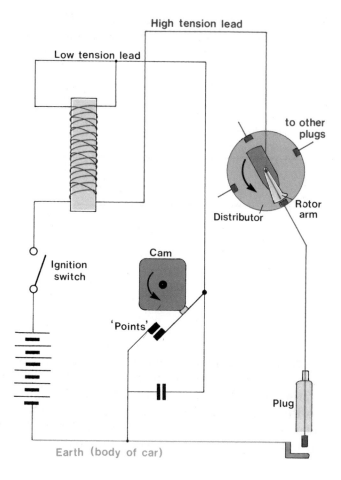

Fig. 25.20 Ignition system of car

The Transformer

A transformer is a device for changing a low a.c. voltage to a high a.c. voltage, or *vice versa*. It consists of two coils of wire, the primary and the secondary, wound on a common iron core, Fig. 25.21. When an a.c. voltage is applied to the primary an alternating current flows through it and an alternating magnetic field is set up in the core. This changing flux threads the secondary and so an e.m.f. is induced in it. The induced e.m.f. is also an alternating one whose magnitude depends on the rate at which the flux is changing and on the number of turns in the secondary coil.

The p.d. applied to the primary is called the input voltage, V_{in}, while the p.d. developed across the secondary is called the output voltage, V_{out}. The relationship between these two voltages depends on the number of turns in the primary, N_p,

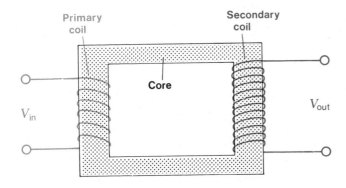

Fig. 25.21 Simple step-up transformer

and the number of turns in the secondary, N_s. It can be shown that, to a reasonable degree of accuracy, if no current is drawn from the secondary,

$$\frac{V_{in}}{V_{out}} = \frac{N_p}{N_s}$$

If the output voltage is greater than the input voltage the transformer is referred to as a step-up transformer. If the output voltage is less than the input voltage the transformer is a step-down transformer.

Example

A transformer is required to convert 220 V a.c. to 4 V a.c. to operate a doorbell. If the primary has 2750 turns how many turns are in the secondary?

$$\frac{V_{in}}{V_{out}} = \frac{N_p}{N_s}$$

$$=> \quad \frac{220}{4} = \frac{2750}{N_s}$$

$$=> \quad N_s = \frac{2750 \times 4}{220}$$

$$= 50$$

Ans. 50

Current in a Transformer

A transformer transfers energy from the primary to the secondary *via* the magnetic field in the core. Ideally, the power obtained from the secondary should be equal to the power fed into the primary, *i.e.*

$$P_{in} = P_{out}$$

Since, in general, $P = VI$

$$P_{in} = V_{in}I_{in}$$

and $\qquad P_{out} = V_{out}I_{out}$

Therefore, in the ideal case,

$$V_{in}I_{in} = V_{out}I_{out}$$

or, $\qquad \boxed{\dfrac{I_{in}}{I_{out}} = \dfrac{V_{out}}{V_{in}}}$

Thus, if the output voltage is ten times the input voltage the output current will be one tenth of the input current. Similarly, if the output voltage is one tenth of the input voltage the output current will be ten times the input current. Because of this the low voltage coil of a transformer must be made of thick wire (*see next section*).

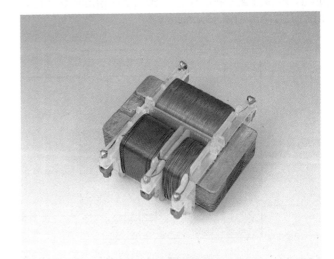

Fig. 25.22 Door-bell transformer. The input voltage is 220 V and there are two output voltages of 3 V and 5 V, respectively. Note the different numbers of turns on the three coils

273

Energy Losses in a Transformer

In practice, the power obtained from the secondary of a transformer is always less than the power fed into the primary, the difference being accounted for by the fact that some of the electrical energy supplied is converted into internal energy in the coils and the core. There are three main reasons for this.

Firstly, electrical energy is converted into internal energy in the wires of the coil at a rate given by $P = RI^2$ *(see p.234)*. The energy "lost" in this way is reduced by making R, the resistance of the coils, as small as possible. This is especially important if the coil must carry a large current *(cf. last section)*.

Secondly, since the core is a conductor situated in a changing magnetic field currents are induced in it. (These currents flow in the core itself and are called eddy currents.) Again, these result in the conversion of electrical energy to internal energy. This effect is reduced by using a laminated core, *i.e.* one which is made up of thin sheets of iron insulated from each other. This increases the resistance of the core and so reduces the magnitudes of the eddy currents.

Thirdly, energy is required to magnetise and demagnetise the core as the current in the primary changes direction. This loss is minimised by using a core made of iron which is easily magnetised and demagnetised.

By these means, and by using a suitably shaped core, *e.g.* as in *Fig. 25.23*, the efficiency of a transformer may be made as high as 99%, *i.e.* 99% of the energy supplied to the primary is available from the secondary.

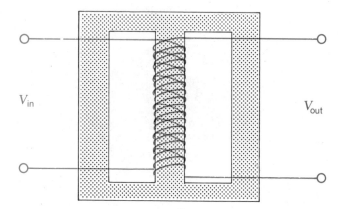

V_{in} V_{out}

Fig. 25.23 Step-up transformer. The shape of the core is designed for maximum efficiency

Uses of Transformer

Transformers play a vital role in the transmission of electrical energy around the country since, in order to minimise losses in the cables, electrical energy must be transmitted at high voltages *(see p.236)*. At the generating stations transformers increase the voltage from 10 kV to 100 kV, 220 kV or even 440 kV. Other transformers reduce these voltages to 500 V for use in factories and 220 V for use in homes and offices. It is the simplicity and high efficiency of transformers which make the use of alternating current more economical than direct current for the transmission of electrical energy, at least over relatively short distances.

Transformers are used in television sets to change the 220 V of the mains supply to the voltages, ranging from a few volts to several kilovolts, required by the various components. In the school laboratory transformers are used in power supplies which, when connected to the mains (220 V) give the various voltages required. Low tension (L.T.) supplies give voltages up to about 24 V; high tension (H.T.) supplies give up to a few hundred volts and extra high tension (E.H.T.) supplies give up to about 6 kV.

Fig. 25.24   E.S.B. transformer

25.5 The Induction Motor

If a magnet is rotated above an aluminium disc, *Fig. 25.25*, the disc will rotate in the same direction as the magnet. This phenomenon was discovered by the French physicist, Dominique Arago (1786 — 1853) and the arrangement is now known as Arago's disc. It is the principle on which the modern induction motor is based and it is explained as follows.

When the magnet moves, the disc is in a changing magnetic field so e.m.f.'s are induced in it. Since the disc is a conductor currents will flow in it and the direction in which these currents flow will be such that their magnetic fields will oppose the motion of the magnet (Lenz's Law). The forces due to these currents act equally on the magnet and the disc (Newton's Third Law). Since the magnet cannot move backwards the disc moves forward, *i.e.* in the same direction as the magnet.

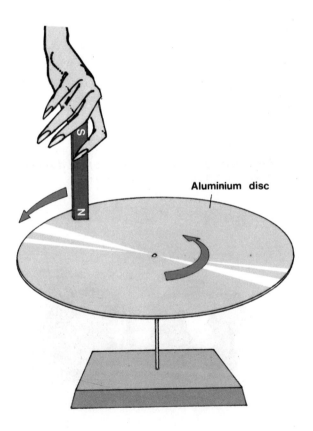

Aluminium disc

Fig. 25.25 Arago's disc

Nicholas Callan (1799 — 1864)

Nicholas Callan was born in Co. Louth in 1799, of a well-to-do family. He was educated at Dundalk Academy and Maynooth College, where he was ordained in 1823. He was appointed professor of natural philosophy (physics) at the college in 1826.

Callan did pioneering work in the study of electricity. He constructed large electromagnets, one of which could lift more than 1 t. He also developed a form of electric motor, the induction motor, and proposed that such a motor could be used in the electrification of the Dublin/Dun Laoghaire railway line. For his work on electricity he experimented with improved forms of battery. (He is said to have judged the power of his batteries by administering shocks from them to his students and observing their reactions!) Callan's greatest claim to fame is his invention, in 1836, of the induction coil. He built a number of such coils, one of which, built in 1857, could produce a spark 38 cm long. The extent of the work involved can be appreciated when it is realised that his secondary coils consisted of several kilometres of fine wire, all insulated by hand with a mixture of beeswax and guttapercha.

Callan gained considerable recognition for his work as a scientist. He was also a highly respected teacher and priest. He died on January 10, 1864.

Senior Physics

The Practical Induction Motor

Arago's disc is not a practical motor since it simply uses kinetic energy to produce kinetic energy. We need a method of producing a rotating magnetic field without using kinetic energy. One way of achieving this is with the arrangement shown in *Fig. 25.26*. A is a metal cylinder which is free to rotate in the direction shown. B_1 and B_2 are two similar electromagnets connected in parallel with each other to an a.c. power supply. C is a capacitor of about 5 μF which is connected in series with B_1. The effect of the capacitor is that the current in B_1 is approximately 90° out of phase with the current in B_2. Consequently, when the magnetic field due to B_1 is a maximum that due to B_2 is zero; when the field due to B_1 is zero that due to B_2 is a maximum, and so on. *Fig. 25.27* shows how the total magnetic field due to both electromagnets varies over one complete cycle of the applied p.d., *V*. From this diagram we see that the direction of the resultant magnetic field rotates through 360° for each cycle of the applied p.d. Thus, on the principle of Arago's disc, the cylinder turns in the same direction as the field.

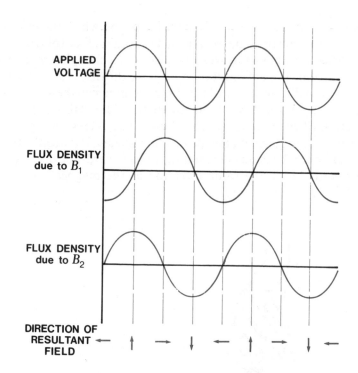

Fig. 25.27 *Production of a rotating magnetic field due to phase difference between currents in two coils*

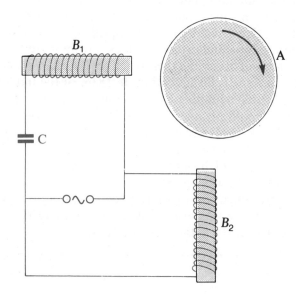

Fig. 25.26 *Principle of the induction motor*

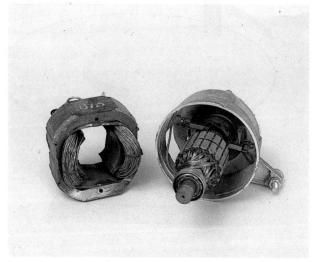

Fig. 25.28 *Induction motor*

276

SUMMARY

Magnetic flux ϕ, is defined by

$$\phi = BA$$

where ϕ is the flux passing through area A at right angles to the flux density, B which is assumed to be uniform. Flux is a scalar quantity and its unit is the weber.

When there is a change in the magnetic flux threading any closed loop an e.m.f. is induced in the loop. The magnitude of the induced e.m.f. is proportional to the rate at which the flux is changing (Faraday's Laws). The direction in which an induced current flows is such that its effects oppose the change which caused it (Lenz's Law).

An alternator is a device for converting kinetic energy to electrical energy. It consists of a magnet rotating beside a coil so that the flux threading the coil changes continuously, thus causing an e.m.f. to be induced in it.

An alternating current is one which continually changes direction, flowing first in one direction and then in the opposite direction. The frequency of the mains supply in this country is 50 Hz. Alternating currents are measured by converting them to direct currents which are then measured using moving coil, or other, meters. An oscilloscope may also be used to measure alternating voltages and hence currents.

An induction coil converts a low d.c. voltage to a high d.c. (for practical purposes) voltage. It consists of two coils, one (the primary) having a small number of turns and the other (the secondary) a very large number of turns, wound on an iron core. When the current in the primary is switched off a large e.m.f. is induced in the secondary.

A transformer converts a low a.c. voltage to a high a.c. voltage or *vice versa*. It consists of two coils wound on an iron core. When an alternating p.d. is applied to the primary an alternating e.m.f. is induced in the secondary. The ratio of the input voltage to the output voltage is given by

$$\frac{V_{\text{in}}}{V_{\text{out}}} = \frac{N_{\text{p}}}{N_{\text{s}}}$$

where N_{p} and N_{s} are the numbers of turns in the primary coil and the secondary coil respectively. Energy losses in a transformer are kept to a minimum by using thick wire for the low voltage coil, by using a suitably shaped core, and by having the core made of iron and laminated.

A capacitor blocks d.c. and conducts a.c. An inductor has a high effective resistance to a.c. and a low effective resistance to d.c.

The effective value of a sinusoidally alternating current is the root-mean-square value, $I_{\text{r.m.s.}}$, which is given by

$$I_{\text{r.m.s.}} = \frac{I_0}{\sqrt{2}}$$

where I_0 is the maximum, or peak, value of the current. Similarly, the root-mean-square voltage is given by

$$V_{\text{r.m.s.}} = \frac{V_0}{\sqrt{2}}$$

Questions 25

Section I

1. Define magnetic flux...

....

2. Magnetic flux is a...

   quantity. Its unit is the.....................................

3. State the laws of electromagnetic induction..............

....

....

4. What is meant by flux linkage?...........................

....

5. Sketch a graph to show how the magnitude of an alternating current varies with time.

6. Name the principal parts of an alternator................

....

....

7. What is the basic principle of an alternator?............

....

....

8. What is meant by back e.m.f. in a motor?..............

....

....

*9. Give the relationship between the root-mean-square value of a sinusoidally alternating voltage and its peak value.

....

*10. A capacitor blocks .... but conducts.............

*11. What is meant by mutual induction?......................

....

....

*12. What is meant by self induction?...........................

....

....

*13. A coil has a............ resistance to a.c. but a............ .... to d.c.

14. What is the function of an induction coil?..............

....

....

15. What is the function of a transformer?...................

....

....

16. What is the basic principle of both the induction coil and the transformer?...

....

....

17. Why does the low voltage winding of a transformer have thicker wire than the high voltage winding?.............

....

....

18. Give three factors which affect the efficiency of a transformer..

....

....

19. If a current is drawn from the secondary of a transformer the output voltage drops. Suggest a reason for this...

....

....

*20. What is the basic principle of the induction motor?...

....

....

Section II

1. A piece of wire, in the form of a rectangle of area 3.0×10^{-2} m^2, is hanging vertically with its plane perpendicular to the earth's magnetic field. If the horizontal component of the earth's magnetic flux density is 1.7×10^{-5} T, calculate the flux threading the loop. If the loop is turned through 90° in 0.2 s calculate the average e.m.f. induced in it.

2. A circular coil of wire has 200 turns of average diameter 0.07 m and is placed in the magnetic field of an electromagnet. If the plane of the coil is perpendicular to the flux density and if the flux density, assumed uniform over the plane of the coil, is 0.2 T, calculate the flux linking the coil. If, when the current to the electromagnet is switched off, the flux density falls to zero over a period of 1.0 s calculate the average e.m.f. induced in the coil.

3. A circular coil of wire of radius 15 cm and having 50 turns is withdrawn from a uniform magnetic field in 0.15 s, the direction of the field being perpendicular to the plane

of the coil. The resistance of the coil is 22 Ω and the average current flowing in it during the time it is being withdrawn from the field is 16 mA. What is the magnitude of the field's flux density?

4. A rectangular coil, 20 cm by 30 cm and having 200 turns, is free to rotate about a vertical axis parallel to its longer side. It is arranged so that its plane is perpendicular to the horizontal component of the earth's magnetic flux density. When the coil is turned quickly through 90° a total charge of 4.6 μC passes through a meter attached to it. Given that the total resistance of the coil is 24 Ω calculate the magnitude of the horizontal component of the earth's magnetic flux density at the position of the coil. If the total value of the earth's flux density is 50 μT what is the value of the angle of dip at the location of the coil?

5. A coil of wire containing 100 turns is in the shape of a square of side 15 cm. The coil is moving at a steady speed of 2.2 m s^{-1} in a direction parallel to the plane of the coil when it enters a magnetic field of uniform flux density 46 mT at right angles to its velocity and to the plane of the coil. Given that the total resistance of the coil is 16 Ω calculate the work done in moving the coil completely into the field. Neglecting the effect of friction, is any work done in moving the coil when it is completely within the field? Explain.

6. A transformer has 2000 turns in the primary coil and 50 turns in the secondary. If the input voltage is 220 V what is the output voltage?

7. A bell transformer is required to produce an output voltage of 8 V when its primary is connected to the mains. If the primary has 1100 turns how many turns must be on the secondary?

8. A step-up transformer in a TV is required to give an output of 3300 V when connected to the mains. If the primary has 1100 turns how many turns must be on the secondary?

9. If the power drawn from the secondary of the transformer in the previous question is 100 W and if the efficiency of the transformer is 90%, what is the power input into the primary? Calculate the current flowing in the primary and in the secondary.

*10. What is the r.m.s. value of a sinusoidally alternating p.d. whose peak-to-peak value is 20 V?

REVISION EXERCISES D

Section I

*1. In an experiment to determine the e.m.f. of a cell using a potentiometer a cell of e.m.f. 1.50 V was first connected to the potentiometer. When the galvanometer read zero the moving contact was on the 64.6 cm mark of the metre scale. With the cell of unknown e.m.f. connected to the potentiometer the galvanometer read zero when the moving contact was on the 61.8 cm mark of the scale. Using these values calculate the e.m.f. of the second cell. State how the accuracy of the result might have been improved. Assuming that the driver cell was connected directly to the potentiometer wire estimate its e.m.f.

2. The resistivity of nichrome was determined from a sample in the form of a wire. The diameter of the wire was measured at five positions on the wire and the following results were obtained: 0.21 mm; 0.19 mm; 0.21 mm; 0.20 mm; 0.21 mm. The length of the wire was found to be 1.22 m and its resistance was found to be 48.2 Ω. Using these data, calculate the resistivity of nichrome. Explain how the length of the wire should be measured in this experiment. Describe the instrument you would use to measure the diameter of the wire.

3. The resistance of a length of copper wire was determined for a series of values of its temperature and the following results were obtained.

$\theta/°C$	15	25	35	45	55	65	75	85
R/Ω	4.1	4.3	4.6	4.9	5.1	5.3	5.6	5.9

Plot a graph to show how the resistance of the wire varies with its temperature. From the graph, estimate the resistance of the wire at 0 °C. Describe, with the aid of a diagram, the apparatus you would use to carry out this experiment.

*4. The resistance of a thermistor was determined for a series of values of its temperature and the following results were obtained.

$\theta/°C$	20	30	40	50	60	70	80	90
R/Ω	1300	900	640	460	340	260	200	150

Plot a graph to show how the resistance of the thermistor varies with its temperature. From the graph, estimate the resistance of the thermistor at 35 °C.

5. In an experiment to verify Joule's law a known mass of water was placed in a calorimeter and a current was passed through the coil for a certain time. The rise in temperature of the water and the calorimeter was recorded. The process was repeated for different values of the current, the mass of water and the time being the same in each case. The following results were obtained.

I/A	0.4	0.6	0.8	1.0	1.2	1.4	1.6	1.8	2.0
$\Delta T/K$	2	4	7	13	17	25	31	40	51

Plot a suitable graph from these data and use it to explain how this experiment verifies Joule's law.
Would it be possible to use the same water throughout the experiment? Discuss the advantages and disadvantages of this as opposed to using fresh water for each value of the current.

6. In an experiment to verify Faraday's first law of electrolysis a current was passed through a copper voltameter and the mass of copper deposited in a given time for a series of values of the current was determined. The following results were obtained.

I/A	0.4	0.6	0.8	1.0	1.2	1.4	1.6	1.8	2.0
m/mg	83	120	164	198	235	278	318	352	390

Plot a graph of m against I and explain how this verifies Faraday's law. Given that the time for which the current flowed in each case was 10 minutes use the graph to calculate the electrochemical equivalent of copper. Mention two precautions which should be taken when carrying out this experiment to ensure a more accurate result.

*7. A high impedance voltmeter was used to measure the p.d. between the terminals of a cell. The measurement was taken firstly with the cell on its own, then with a number of different resistors connected across the terminals of the cell. The reading on the voltmeter was recorded for each value of the resistance and the following results were obtained.

R/Ω	0	50	25	15	10	5
V/V	1.5	1.47	1.44	1.40	1.35	1.23

Use these data to determine the internal resistance of the cell. Why is it important to use a high impedance voltmeter in this experiment? What alternative apparatus could have been used in this experiment?

Section II

1. State Coulomb's law. Calculate the magnitude of the force between two point charges each of 2.2 μC a distance 40 mm apart. (Permittivity of free space, $\varepsilon_0 = 8.9 \times 10^{-12}$ F m^{-1}.)

*2. Define electric field intensity and electric field flux. Three point positive charges, each of 4.4 pC are situated at the corners of an equilateral triangle, ABC, of side 40 cm. Calculate the resultant field intensity at the mid-point of BC. ($\varepsilon_0 = 8.9 \times 10^{-12}$ F m^{-1}.)

3. A capacitor of capacitance 4.5 μF is connected in series with a battery and a resistor. The p.d. across the capacitor rises from 0 to 10 V in 15 s. Calculate (i) the charge on the capacitor at the end of the 15 s, (ii) the average current flowing in the circuit during the 15 s.

4. State Ohm's Law and describe how you would verify it experimentally for a piece of iron wire.

*5. Referring to *Fig. I*, calculate (i) the total resistance of the circuit, (ii) the total current flowing in the circuit, (iii) the p.d. across the 10 Ω resistor, (iv) the current flowing through the 20 Ω resistor.

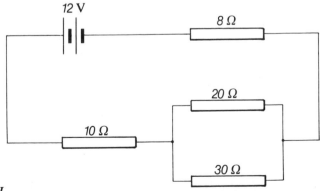

Fig. I

*6. Referring to *Fig. II*, calculate (i) the total resistance of the circuit, (ii) the total current flowing in the circuit, (iii) the p.d. across the 20 Ω resistor.

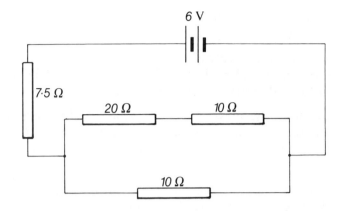

Fig. II

7. Describe how you would determine the resistance of a length of wire. What steps would you take to ensure an accurate result?

8. Describe an experiment to investigate how the resistance of a length of wire varies with its temperature.

9. Define power. How much electrical energy is converted in a 10 Ω resistor when a current of 2.5 A flows through it for 10 minutes? Into what form is the electrical energy converted?

10. A p.d. of 24 V is maintained between the ends of a length of wire. If the current flowing in the wire is 2.5 A calculate the amount of energy converted in 5 minutes. What is the total charge passed through the wire in this time?

11. Electrical energy is to be transmitted through cables having a total resistance of 40 Ω at a rate of 100 kW. If the energy is transmitted at a potential of (a) 20 kV (b) 100 kV, calculate, in each case, (i) the current flowing, (ii) the power lost in the cables.

12. A car head lamp bulb is rated as 12 V, 45 W. What is the resistance of the filament of the bulb? What is the total current flowing in a circuit in which two such bulbs are connected in parallel?

13. Explain why fuses are used in electrical circuits and how they work. Why is it dangerous to replace a fuse with a piece of aluminium foil or with wire other than fuse wire of the correct rating?

14. Draw a circuit diagram showing four "power" sockets connected in a ring main circuit.

15. What is an electrolyte?
State Faraday's First Law of Electrolysis and explain how you would verify it experimentally.

16. Calculate the mass of copper deposited in a copper voltameter when a steady current of 2.5 A flows through it for 20 minutes. (e.c.e. of copper $= 3.3 \times 10^{-7}$ kg C^{-1}.)

17. Calculate the electrochemical equivalent of hydrogen if 25 cm^3 of hydrogen are liberated by a current of 0.33 A in 10 minutes. (Take the density of hydrogen to be 8.5×10^{-2} kg m^{-3}.)

18. Draw a labelled diagram of a primary cell. What is polarisation and what effect does it have? In what respect(s) does a primary cell differ from a secondary cell?

19. How would you demonstrate the magnetic effect of a current? Draw a diagram showing the magnetic field due to a current in (i) a straight conductor, (ii) a coil, (iii) a solenoid.

*20. Describe how you would demonstrate that a current-carrying conductor in a magnetic field experiences a force. Calculate the magnitude of the force on a wire of length 10 cm carrying a current of 3.5 A at an angle of 30° to a magnetic field of flux density 0.45 T.

21. Draw a labelled diagram of a simple d.c. motor and explain how it works. What is the function of the split-ring commutator?

22. Draw a labelled diagram of a moving coil galvanometer and explain how it works. How may the instrument be made less sensitive?

23. Use diagrams to show how a moving coil galvanometer may be used as (i) an ammeter, (ii) a voltmeter.

24. A moving coil galvanometer has an f.s.d. of 1 mA and an internal resistance of 100 Ω. What is the resistance of the shunt required to convert it into an ammeter of f.s.d. (i) 1 A, (ii) 5 A?

25. Calculate the resistance of the multiplier required to convert the galvanometer of the previous question into a voltmeter of f.s.d. (i) 5 V, (ii) 50 V.

26. State the laws of electromagnetic induction and describe experiments to demonstrate them.

27. Draw a labelled diagram of a simple a.c. generator. In what respect(s) does an a.c. generator differ from a d.c. motor?

28. Draw a labelled diagram of an induction coil and explain how it works. Give one use of an induction coil.

29. Draw a labelled diagram of a transformer and explain how it works. State the steps taken to reduce energy losses in a transformer.

30. A transformer is to be used to give an output of 3.3 kV from a mains input. If there are 1200 turns on the primary coil how many must there be on the secondary?

26 Electron Beams

26.1 The Electron

Early experiments (c. 1895) on cathode rays *(see below)* established that they were negatively charged. It was also shown that they had momentum, suggesting that they consisted of a stream of particles. The English physicist, J. J. Thompson (1856 — 1940), determined the ratio of the charge on these particles to their mass (this ratio, e/m, is called the **specific charge**) and found that the result was the same no matter what material was used for the cathode. This led Thompson to the conclusion that these particles were constituents of all matter. Earlier it had been suggested, by Faraday and others, that charge was not continuous but was made up of discrete units, in the same way as matter is made up of discrete units which we call atoms. The Irish physicist, G. J. Stoney (1826 — 1911), suggested, in 1891, that the name **electron** be given to the unit of charge. This **electronic charge** seemed to be the charge carried on a monovalent ion. From Faraday's work on electrolysis this charge was shown to be equal to 1.6×10^{-19} C. Thompson assumed that the charge on the particles of cathode rays was equal to that on a monovalent ion and as a result the particles came to be known as electrons. Knowing the charge, and also the charge to mass ratio, Thompson calculated the mass of the electron, arriving at a value of 9.0×10^{-31} kg. (The currently accepted value is 9.1×10^{-31} kg.) While this calculation was based on an assumption there is now no doubt that the assumption was correct.

In 1906, the American physicist, R. A. Millikan started a series of experiments which verified that charge was indeed made up of units of 1.6×10^{-19} C. The apparatus for this experiment (which is now known as Millikan's oil drop experiment) is shown in *Fig. 26.1*. It consists of a pair of horizontal parallel metal plates a few millimetres apart, the top one having a small hole in it. The plates are connected, through a reversing switch, to a high voltage power supply. A spray of oil is produced above the plates, resulting in a few tiny drops of oil, most of which are charged, falling through the hole in the upper plate. The space between the plates is illuminated from one side and the drops are viewed through

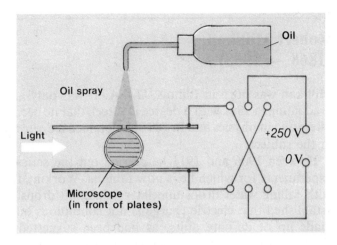

Fig. 26.1 Millikan's oil drop experiment

a microscope. When a small drop falls freely it does so with a constant speed due to the fact that it is subject to two equal and opposite forces — its weight and the frictional force between it and the air through which it is falling. This speed is determined by noting how long it takes to fall a measured distance. When a p.d. of the correct polarity (the top plate must have the opposite charge to the charge on the drop) is applied to the plates the drop will move upwards, again at a constant speed, which is determined as before. The forces on the drop are now its weight, the frictional force and the force due to the electric field ($F = Eq$, *p. 202*). Knowing these two speeds it is possible to calculate the value of q, the charge on the drop. It is found that q is always a whole-number multiple of 1.6×10^{-19} C.

Experiments by Rutherford *(see Chapter 29)* showed that electrons occupy the space around the nuclei of atoms. The properties of electrons may therefore be summarised as follows.

> **Electrons are very small particles which move around the nuclei of atoms. They have a mass of 9.1×10^{-31} kg and a negative charge of 1.6×10^{-19} C.**

Robert Millikan (1868 — 1953)

Millikan was born in Illinois, U.S.A. It was only after graduating in 1891 with a degree in Greek that he became interested in physics and in 1895 he obtained a doctorate in the subject.

Between 1906 and 1911 he carried out the series of experiments for which he is now famous. Working first with falling water drops and later falling oil drops, he established that electric charge is not continuous but is made up of definite units, as had been suggested by Faraday's work on electrolysis a century before. He also verified experimentally Einstein's photoelectric law and used his results to determine the value of Planck's constant.

In 1921 Millikan moved from the University of Chicago to the Californian Institute of Technology where he remained until his retirement. While there, he studied the radiation which reaches earth from outer space. He named this radiation "cosmic rays" and believed it to be a form of electromagnetic radiation, although others showed it to be mostly protons. He was awarded the Nobel Prize in Physics for his work on charge and the photoelectric effect. He died on December 19, 1953.

26.2 Thermionic Emission

Fig. 26.2 shows an evacuated glass bulb into which has been sealed a metal filament, K, and a metal plate, A. When the low voltage circuit is switched on, the filament glows red-hot and the galvanometer registers a current. The direction in which the current flows, as indicated by the galvanometer, is as shown, indicating that electrons are crossing from K to A inside the bulb. (Remember that electrons flow in the opposite direction to conventional current.) When the low voltage circuit is switched off the current stops flowing. Also, if a p.d. of a few volts is applied between K and A, making A negative with respect to K, no current flows. From these observations we conclude that electrons are emitted from the metal filament when it is hot. This phenomenon, which was discovered by Thomas Edison, is known as **thermionic emission**.

> **Thermionic emission is the emission of electrons from the surface of a hot metal.**

Thermionic emission may be explained as follows. As we have already seen, some of the electrons in metals are only weakly attracted to the nuclei of the atoms. When the metal is heated these electrons gain enough energy to escape from the metal entirely — the higher the temperature of the metal the more electrons are released.

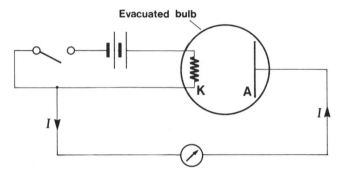

Fig. 26.2

Cathode Rays

If the metal plate (the anode) is made positive with respect to the filament (the cathode), *Fig. 26.3*, and has a hole cut in it the glass behind it begins to glow with a greenish light. (The brightness of the light may be increased by coating the inside of the glass with special substances called phosphors.) If an obstacle of a distinctive shape is placed between the plate and the glass, a sharp shadow of the obstacle is seen on the glass. This indicates that the electrons travel out from the filament in straight lines and led to their being called **cathode rays** before their true identity was known.

The above experiments, together with other similar

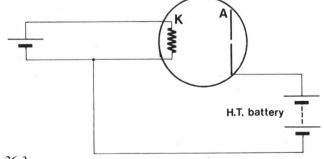

Fig. 26.3

experiments, lead to the following conclusions regarding the nature of electrons.

1. They travel in straight lines.
2. They transport energy (light is produced when they strike the glass).
3. They transport charge (a current flows in the external circuit, the direction in which the current flows indicating that the charge carried by the electrons is negative).

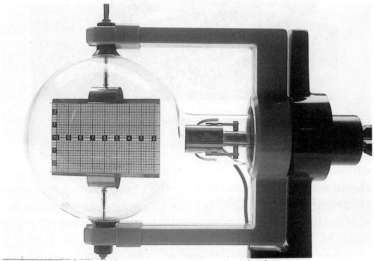

Fig. 26.4(a) Producing a narrow beam of cathode rays

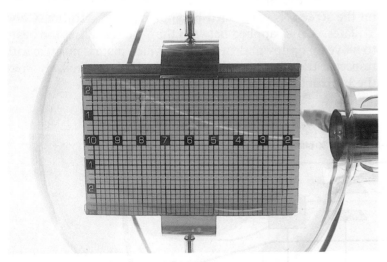

Fig. 26.4(b) Deflection of cathode rays in an electric field

Deflection of Cathode Rays

The behaviour of cathode rays in electric and magnetic fields may be investigated using the apparatus shown in *Fig. 26.4(a)*. Using a suitably shaped anode, the electrons from the cathode are collimated into a narrow beam which them strikes a phosphor-coated screen set at an angle to the beam. An electric field is set up by applying a p.d. of a few kilovolts between the horizontal metal plates. The resulting deflection of the beam is shown in *Fig. 26.4(b)*. To investigate the deflection of the beam in a magnetic field, the apparatus is placed between two coils which are connected to a low voltage power supply, *Fig. 26.4(c)*.

Fig. 26.4(c) Beam of cathode rays in an electric field and a magnetic field. The forces due to the two fields are equal in magnitude but opposite in direction

This apparatus may be used to determine the specific charge on the electron. The electric and magnetic fields are both switched on at the same time and the strength of the magnetic field is adjusted until the beam is again undeflected, *Fig. 26.4(c)*. This means that the force on the electrons due to the magnetic field is equal in magnitude and opposite in direction to the force on the electrons due to the electric field, *i.e.*

$$evB = Ee$$

$$=> \qquad v = \frac{E}{B} \qquad (i)$$

The work done in accelerating the electrons is eV, where V is the p.d. between the cathode and anode. Since the work done must be equal to the kinetic energy gained, we have

$$eV = \tfrac{1}{2} mv^2$$

$$\Rightarrow \quad \frac{e}{m} = \frac{v^2}{2V}$$

Substituting for v from *Eqn. (i)* gives

$$\frac{e}{m} = \frac{E^2}{2B^2V}$$

All the quantities on the right hand side of this equation may be determined so e/m may be calculated. Its value is found to be 1.76×10^{11} C kg^{-1}.

The Cathode Ray Tube (CRT)

The tube referred to in the previous section is essentially a cathode ray tube, *i.e.* a tube in which a beam of cathode rays can be produced and controlled. *Fig. 26.5* shows a cathode ray tube as used in a cathode ray oscilloscope.

A beam of electrons is produced by thermionic emission at the cathode. The beam passes through the grid and on to a pair of cylindrical anodes. From the anodes a narrow beam passes between two pairs of deflecting plates, X and Y, at right angles to each other, and onto the screen. At the screen some of the kinetic energy of the electrons is converted to light, the colour of the light depending on the phosphor with which the screen is coated.

Fig. 26.6 Cathode ray tube in television

The grid is kept at a negative potential with respect to the cathode. Increasing the negative potential of the grid reduces the number of electrons reaching the screen and hence reduces the brightness of the light emitted from the screen. The anodes accelerate the electrons leaving the cathode (to speeds of the order of 10^7 m s^{-1}). The electric field between the anodes is designed to focus the electrons onto a small spot on the screen.

The position of the beam, and hence of the spot of light on the screen, is controlled by applying p.d.'s to the X and Y plates. A p.d. applied between the Y plates causes the beam to move up or down, depending on which plate is positive with respect to the other. A p.d. between the X plates similarly alters the horizontal position of the beam.

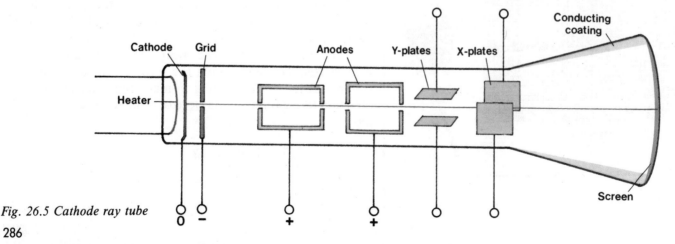

Fig. 26.5 Cathode ray tube

The oscilloscope can be used for measuring p.d.'s since the position of the spot on the screen depends on the p.d. between the deflecting plates. As we noted in the previous chapter it is particularly useful for measuring alternating voltages. It can also be used to display signals from, for example, a microphone and to measure the frequency of a.c. signals.

The CRT in a television is essentially the same as that in an oscilloscope. The main difference is that magnetic fields, rather than electric fields, are used to deflect the beam. Also, in a colour television there are three beams and three different types of phosphor on the screen, one for each of the primary colours (*see p. 137*).

26.3 X-Rays

If a beam of electrons is accelerated to a very high speed and then allowed to strike a metal target, electromagnetic radiation of very short wavelength is produced. This radiation is called X-rays and was discovered by Wilhelm Röntgen in 1895.

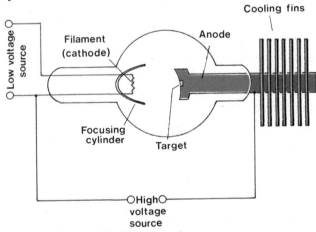

Fig. 26.7 X-ray tube (shielding not shown)

A diagram of a modern X-ray tube is shown in *Fig. 26.7*. A filament is sealed in a highly evacuated tube and is heated by a current from a low voltage winding of a transformer. Electrons are emitted from the hot filament by thermionic emission and are accelerated through a p.d. of the order of 100 kV. (An a.c. supply is normally used, so the electrons flow only on the half cycles when the anode is positive with respect to the cathode.) The electrons strike a target which is made of tungsten and set in a block of copper. When the electrons, travelling at speeds of the order of 10^8 m s^{-1}, strike the target some of their kinetic energy is converted into X-rays. However,

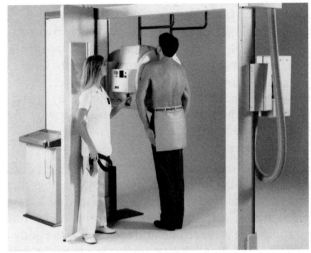

Fig. 26.8 X-ray machine (Courtesy of Siemens)

most of the kinetic energy (up to 99.9%) is converted into internal energy in the target. The target, consequently, becomes very hot and it is for this reason that tungsten, with a melting point of 3380 °C, is used. Even so, the target must be cooled. In the type of tube shown in the diagram cooling is achieved by blowing cold air over the fins attached to the anode. In another type, cold water is pumped through pipes embedded in the anode. Since over-exposure to X-rays is dangerous, the tube is surrounded by a lead shield. The lead absorbs the X-rays and so people who must work in the vicinity of the tube are protected from exposure to the X-rays. A "window" in the shield allows a narrow beam of X-rays to be emitted.

Properties of X-rays

As we have already seen, X-rays are electromagnetic radiation of short wavelength (*see The Electromagnetic Spectrum, p. 138*). The actual wavelengths produced by a given tube depend on the p.d. across the tube and the material of the target, but are typically of the order of 10^{-10} m. The properties of X-rays may be summarised as follows.

1. They are not deflected in electric or magnetic fields. This means that they carry no charge.
2. They show diffraction and interference effects. This indicates that they have a wave nature. Crystals are used for this purpose, since the spacing between atoms in a crystal is approximately of the same order as the wavelength of X-rays.
3. They affect photographic emulsions in the same way as light.

4. They cause certain substances to fluoresce.
5. They can penetrate many substances which are opaque to light. In general, X-rays are absorbed by dense substances containing elements of high atomic number (*see p. 318*), *e.g.* lead, calcium, *etc.*
6. They ionise materials through which they pass. This provides a method of detecting them and is why they are harmful to the human body.

Uses of X-Rays

X-rays are used in medicine to "photograph" bones and internal organs of the body, with a view to detecting fractures, cancers, etc. This is made possible by the fact that X-rays pass through skin and soft tissue but are absorbed by bone. A photographic film is placed behind the part of the body being investigated and a small X-ray source is placed in front. In this way a shadow photograph, *Fig. 26.9,* is taken of the bones. In practice, the photographic film is placed between two fluorescent screens and it is the light from these which affects the film rather than X-rays directly. This allows the film to be developed after a much shorter exposure than the direct method.

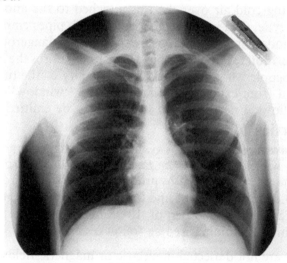

Fig. 26.9 X-ray photograph (Courtesy of Siemens)

X-ray photographs of lungs can reveal disease due to the fact that damaged tissue is more dense than healthy tissue. Damaged tissue therefore shows up as a "shadow" on an X-ray photograph. In order to photograph some organs it is necessary that a dense substance is first introduced into them. For example, in order to photograph the stomach, the patient is first given a "meal" of barium sulphate. This results in a shadow of the stomach being produced on the X-ray photograph.

Other uses of X-rays include the treatment of cancer, detection of flaws in metal castings and monitoring the thickness of various materials at the production stage, *e.g.* the tread for vehicle tyres.

Wilhelm Röntgen (1845 — 1923)

Röntgen was born in Prussia and educated in Holland and Switzerland. After receiving his doctorate in 1869 he worked as an assistant to the German physicist, August Kundt.

On November 5, 1895, while investigating the properties of cathode rays, he noticed that a piece of paper coated with barium platinocyanide was glowing even though the cathode ray tube was enclosed in black cardboard and the room was in darkness. When he switched off the tube, the paper stopped glowing. He concluded that invisible rays, capable of penetrating cardboard, were coming from the cathode ray tube. Since the nature of the rays was unknown he called them X-rays. After his discovery, Röntgen worked furiously to establish the properties of X-rays and, on December 28, 1895 he published his results.

The discovery of X-rays opened up a whole new field in the study of physics at a time when many believed that, apart from incidental details, all that was to be known had already been discovered. In 1901, Röntgen was awarded the first Nobel Prize in Physics. He made no attempt to profit from his discovery and he died, quite poor, in Munich in 1923.

26.4 Photoelectric Emission

Fig. 26.11 shows a freshly cleaned piece of zinc on the cap of a negatively charged electroscope. When ultraviolet radiation is directed onto the zinc the leaves are seen to collapse. Since the electroscope was negatively charged this means that electrons are being removed from the cap and leaves. This leads us to the conclusion that the ultra violet radiation causes electrons to be emitted from the zinc. The fact that the leaves will not collapse if the electroscope were positively charged to

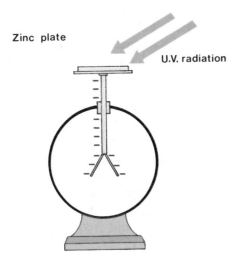

Fig. 26.10

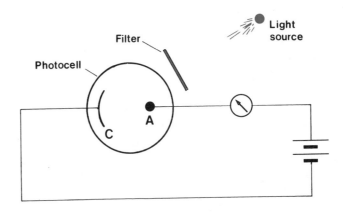

Fig. 26.11

start with supports this conclusion. The removal of electrons in this way is known as the **photoelectric effect**.

> **The photoelectric effect is the emission of electrons from the surface of a metal as a result of electromagnetic radiation falling on it.**

This phenomenon may be further investigated using the circuit shown in *Fig. 26.11*. PC is an evacuated glass bulb containing a semi-cylindrical cathode, C, and a wire anode, A. When light falls on the cathode electrons are emitted from it. PC is called a photoelectric cell. The anode is kept at a positive potential with respect to the cathode, so the electrons emitted from the cathode are attracted to the anode. From the anode, the electrons flow around the external circuit back to the cathode. The frequency of the light falling on the cathode may be controlled using suitable filters.

It is found that, below a certain frequency, called the **threshold frequency**, no current flows however bright the light may be. Further, it may be shown that, if the brightness, *i.e.* the **intensity**, of the light is kept constant, the current does not vary with changing frequency, *i.e.* the number of electrons emitted per second is independent of the frequency.

Above this minimum frequency, the magnitude of the current is proportional to the intensity of the light, *Fig. 26.12*. This relationship may be investigated using the circuit shown in *Fig. 26.11*. A given filter is used, thus keeping the frequency of the light falling on the cathode constant, while the brightness

of the light is varied by changing the distance between the cell and the lamp. Since the intensity of the light varies inversely with the square of the distance a graph of current against the inverse of the square of the distance should be a straight line through the origin.

Photoelectric cells are used in television cameras and in certain types of burglar alarm. *(See also The Light-Dependent Resistor, p. 298).*

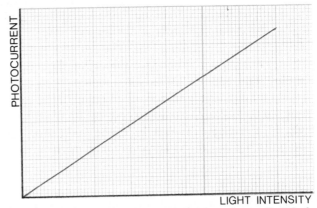

Fig. 26.12 *Variation of the current through a photocell with the intensity of the light falling on it*

The Quantum Theory

The photoelectric effect was discovered by the German physicist Heinrich Hertz (1857 — 1894) in 1887 and it soon became clear that it could not be explained in terms of the wave theory of electromagnetic radiation. Around the same time, the wave theory was proving inadequate in dealing with some other aspects of the nature of electromagnetic radiation, in particular the emission of radiation from hot surfaces. In 1900 another German physicist, Max Planck (1858 — 1947), proposed that radiation was emitted from a hot surface, not as a continuous stream of energy but as a series of "bundles" or "packets" of energy which he called **quanta** or **photons**. In other words, just as apparently continuous matter is made up of discrete particles called atoms, so radiant energy is emitted in discrete "particles" called quanta or photons. Planck further proposed that the amount of energy in a photon was proportional to the frequency of the wave. The constant of proportionality, *h*, has come to be known as **Planck's constant** and it is one of the most fundamental constants in physics. Thus, the energy of a photon is given by

$$E = hf$$

The accepted value of *h* is 6.6×10^{-34} J s. The fact that *h* is so small explains why we do not notice the particle nature of electromagnetic radiation in everyday life anymore than we notice the atomic nature of matter.

It is clear that two theories are required to explain the behaviour of electromagnetic radiation, including light. Neither theory, the Wave Theory or the Quantum Theory, is a complete description of all aspects of the nature of electromagnetic radiation. Neither theory is "wrong"; each is simply limited to particular characteristics of radiant energy.

Just as radiation is shown to have a particle nature, so particles have been shown to have a wave nature. For example, a beam of electrons exhibits diffraction and interference effects when passed through a very thin sheet of gold. In the electron microscope electrons are used instead of light. Since the wavelength of the electrons can be made very short much greater magnification can be achieved than with an optical microscope.

The Photoelectric Law

In 1905, Albert Einstein extended the ideas of the Quantum Theory to include the absorption of electromagnetic radiation, with particular reference to the photoelectric effect. He assumed that radiation was also absorbed in discrete quanta and that, when a quantum was absorbed by a metal, all of its energy was given to one electron. Thus, if the energy of the photon is greater than the energy required to remove an electron from the surface of the metal, photoelectric emission takes place. Since the energy of a photon is proportional to

the frequency of the wave this explains why no electrons are emitted by radiation below a certain frequency. The relationship between the frequency of the wave and the energy of an emitted electron (electrons emitted by photoelectric emission are sometimes called photoelectrons) is summarised in an equation known as **Einstein's Photoelectric Law:**

$$hf = \phi + \tfrac{1}{2}mv^2 \qquad \text{(i)}$$

where ϕ, called the **work function** of the metal, is the energy required to remove an electron from the surface, and $\tfrac{1}{2}mv^2$ is the maximum kinetic energy of the emitted electron. From *Eqn. (i)* the minimum frequency which will cause an electron to be emitted, *i.e.* the threshold frequency, f_0, is given by

$$hf_0 = \phi$$

Thus, *Eqn. (i)* may be written as

$$hf = hf_0 + \tfrac{1}{2}mv^2$$

In 1916 Millikan carried out an experiment in which he determined the maximum kinetic energy of the electrons emitted by a metal due to radiation of different frequencies. He found that a graph of frequency (*f*) against maximum kinetic energy (E_k) was a straight line as shown in *Fig. 26.13*. He also found that the intercept, f_0, on the frequency axis depended on the type of metal from which the electrons were being emitted. These results are exactly as predicted by Einstein's photoelectric equation and are therefore verification of the equation and of the theory on which it is based.

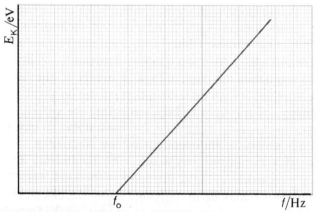

Fig. 26.13 Variation of kinetic energy of electrons with frequency of radiation

The relationship between the kinetic energy of the electrons and the frequency of the incident light may be investigated using the circuit shown in *Fig. 26.14*. Note that the anode is negative with respect to the cathode. The p.d. between the anode and cathode is gradually increased until the galvanometer just reads zero. The fact that there is no current flowing

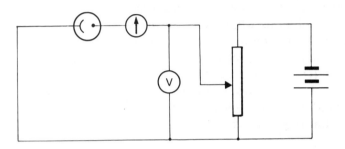

Fig. 26.14

indicates that the most energetic electrons are unable to cross from the cathode to the anode against the p.d. If the p.d. between the anode and cathode is V then the work done in bringing a charge e (the charge on an electron) from one to the other is eV. Since the most energetic electrons are just not able to reach the anode this means that the maximum kinetic energy of the electrons is equal to eV. Thus, V, the p.d. required to *just* stop the electrons reaching the anode, is proportional to the maximum kinetic energy of the electrons. A graph of V against f should therefore be a straight line similar to the graph of *Fig. 26.13*.

The intensity of the radiation falling on a metal is a measure of the amount of energy striking its surface per second. According to the Quantum Theory, each photon carries a definite amount of energy, the amount depending only on the frequency. It follows that the intensity is proportional to the number of photons arriving per second. Each photon causes one electron to be emitted, so the Quantum Theory predicts that the number of electrons emitted per second depends on

the intensity of the radiation. This agrees with experimental observation, as we noted earlier.

26.5 Spectra and Electron Energies

The photoelectric effect tells us that when a body absorbs radiation its electrons gain energy. It is logical to assume therefore that when a body emits radiation its electrons lose energy. Since, according to the Quantum Theory, radiation is emitted in discrete quanta, it follows that the energies of electrons in atoms are likely to be quantised also.

It is useful to imagine the electrons in an atom being arranged on a series of "shelves", one above the other, *Fig. 26.15*. An electron on the highest "shelf" has the greatest energy, one on the lowest "shelf" has the least energy. If an electron falls from one "shelf" to another it loses a certain, definite amount of energy — a quantum. These imaginary "shelves" are called **energy levels**.

Fig. 26.15 Energy level diagram

Electrons tend to occupy the lowest energy levels available. Thus, if an atom has two electrons (the neutral helium atom) they will normally be in the two lowest energy levels while the other energy levels will be empty. If energy is supplied to the atom (for example, by heating the gas) one of the electrons may be raised to a higher level. But this is a very unstable situation — like kicking a football onto the crossbar — and the electron immediately falls back to its original level, emitting its extra energy as a quantum of electromagnetic radiation.

We can now see how line emission spectra (*p. 140*) are produced. Electrons which have been "excited" to higher energy levels fall back, either directly or in a series of stages, to their original levels, emitting the excess energy as light of definite frequencies ($E = hf$, where h is Planck's constant) in the process. *Fig. 26.16* gives a picture of what might happen when an electron in a particular atom has been excited to a higher level, perhaps by the passage of an electric current or by heating.

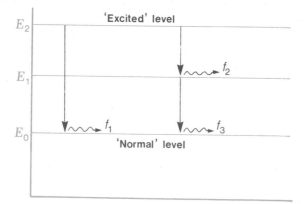

Fig. 26.16 When an electron falls to a lower level a photon is emitted

If the electron goes directly from the "excited" level to the "normal" level, the energy of the emitted photon will be

$$E = E_2 - E_0$$

and the frequency of the radiation emitted will be

$$f_1 = \frac{E_2 - E_0}{h}$$

If the electron stops first at the intermediate level before going on to the lowest level two photons will be emitted, one of energy $E_2 - E_1$ and one of energy $E_1 - E_0$. The corresponding frequencies would be given by

$$f_2 = \frac{E_2 - E_1}{h}$$

and

$$f_3 = \frac{E_1 - E_0}{h}$$

The emission spectrum of a gas consisting of this type of atom would have three bright lines corresponding to frequencies of

292

f_1, f_2 and f_3.

The formation of an absorption spectrum is the reverse of this process. When white light is passed through the gas, photons of frequencies f_1, f_2 and f_3 may be absorbed, raising electrons to higher energy levels. When the electrons fall back to the "normal" level they re-emit these photons. However, since these photons are emitted in random directions the majority will not be emitted in the same direction as that of the white light. So, when the spectrum of the transmitted light is examined it is found to contain dark lines corresponding to those photons which were absorbed.

When two or more atoms join together to form a molecule many more energy changes are possible. Thus, gases like oxygen (O_2) and hydrogen (H_2) give band spectra. In liquids and solids a large number of atoms are very close together and virtually all energy changes are possible, so a continuous spectrum is produced.

26.6 Conduction in Gases

Since gases do not normally contain free charge carriers they do not conduct electricity. For a gas to become conducting it is necessary for some of its molecules to be ionised. This can be achieved by radiation falling on the gas or by collisions between neutral molecules and ions or electrons which have been accelerated in an electric field.

Fig. 26.17 Discharge tube

Fig. 26.17 shows a sealed tube containing two metal electrodes and a gas at low pressure. If the current flowing through the tube is measured for different values of the applied p.d., a graph similar to that shown in *Fig. 26.18* is obtained. At low voltages a small current flows. This current is due to a small number of ions and electrons produced either deliberately by, for example, X-rays, or accidently by cosmic rays (high energy particles, mainly protons, reaching earth from space). Electrons may also be produced by photoelectric emission from the cathode.

In the electric field between the electrodes, the positive ions move towards the cathode while the negative ions and electrons move towards the anode. At low voltages, some of the ions

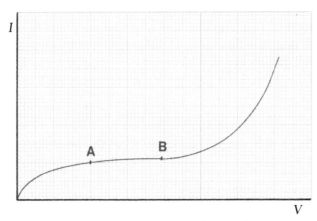

Fig. 26.18 Variation of current with applied p.d. for a gas

recombine before reaching the electrodes. As the p.d. between the electrodes is increased, more of the ions reach the electrodes and the current increases. At A (*Fig. 26.18*), all the ions produced in the tube are reaching the electrodes and the current becomes independent of the p.d. If the p.d. is increased further to B, the current starts to increase again. At this point the ions gain enough energy in the electric field to ionise other molecules by collision, thus producing more ions. These ions in turn may produce more ions and so an **avalanche** effect takes place. Beyond B the current rises very rapidly, if not limited by an external resistor connected in series with the tube.

A current through a gas is referred to as a **discharge**. Such currents consist of positive ions moving in one direction and negative ions moving in the opposite direction. *(Compare this with conduction in solids, p. 193, and conduction in electrolytes p.241.)* If the pressure is low enough or the voltage is high enough (*e.g.* about 10^2 Pa and 10^3 V) the gas emits electromagnetic radiation (usually light and ultra violet) as the current flows. Such a discharge is called a **glow discharge**. The frequency of the radiation (and hence the colour of the light) emitted depends on the nature of the gas *(see previous section)*. Thus neon gives a purplish light, sodium gives a yellow light, *etc.* In fluorescent lamps most of the radiation emitted is ultra violet. This is absorbed by a coating (a phosphor) on the inside of the tube and the energy is re-emitted as light.

SUMMARY

Thermionic emission is the emission of electrons from the surface of a heated metal. A beam of electrons, also called cathode rays, may be produced by accelerating the electrons through a specially shaped anode. Since cathode rays are (negatively) charged they may be deflected in electric and magnetic fields.

X-rays are produced when a beam of electrons is accelerated to very high speeds and allowed to strike a metal target. X-rays are electromagnetic radiation of very short wavelength ($\sim 10^{-10}$ m). They are used in medicine and in industry because of their ability to pass through materials which are opaque to light.

Photoelectric emission is the emission of electrons from the surface of a metal by incident radiation. The number of electrons emitted per second is proportional to the intensity of the incident radiation, while the maximum kinetic energy of the electrons depends on the frequency of the radiation.

The quantum theory holds that electromagnetic radiation is made up of discrete quanta or photons. The energy, E, of a quantum is related to the frequency, f, of the radiation by

$$E = hf$$

where h is Planck's constant. Einstein's photoelectric law states that

$$hf = \phi + \tfrac{1}{2}mv^2$$

where ϕ is the work function of the metal and $\tfrac{1}{2}mv^2$ is the maximum kinetic energy of the emitted electron.

In atoms, electrons can have only certain energies. When an electron falls from a higher energy level to a lower one the energy is emitted as a quantum of radiation; conversely, an electron will absorb a quantum of radiation if the energy of the quantum is exactly that needed to lift the electron to a higher energy level. Since only certain energy levels exist only certain frequencies of radiation are emitted or absorbed. The spectra of different elements thus reflect the electronic structure of their atoms and so can be used to identify the elements.

Gases do not normally conduct. If some of the gas molecules are ionised a current may flow. Such a current is called a discharge and it consists of positive ions moving in one direction and negative ions and electrons moving in the opposite direction. In gases at low pressure a glow discharge may be produced — electromagnetic radiation is emitted from the gas as the current flows through it.

Questions 26

DATA:
$$h = 6.6 \times 10^{-34} \text{ J s}$$
$$c = 3.0 \times 10^8 \text{ m s}^{-1}$$
$$e = 1.6 \times 10^{-19} \text{ C}$$
$$m_e = 9.1 \times 10^{-31} \text{ kg.}$$

Section I

1. What is meant by the specific charge of a particle?
   ....
   ....

2. Name the physicist who first determined the value of the charge carried by the electron........................
   ....

3. Give two properties of electrons.........................
   ....
   ....

4. What is meant by thermionic emission?................
   ....
   ....

5. Give two properties of cathode rays.....................
   ....
   ....

*6. In a cathode ray tube the..........................controls the speed of the electrons and hence the...............
   ....of the trace on the screen.

*7. What are the functions of the anodes in a CRT?
   ....

8. State two practical applications of the cathode ray tube.
   ....

9. What are X-rays?...
   ....

10. Why does the target in an X-ray tube become very hot?
    ....
    ....

11. Give three properties of X-rays...........................
    ....
    ....

12. Give two uses of X-rays.....................................
    ....
    ....

13. What is the photoelectric effect?.........................
    ....
    ....

14. What determines the current flowing through a photocell? ....
    ....
    ....

*15. In the equation $E = hf$, what does each of the symbols represent?...
    ....
    ....

*16. Write down an expression for Einstein's photoelectric law ....

....

....

*17. What is the relationship between the threshold frequency for a metal and its work function? ....

....

18. What is an emission spectrum? Name the three types of emission spectrum...

....

....

19. Give an example of an absorption spectrum............

....

20. The charge carriers in a gas are...........................

....

Section II

1. In the experiment to demonstrate photoelectric emission with the gold leaf electroscope the zinc must be freshly cleaned. Suggest a reason why this is necessary. Why would the leaves not collapse if (i) the zinc were covered with a piece of ordinary glass, (ii) the electroscope were charged positively initially? Explain.

*2. Calculate the energy of a photon of u.v. radiation, the wavelength of which is 3.0×10^{-7} m. If u.v. radiation of this wavelength falls on a metal which has a work function of 4.0×10^{-19} J, what is the maximum kinetic energy of the electrons emitted?

*3. A metal has a work function of 2.0 eV. What is the maximum wavelength of radiation which will remove electrons from it?

*4. Radiation of wavelength 2.5×10^{-7} m falls on a metal with a work function of 4.0 eV. Calculate (i) the maximum kinetic energy, (ii) the maximum speed, of the emitted electrons.

*5. When radiation of wavelength 1.6×10^{-7} m falls on a metal surface the maximum kinetic energy of the emitted photoelectrons is 3.2 eV. Calculate (i) the work function of the metal, (ii) the threshold frequency for the metal.

*6. *Fig. I* shows a graph of kinetic energy of emitted photoelectrons as a function of the frequency of the incident radiation. Use this graph to determine (i) a value for Planck's constant, (ii) the minimum voltage required to stop the electrons when the frequency of the incident radiation was 1.84×10^{15} Hz.

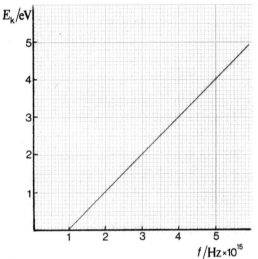

Fig. I.

7. What is the kinetic energy acquired by electrons in an X-ray tube when the p.d. across the tube is 100 kV? What is the speed of these electrons?

*8. In an X-ray tube operating at 50 kV all the energy of one electron appears as an X-ray photon. Calculate the frequency and wavelength of this photon.

*9. When the p.d. across an X-ray tube is 100 kV the current flowing from anode to cathode is 24 mA. If 0.5% of the electrical energy supplied is emitted as X-rays of wavelength 0.15 nm, and given that the energy of each photon comes from one electron only, calculate the number of photons, of this particular wavelength, which are emitted per second.

10. "X-ray production is the converse of the photoelectric effect." Discuss this statement with particular reference to the wavelengths/frequencies of the radiations and the energies/speeds of the electrons involved in each case.

27 Semiconductors I

27.1 Semiconductors

Previously *(Chapter 18)* we divided materials into two classes, *viz.* conductors and insulators, *i.e.* materials which allow charge to flow through them and materials which do not. We say that conductors have a low resistance and insulators a very high resistance. **Semiconductors** are materials which have resistivities between those of conductors and those of insulators, *Table 27.1*. The most common examples of semiconductors are silicon and germanium.

Material	Resistivity/ $\Omega\,m$
Copper	2×10^{-8}
Aluminium	3×10^{-8}
Carbon	3×10^{-5}
Germanium	6×10^{-1}
Silicon	2×10^{3}
Glass	10^{12}
Polystyrene	10^{15}

Table 27.1

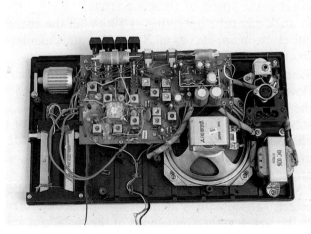

Fig. 27.1 Transistor radio is based on semiconductor materials

Conduction in semiconductors

In *Chapter 18* we noted that conduction in a metal is due to the movement of loosely held electrons. These are electrons which are only weakly attracted to the nuclei of the atoms and so will move under the influence of an applied electric field. In the case of metals approximately one electron per atom is free to move in this way. In a semiconductor material, *e.g.* silicon, at a temperature of 0 K, there are no free electrons. If the temperature is increased a small proportion of the electrons gain enough energy to break free of the nuclei and become available for conduction. In silicon, at room temperature, there is approximately one free electron for every 10^{10} atoms.

A silicon atom has four electrons in its outer "shell". These are the electrons which are involved in forming bonds between the silicon atom and other atoms. In a crystal of silicon each atom is bonded to four other silicon atoms, each bond consisting of two shared electrons, *Fig. 27.2*. At temperatures above absolute zero some of these electrons leave the bond and become free to move through the crystal.

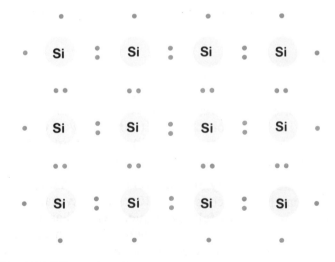

Fig. 27.2 Silicon crystal

When a silicon atom loses one of its four outer electrons the "space" left behind behaves as if it had a positive charge (it has a strong attraction for a negative electron), *Fig. 27.3*. Such spaces are therefore called **positive holes**. When a piece of silicon is placed in an electric field the free electrons drift towards to the positive, just as they do in a metal. At the same time, a bound electron near a positive hole may gain enough

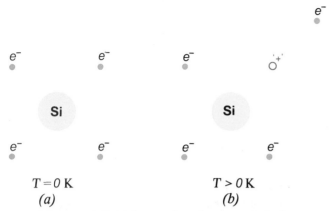

Fig. 27.3 Silicon at 0 K (a) has no free electrons. At higher temperatures (b) some electrons are freed, leaving positive holes

energy to move away from its own atom and into the positive hole, thus creating a positive hole at its original position. As this process is repeated the positive hole, in effect, moves towards the negative, *Fig. 27.4.* (It is like a row of ten seats with nine people sitting on them, leaving an empty seat at one end. If everyone moves down one place in turn, the empty seat effectively "moves" up the row.) Thus, in semiconductors,

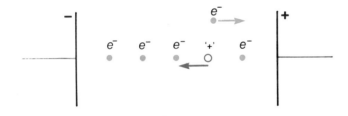

Fig. 27.4 In an electric field electrons move one way, holes move the opposite way

conduction is due to both positive holes and negative electrons, moving in opposite directions. In a pure semiconductor there are equal numbers of holes and electrons and the process is then called **intrinsic** conduction. The higher the temperature of a piece of semiconductor material the more free electrons and holes are available for conduction. In other words the resistance of a semiconductor decreases with increasing temperature, *Fig. 27.5.* Note that the change in resistance with temperature is not linear.

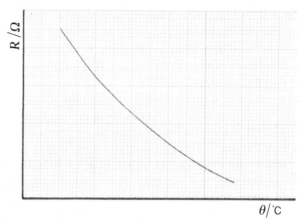

Fig. 27.5 Variation of resistance of a semiconductor with temperature

Doping

The conductivity of a pure semiconductor can be greatly increased by the addition of a very small quantity of another element, *e.g.* boron or phosphorus. This process is called **doping** and the element added is referred to as the impurity element.

Phosphorus has five electrons in its outer "shell". When a phosphorus atom is added to a crystal of silicon, there is an extra electron which does not fit into the crystal structure, *Fig. 27.6.* These extra electrons are available for conduction and so the conductivity is increased, *i.e.* the resistance of the piece of material is reduced. Only a very small quantity of phosphorus is required to produce a substantial reduction in the resistance. In fact, the addition of one atom of phosphorus for every 10^7 atoms of silicon reduces the resistance by a factor of about 10^3.

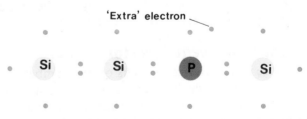

Fig. 27.6 Doping silicon with phosphorus increases the number of free electrons

In a piece of silicon doped with phosphorus there are more electrons than holes. In fact, using the sample values from the previous paragraph, there are approximately a thousand times more electrons than holes. So, in this type of material conduction is due mainly to negative electrons. We say that electrons are the **majority charge carriers**, while the positive holes are the **minority charge carriers**. A semiconductor in which electrons are the majority carriers is called a negative-type, or **n-type**, semiconductor.

The resistance of a piece of silicon can also be reduced by doping it with boron. Boron has only three electrons in its outer "shell". The addition of a boron atom to a silicon crystal thus introduces a "space" into the crystal which would normally be occupied by an electron. In other words, the addition of boron to a silicon crystal increases the number of holes, *Fig. 27.7*. Once again, there are more charge carriers available for conduction so the resistance of the piece of semiconductor is reduced. In this case, holes are the majority charge carriers and electrons are the minority carriers. A semiconductor in which positive holes are the majority carriers is called a positive-type, or **p-type**, semiconductor.

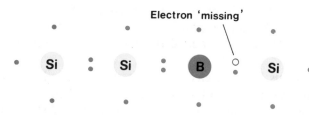

Fig. 27.7 *Doping silicon with boron increases the number of positive holes*

It is important to realise that doping does not introduce an overall charge to the crystal. The impurity atoms have different numbers of electrons in their outer shells compared with the silicon atoms but they still have equal numbers of protons and electrons, *i.e.* they are electrically neutral. Since conduction in a doped semiconductor is due mainly to the impurity elements the process is referred to as **extrinsic** conduction.

The Light Dependent Resistor (LDR)

The resistance of any device depends on the number of free electrons and/or holes which are available for conduction. We learned in the last section that the number of free charge carriers in a semiconductor may be increased by supplying energy to it. One way of doing this is to direct a beam of light, or other form of electromagnetic radiation, on to it.

An l.d.r. consists of a thin layer of semiconductor, usually cadmium sulphide (CdS), to which connection is made with

Fig. 27.8 *Electrodes in a CdS cell*

suitably shaped electrodes, *Fig. 27.8*. The electrodes are made in this way to reduce the total resistance of the resistor. (This type of resistor is sometimes referred to as a CdS *cell*.) If the resistor is connected to an ohmmeter, *Fig. 27.9*, and the brightness (intensity) of the light falling on it is increased by bringing the lamp nearer to it, the resistance is seen to fall from several megohms to less than 100 Ω.

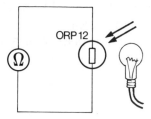

Fig. 27.9

Fig. 27.10 *CdS cell*

CdS cells may be used, in conjunction with transistors and/or relays (*see p. 254*), to switch on street lights, security lights, *etc.*, in the evening and switch them off again in the morning. They are also used in meters which measure light intensity, *e.g.* in a camera.

The Semiconductor Thermistor

We noted earlier that the resistance of a semiconductor falls with increasing temperature. A semiconductor thermistor is a device whose resistance falls very rapidly with increasing temperature. Such thermistors are usually made of a mixture of the oxides of iron, nickel and cobalt and may be used as temperature-controlled switches (*see p. 312*) and as thermometers (*see p. 229*). Thermistors are sometimes referred to as temperature dependent resistors. (The resistance of all resistors depends on their temperature. Thermistors differ in that their resistance is very sensitive to changes in temperature.)

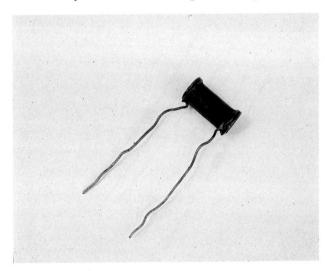

Fig. 27.11 Thermistor

Fig. 27.12 shows part of a circuit in which a thermistor is being used to control the temperature of a piece of electrical equipment. If the temperature of the equipment rises the

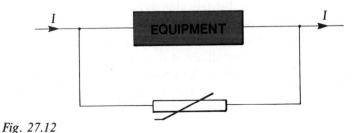

Fig. 27.12

resistance of the thermistor, which is placed close to the equipment, falls. This allows more current to flow through the thermistor, so less current flows through the equipment and its temperature falls. (This assumes that the total current in the circuit is kept constant. It is possible to construct a circuit in which a constant current will be maintained.) Thus, by a suitable choice of thermistor, the temperature of the equipment may be maintained at any desired level.

27.2 The P-N Junction

A p-n junction is a junction between a piece of p-type semiconductor and a piece of n-type. In practice, it is made by taking a single crystal of silicon (or germanium) and, by doping with the appropriate elements, making one part of it a p-type semiconductor and the remainder an n-type.

When a piece of p-type material is brought in contact with a piece of n-type material electrons near the junction move into the p-type material where they combine with positive holes. At the same time holes move into the n-type material where they combine with free electrons. As a result, a layer is formed on either side of the junction in which there are virtually no free electrons or holes. This region is called the **depletion layer**. Since there are, for most practical purposes, no free charge carriers in this region it acts as an insulator between the p-type and the n-type material. Since there has been a transfer of electrons to the p-type material this side of the junction now has a net negative charge. Similarly, due to the movement of holes into the n-type this side of the junction has a net positive charge, *Fig. 27.13*. There is therefore a potential difference across the junction. This p.d. is referred to as the **junction voltage**. For silicon its value is about 0.6 V while for germanium it is about 0.2 V.

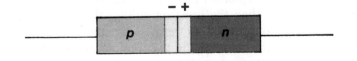

Fig. 27.13 P-n junction

The Junction Diode

A junction diode consists of a piece of p-type semiconductor and a piece of n-type, joined together as described in the previous section. If a diode is connected in a circuit as shown in *Fig. 27.14(a)*, so that the n-type material is positive with respect to the p-type, the lamp does not light, *i.e.* no current flows through the diode. This is explained as follows. When the n-type region is made positive the free electrons in it are attracted to the positive of the battery and are therefore drawn back from the junction. At the same time the positive holes in the p-type region are attracted towards the negative of the battery and away from the junction. The result of this movement of charge carriers is that the depletion layer is widened. Since the depletion layer is effectively an insulator no current flows through the device. When a diode is connected in a circuit in this way, *i.e.* with the n-type positive with respect to the p-type, it is said to be **reverse biased**.

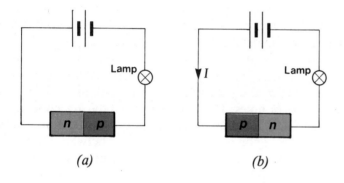

(a)   (b)

Fig. 27.14
(a) Lamp does not light   (b) Lamp lights

Now consider what happens if the connections to the diode are reversed, *Fig. 27.14(b)*, so that the p-type is now positive with respect to the n-type. The free electrons in the n-type are now repelled from the negative terminal of the battery and driven into the depletion layer. At the same time holes are driven into the depletion layer from the other side. As a result the width of the depletion layer is reduced. If the applied voltage is greater than the junction voltage the width of the depletion layer is reduced to zero. Since there is now no insulating material between the two regions of the diode a current can flow through it. When a diode is connected in a circuit in this way, *i.e.* with the p-type positive with respect to the n-type, it is said to be **forward biased**.

300

A diode is thus a device which allows current to flow in one direction but not in the other. *Fig. 27.15* shows a diagram of a diode *(a)*, with its symbol *(b)*; *(c)* shows some common diodes.

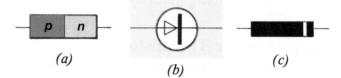

(a)   (b)   (c)

Fig. 27.15 Semiconductor diode. (a) Structure; (b) Symbol; (c) Common shape with band to indicate cathode

The coloured band indicates the n-type end of the diode, *i.e.* for forward bias this end should be connected to the negative of the battery or power supply. It should be noted that *(a)* is a schematic diagram. In practice, neither the crystal itself nor the junction between the p-type and n-type regions is regular in shape.

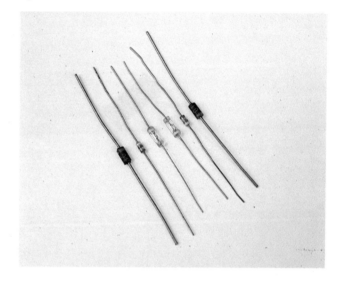

Fig. 27.16 Semiconductor diodes

If the p.d. applied to a diode is increased above the junction voltage the current through the device increases rapidly with increasing p.d. It is usually desirable to have a resistor in series with the diode to ensure that the current through it cannot exceed the maximum rated value. The variation of current with applied p.d. for a silicon diode is shown in *Fig. 27.17*. This graph is called the **characteristic curve** of the diode. Note that, when the diode is forward biased, virtually no current flows until the applied p.d. exceeds about 0.6 V. When the diode

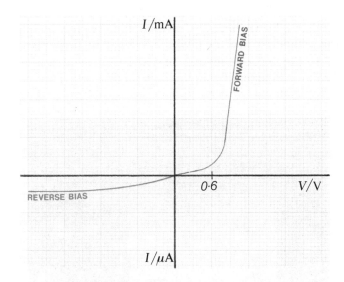

Fig. 27.17 Variation of current with p.d. for a diode

is reverse biased a very small current flows (typically less than 1 μA for a silicon diode at room temperature). The reason for this may be explained as follows. At 0 K there are no free charge carriers in the depletion layer. At temperatures above 0 K some electrons may gain enough energy to break free of their atoms and become available for conduction. It is these electrons, and the holes which they leave behind, which carry the current when the diode is reverse biased. The higher the temperature the more electrons will gain enough energy to become available for conduction and so the larger will be the reverse current. For small values of the applied p.d. the reverse current is essentially independent of the voltage. However, if the applied p.d. is increased beyond a certain value the depletion layer will break down and a very large current will flow.

EXPERIMENT 27.1
To Plot the Characteristic Curve of a Diode

Apparatus: Diode, battery or low voltage power supply, voltmeter, milliammeter, microammeter, rheostat, 100 Ω resistor.

Procedure:

1. Set up the circuit as shown in *Fig. 27.18*. Adjust the rheostat to give a p.d. of 0.1 V across the diode. Record the current flowing.

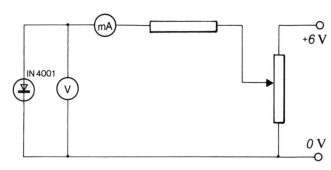

Fig. 27.18

2. Using the rheostat, increase the p.d. in steps of 0.1 V, noting the current at each step, until the current reaches about 50 mA. Plot a graph of I against V.

RESULTS

I/A	V/V

4. Replace the milliammeter with the microammeter and reverse the connections to the diode. Change the position of the voltmeter so that it measures the p.d. across both the diode and the microammeter.

5. Repeat the procedure given above and again plot the values of I and V on the graph obtained in the first part of the experiment.

RESULTS

I/A	V/V

Questions

1. What is the purpose of the 100 Ω resistor?
2. Why is it necessary to change the position of the voltmeter for the second part of the experiment? How might this be avoided?

The Rectifier

One common application of the diode is the rectifier — a device which converts alternating current to direct current. *Fig. 27.19* shows a simple half-wave rectifier. The input and output

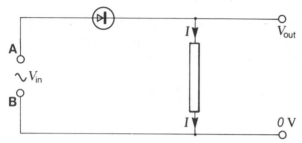

Fig. 27.19 Half-wave rectifier

voltages are shown in *Fig. 27.20(a)* and *Fig. 27.20(b)*, respectively. On the first half of the cycle, when A is positive

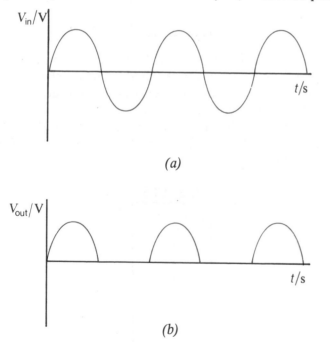

(a)

(b)

Fig. 27.20 Half-wave rectifier: (a) Input voltage;(b) Output voltage

with respect to B, the diode is forward biased and so it conducts and a current flows through the resistor in the direction shown. On the second half of the cycle, A is negative with respect to B and so the diode is reverse biased and therefore it cannot conduct — no current flows through the resistor. Thus current flows through the resistor in one direction only.

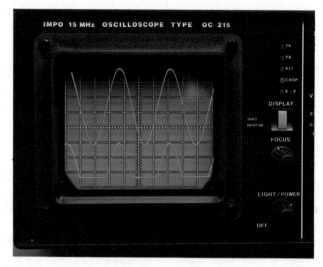

Fig. 27.21 Using an oscilloscope to display half-wave rectification

While the output from a half-wave rectifier is direct in the sense that the direction in which the current flows does not change, it is far from constant. Indeed, for 50% of the time the output is zero! A more satisfactory arrangement is the full-wave rectifier, *Fig. 27.22*. On the first half of the cycle, when

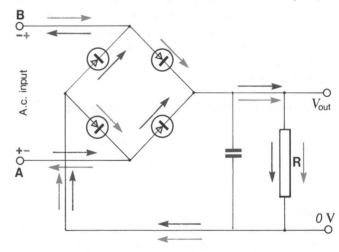

Fig. 27.22 Full-wave rectifier

A is positive with respect to B, current flows along the path indicated by the red arrows. When A is negative with respect to B the current flows along the path shown by the blue arrows. Note that, in both cases, the direction in which the current flows through the resistor, R, is the same. The input and output of a full-wave rectifier are shown in the photograph of *Fig. 27.23(a)*.

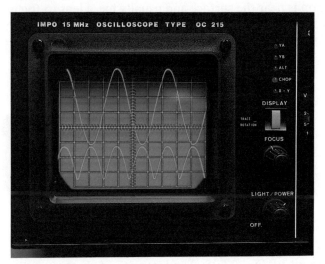

Fig. 27.23(a) Full-wave rectification

To obtain a more constant output from the rectifier a smoothing circuit is required. In its simplest form this consists of a capacitor connected in parallel with the output, *Fig. 27.22*.

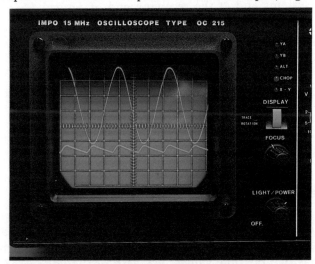

Fig. 27.23(b) Full-wave rectification with smoothing

The smoothed output is shown in *Fig. 27.23(b)*. The output of a half-wave rectifier may be similarly smoothed by connecting a capacitor in parallel with the output. The larger the capacitance of the capacitor the less variation there is in the output voltage.

The Light-Emitting Diode (l.e.d.)

We noted earlier that, when a p-n junction is forward biased, electrons and holes combine at the junction. As a result, the electrons lose energy. This energy is converted to internal energy in the junction and so the temperature of the junction rises. However, in certain semiconducting materials, the energy lost by an electron may be converted to a photon, *i.e.* light is produced at the junction. If the diode is constructed so that the junction is near to the surface this light may escape and the diode is then known as a **light-emitting diode (l.e.d.)**. The most commonly used material for l.e.d.'s is gallium arsenide phosphide, the colour of the light emitted depending on the exact composition of the diode. Red, green and yellow l.e.d.'s are currently available. *Fig. 27.24* shows some typical l.e.d.'s together with their symbol.

 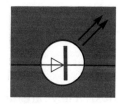

Fig. 27.24   Light emitting diodes and symbol

L.e.d.'s are widely used in the displays of calculators, digital clocks, *etc.* Each digit is formed from a seven segment display, arranged as shown in *Fig. 27.25*. Each segment in the display is a bar-shaped l.e.d. For example, the number 7 is formed by lighting the top diode and the two diodes on the right hand side. Since the maximum current through the diode must not exceed a few milliamps, each diode has a resistor connected in series with it. *Fig. 27.26* shows a typical circuit for a single l.e.d.

Light-emitting diodes are also used as indicators, showing whether a particular circuit or component is switched on or not. As indicators, l.e.d.'s have several **advantages** over

Fig. 27.25 Seven segment display

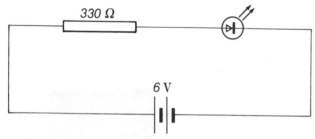

Fig. 27.26

filament bulbs. They are smaller, require much smaller currents, are more reliable and are capable of operating at much higher speeds. L.e.d.'s are also used for transmitting information along optical fibres (*p. 30*).

In many instruments, l.e.d.'s are being replaced by liquid crystal displays (l.c.d.'s). These have the advantage that they draw even smaller currents than l.e.d.'s. However, they are more expensive and have to be illuminated by a separate light source in the dark.

The Photodiode

In a reverse biased p-n junction a negligible current flows due to the virtual absence of free charge carriers, as explained on *p. 301*. We noted earlier that shining light on a semiconductor increases the number of charge carriers available for conduction

(*cf. l.d.r., p. 298*). If a reverse biased p-n junction is illuminated with light, free electron-hole pairs are created in the depletion layer and so a small current will flow. Since each photon of incident light creates one electron-hole pair the size of the current flowing is proportional to the intensity of the light falling on the junction. A junction diode intended for use in this way is called a **photodiode**. It is constructed so that the junction is near the surface and usually has a lens to focus light onto the junction. A photodiode, with its symbol, is shown in *Fig. 27.27*.

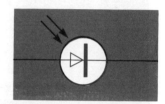

Fig. 27.27 Photodiode and symbol

The behaviour of a photodiode may be investigated using the circuit of *Fig. 27.28*. The intensity of the light falling on the diode is increased by moving the lamp closer to it. Note that the diode is reverse biased and that the current is of the order of a few microamps.

Photodiodes are used in photometers, high speed counters, alarm systems and at the receiving end of optical fibres.

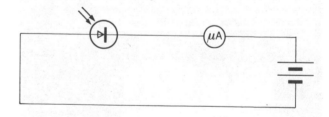

Fig. 27.28   Photodiode circuit — note polarity of battery

Logic Gates

Logic gates are circuits used in calculators, computers, *etc.* Their importance lies in the fact that their output voltages depend in a particular way on their input voltage(s). Each of these has two possible values, *viz.* zero or the voltage of the battery or power supply. (In practice, the input and output voltages may not be exactly equal to these values.) When a voltage is at, or near, zero it is said to be "low" and is represented by the number 0. When a voltage is at, or near, the supply voltage is said to be "high" and is represented by the number 1.

There is a number of different types of gate. Here we shall consider only two — the **AND gate** and the **OR gate**. In practice these circuits would usually be constructed using transistors and incorporated into integrated circuits (*see p. 316*). We shall consider only the diode equivalents.

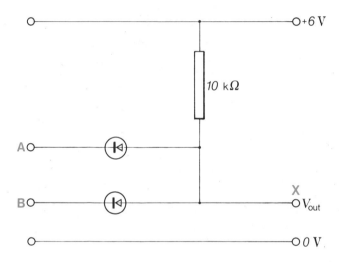

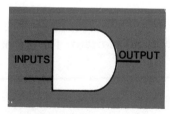

Fig. 27.29 AND gate and symbol

Fig. 27.29 shows an AND gate with its symbol. The gate has two inputs, A and B. If A and B are both connected to the +6 V line no current can flow through the resistor, so the p.d. across the resistor is zero. Therefore the potential at X is also 6 V, *i.e.* the output is 6 V, the same as the supply voltage. In logic terms we say that when both inputs are "high" the output is also "high". If B is now connected to the 0 V line a current flows through the resistor since the diode is now forward biased. In practice the resistance of the resistor is very

much greater than the resistance of the diode, so the p.d. across the resistor is almost 6 V and so the potential at X is almost 0 V, *i.e.* the output is "low". So, when A is "high" and B is "low" the output is "low". Similar reasoning shows that when A is "low" and B is "high" the output is again "low". Also, if A and B are both "low" the output is "low".

The relationships between the input and output voltages are usually summarised in the form of a **truth table**, *Fig. 27.30*. From this we see that the output is "high" only if both A **and** B are "high" — hence the name, AND gate.

A	B	Output
1	1	1
1	0	0
0	1	0
0	0	0

Fig. 27.30 Truth table for an AND gate

The OR gate, with its symbol, is shown in *Fig. 27.31*. If either A or B is connected to the +6 V line the corresponding diode is forward biased and so a current flows through the resistor.

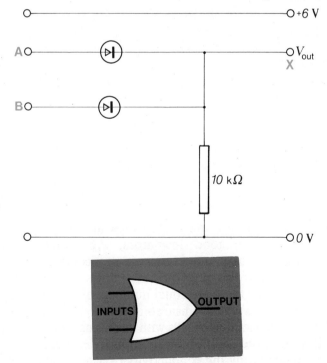

Fig. 27.31 OR gate and symbol

Since the p.d. across the diode is small compared with the p.d. across the resistor, the potential at X is almost 6 V, *i.e.* the output is "high". So, if A or B or both are "high" the output is "high". If A and B are both connected to the 0 V line no current can flow through the resistor and so the p.d. across it is zero, *i.e.* the output is "low". Again, these relationships are summarised in a truth table, *Fig. 27.32*.

A	B	Output
1	1	1
1	0	1
0	1	1
0	0	0

Fig. 27.32 Truth table for an OR gate

EXPERIMENT 27.2
To Establish Truth Tables for AND and OR Gates

Apparatus: 2 diodes, 10 kΩ resistor, voltmeter, battery or low voltage power supply.

Procedure:
1. Set up the circuit shown in *Fig. 27.29*.
2. Connect points A and B to the positive of the battery/power supply. Note the reading on the voltmeter. Record the appropriate logic level, 1 or 0, in the truth table.
3. Disconnect B and connect it to the negative of the battery/power supply. Record the logic level.
4. Reverse the connections of A and B and record the logic level.
5. Connect both A and B to the negative of the battery/power supply and record the logic level.

RESULTS
Truth Table for AND gate

A	B	Output

6. Set up the circuit shown in *Fig. 27.31* and repeat the procedure given in Steps 2 to 5 above.

RESULTS
Truth Table for OR gate

A	B	Output

Questions

1. Explain why the reading on the voltmeter may not be exactly equal to zero or the supply voltage, as appropriate. What type of voltmeter would be most suitable for this experiment?
2. Is the value of the resistor used in this experiment critical? Explain.
3. Suggest how the circuit might be modified so that a l.e.d. might be used as an indicator for the output.

SUMMARY

Semiconductors are materials which are neither good conductors nor good insulators, *e.g.* silicon and germanium. The charge carriers in semiconductors are negative electrons and positive holes. In a pure semiconductor there are equal numbers of each and the process is known as intrinsic conduction. The resistance of a pure semiconductor falls with increasing temperature.

Doping is the addition of an impurity element with a view to increasing the conductivity of a semiconductor.

In a p-type semiconductor the impurity element increases the number of holes — holes are now the majority charge carriers. In an n-type semiconductor the impurity element increases the number of electrons — electrons are now the majority carriers. A doped semiconductor is known as an extrinsic semiconductor.

A light dependent resistor is a device, usually made of cadmium sulphide (CdS), whose resistance decreases as the brightness of the light falling on it increases. CdS cells are used in light-operated switches and in light meters.

A semiconductor thermistor is a device whose resistance falls very rapidly with increasing temperature. Thermistors are used in temperature controlled switches, as thermostats and as thermometers.

A junction diode consists of a piece of n-type material joined to a piece of p-type material. When the p-type is made positive with respect to the n-type — forward bias — current can flow easily. When the diode is reverse biased current cannot flow. Diodes are used in rectifiers —devices which convert a.c. to d.c.

A light-emitting diode (l.e.d.) is a diode which emits light when it is forward biased. L.e.d.'s are used in digital displays, as on/off indicators and as transmitters with optical fibres.

The photodiode is a reverse biased diode which conducts when light falls on the p-n junction. Photodiodes are used as counters, in alarm systems and at the receiving end of optical fibres.

Logic gates are circuits whose output voltage depends on their input voltage(s). The output of an AND gate is high only if both inputs are high. The output of an OR gate is high if either, or both, input(s) is/are high.

Questions 27

Section I

1. What is a semiconductor?..

2. The charge carriers in a semiconductor are...and...................................

3. An .... semiconductor has equal numbers of both types of charge carrier.

4. Doping silicon with phosphorus increases the number of ...., which are then called the................. charge carriers.

5. The .... carriers in silicon doped with .... are positive holes.

6. As the .... of the light falling on a CdS cell increases its .... decreases.

7. Sketch a graph showing how the resistance of a semiconductor thermistor varies with temperature.

8. When a p-n junction is reverse biased the width of the .... increases.

9. You set up a circuit consisting of a diode and milliammeter in series with a variable, low-voltage power supply. You switch on and the milliammeter shows no current. Suggest a reason...

10. What is the function of a rectifier?...

*11. Give two advantages of using l.e.d.'s as indicators. ....

*12. Give two other uses of l.e.d.'s...

*13. A photodiode is normally .... biased.

*14. The output of an AND gate is high when...

*15. The output of an OR gate is low when..

Section II

1. Describe how you would investigate how the resistance of a CdS cell varies with the intensity of the light falling on it. Sketch the graph you would expect to obtain from this experiment.

2. Describe how a depletion layer is formed. Explain how reverse biasing the junction increases the width of the depletion layer.

3. A reverse biased p-n junction may be used as a variable capacitor. Explain how this is so and how the capacitance of such a capacitor could be varied. (Such diodes are called varicap diodes and they are used to tune TV and v.h.f. radio sets.)

4. Consider the circuit shown in *Fig. I*, then sketch a graph to show how the p.d. across the resistor R varies with time.

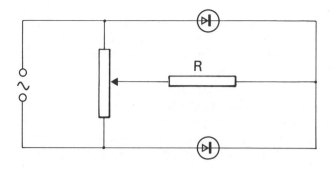

Fig. I

*5. In the AND gate circuit of *Fig. 27.29* a voltmeter is used to measure the output voltage. If the internal resistance of the voltmeter is 50 kΩ what will be the approximate reading on it when (i) A and B are high, (ii) A and B are low?

28 Semiconductors II

28.1 Transistors

A transistor is a three terminal device in which the current flowing between one pair of terminals is controlled by the p.d. between another pair. There are two main types of transistor, *viz.* the **bi-polar transistor** and the unipolar, or field effect, transistor. We shall consider firstly the structure and operation of the **field effect transistor (FET)**.

Fig. 28.1 Selection of transistors

The Field Effect Transistor

The most common type of FET is the junction-gate FET (JUGFET). *Fig. 28.2* shows a schematic diagram of a JUGFET, together with its symbol. The three terminals of the transistor are known as the **source** (S), **gate** (G) and **drain** (D), respectively. The source is connected to the drain by a channel of n-type material laid on a base, or substrate, of p-type material. The gate is connected to a piece of p-type material.

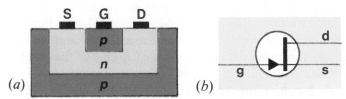

Fig. 28.2 JUGFET (a) Structure, (b) Symbol

The operation of an FET may be investigated using the circuit shown in *Fig. 28.3*[*].

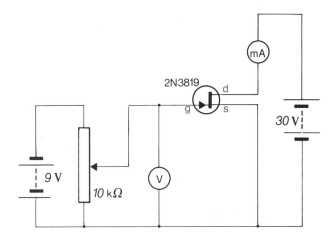

Fig. 28.3

The gate is gradually made more negative with respect to the source by adjusting the rheostat, while the p.d. between the source and the drain is kept constant. For a series of values of the gate-source voltage, V_{GS}, the current flowing from drain to source (conventional current flows from positive to negative) is noted. This current is called the drain current, I_D. From the resulting graph, *Fig. 28.4*, we see that, as the gate becomes more negative with respect to the source, the drain current decreases, reaching zero when V_{GS} is around —5 V.

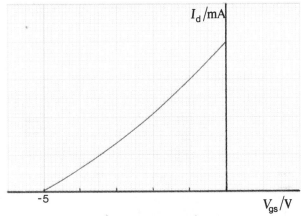

Fig. 28.4 Variation of drain current with gate voltage for fixed source voltage

If the gate-source voltage is kept constant while the drain-source voltage, V_{DS}, is varied it is found that the drain current is largely independent of the drain-source voltage for values of V_{DS} above a few volts, *Fig. 28.5*.

[*]Due to the wide spread of component values, the values given in circuit diagrams in this chapter should be taken as a guide only.

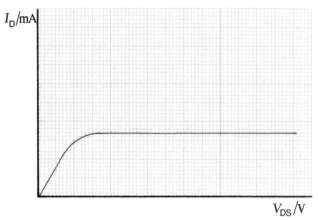

Fig. 28.5 Variation of drain current with source voltage for fixed gate voltage

To understand how a JUGFET works consider the diagram shown in *Fig. 28.6*. This is a simplified version of the circuit shown in *Fig. 28.3*. In the arrangement shown in the diagram the source is common to both circuits. This is the most common arrangement although other arrangements are possible. Since the n-type channel is in contact with p-type material a depletion layer exists along the junction as indicated by the yellow area in *Fig. 28.6*. The gate is negative with respect to the channel, *i.e.* the junction is reverse biased. The extent of the depletion layer depends on the size of the voltage applied to the gate.

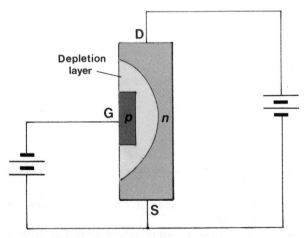

Fig. 28.6 Simple bias circuits for JUGFET (Substrate not shown)

As a result of the p.d. between the source and the drain, electrons travel up through the channel from the source to the drain. The wider the depletion layer the narrower the channel and so the more difficult it is for electrons to pass through,

i.e. the smaller the current flowing. Since the width of the depletion layer depends on the gate voltage it follows that the size of the current flowing in the channel also depends on the gate voltage — the more negative the gate voltage the smaller the current flowing.

In the JUGFET described in the previous paragraphs conduction takes place in a channel of n-type material, *i.e.* the charge is carried by electrons. For this reason this type of FET is called an **n-channel** device. P-channel devices are also available. In these, conduction takes place in a channel of p-type material, *i.e.* the charge carriers are positive holes. Obviously, the bias voltages for p-channel devices must be applied with the opposite polarity to those for n-channel devices.

The Bi-Polar Transistor

A bi-polar transistor consists of two junction diodes joined back-to-back. *Fig. 28.7* shows a schematic diagram of a bi-polar transistor, together with its symbol. In practice, the three layers of semiconducting material are formed on the same crystal. The middle layer, called the **base** (b), is extremely thin ($\sim 10^{-6}$ m) and is only lightly doped. The outer layers are much thicker and are called the **emmiter** (e) and the **collector** (c) respectively. The collector is usually larger than the emitter.

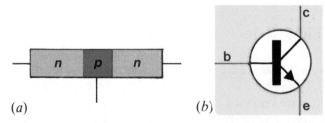

Fig. 28.7 Bi-polar transistor (a) Structure, (b) Symbol

The operation of a bi-polar transistor may be investigated using the circuit shown in *Fig. 28.8*. The current to the base — the base current, I_b — is gradually increased by adjusting the rheostat, while the p.d. between the collector and emitter is kept constant. The resulting values of I_b and of the current flowing into the collector — the collector current, I_c — are noted. A graph is then plotted of I_c against I_b, *Fig. 28.9*. From the graph, it is clear that, over the range of values shown, the collector current is proportional to the base current, *i.e.*

$$I_c \propto I_b$$

$$\Rightarrow \boxed{I_c = kI_b}$$

where k is a constant. (k is normally written as h_{FE} and is called the d.c. current gain.)

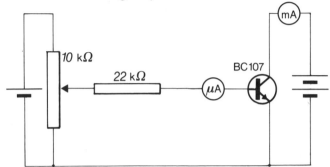

Fig. 28.8

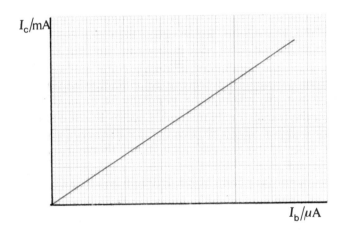

Fig. 28.9 *Variation of collector current with base current for fixed collector voltage*

For the transistor used in this experiment the collector current is approximately 100 times the base current, *i.e.* $k = 100$. Transistors are available with values of k ranging from about 15 to several thousand. Thus, for example, in a typical medium power transistor, a base current of 0.1 mA might give a collector current of 150 mA. It is for this reason that a bi-polar transistor may be referred to as a current amplifier (*p.313*).

By varying the collector-emitter voltage, V_{ce}, while keeping the base current constant it may be shown that the collector current is almost independent of the voltage for values above about 0.6 V, *Fig. 28.10*. Thus, the collector current depends primarily on the base current. A transistor may therefore be thought of as a current-controlled switch — the base current switches on the collector current (*see p.312*).

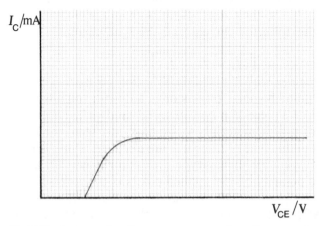

Fig. 28.10 *Variation of collector current with collector voltage for fixed base current*

To understand how a bi-polar transistor works consider the diagram of *Fig. 28.11*, which is a simplified version of the circuit shown in *Fig. 28.8* with the transistor in diagrammatic form. The base is positive with respect to the emitter, *i.e.* the base-emitter junction is forward biased, so electrons can flow easily from the emitter into the base. Since the base is so thin and only lightly doped very few of the electrons combine with holes in the base. Most of them ($\sim 99\%$) pass on into the collector which is positive with respect to the emitter. From the collector the electrons flow through the external circuit back

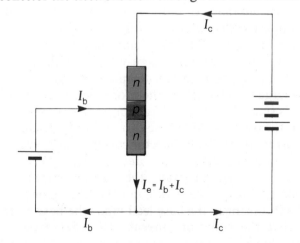

Fig. 28.11

to the emitter. These electrons make up the collector current, I_c. The remaining 1% of the electrons make up the base current, I_b. Thus, the collector current is approximately 100 times the base current. It should be clear that the emitter current, I_e, *i.e.* the current flowing from the emitter, is made up of the collector current and the base current, *i.e.*

$$I_e = I_c + I_b$$

This type of bi-polar transistor is an n-p-n transistor, *i.e.* it consists of a piece of p-type material sandwiched between two pieces of n-type material. As we have just seen, the majority carriers in this type of transistor are electrons. P-n-p transistors are also available. In this type, the majority carriers are holes and the polarity of the applied p.d.'s is the reverse of those for the n-p-n type. The symbol for the p-n-p transistor is the same as that for n-p-n, but with the arrow reversed.

28.2 Transistor Circuits

We have already seen some basic transistor circuits. We shall now look at some applications of these, and other, circuits. We shall consider only circuits based on bi-polar transistors. However, similar circuits could be constructed using field effect transistors.

The Transistor as a Switch

As we have already seen, the collector current is controlled by the base current in a bi-polar transistor. Thus, a change in the base current may be used to switch on, or off, a device connected in the collector circuit. *Fig. 28.12* shows a simple circuit for a fire alarm. TH3 is a semiconductor thermistor, *i.e.* a resistor whose resistance falls very rapidly with increasing temperature (*p.299*). At normal temperature, the resistance of TH3 is large ($\sim 400\ \Omega$) and therefore the current flowing through it is small. Since only a fraction of this current enters the base of the transistor, the base current is very small. Since the base current is small the collector current is also small and so the lamp in the collector circuit does not light. When TH3 becomes warm its resistance falls, the base current increases, so the collector current increases and the lamp lights.

The 22 kΩ resistor in the base lead limits the base current, ensuring that the collector current can never exceed the maximum allowed for the particular transistor (100 mA for the BC107). The 47 Ω resistor, R, provides an alternative path

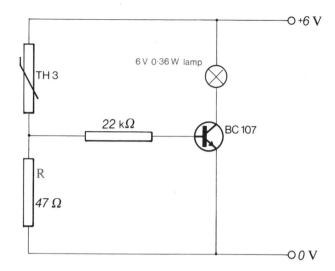

Fig. 28.12 Transistor switch

for the current flowing through the thermistor. The larger the resistance of this resistor the larger the fraction of the current through the thermistor which goes into the base of the transistor. If the resistance of R is large the transistor will switch on even when the thermistor is cold because virtually all of the current through the thermistor will flow through the transistor and this current will generally be large enough to switch on the transistor. If the resistance of R is too small the transistor will not switch on, even when the thermistor is hot, because most of the current through the thermistor will by-pass the transistor.

Although we have described the operation of the transistor in terms of currents, and the bi-polar transistor is in fact known as a current-controlled device, currents only flow as a result of p.d.'s. So the operation of the transistor may also be explained in terms of p.d.'s. The p.d. across R provides the base bias voltage, *i.e.* the voltage across the base-emitter junction. Together, the thermistor and R constitute a potential divider (*see p.221*). When the thermistor is cold its resistance is large and so the p.d. across it is large. Therefore, the p.d. across R is small, no current flows to the base and so the transistor is not switched on. When the thermistor is warm its resistance is small so the p.d. across it is small. Therefore, the p.d. across R is large, a base current flows and the transistor is switched on.

In some cases, the collector current may not be large enough to operate a bell, or other device, in the collector circuit. If a larger transistor is not available, or is not suitable, the solution is to use a relay (*see p.254*). *Fig. 28.13* shows how a relay is incorporated into the simple fire alarm circuit of *Fig. 28.12*. In this case, when the transistor is switched on, the collector current flows through the coil of the relay, causing the contacts to close and switch on the bell. The diode is connected in parallel with the relay coil to prevent induced e.m.f.'s damaging the transistor.

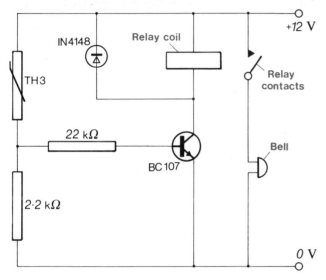

Fig. 28.13 Using a transistor to switch a relay

The Transistor as a Current Amplifier

We have already seen how, in a bi-polar transistor a small change in the base current can produce a large change in the collector current, *i.e.* the transistor acts as a current amplifier. Thus, a transistor can be used to amplify a small, varying current, *e.g.* the current produced in a radio aerial or a microphone, into a current large enough to operate a loudspeaker. *Fig. 28.14* shows a simple circuit which may be used to demonstrate current amplification. The current from the microphone is fed into the base while the collector current operates a small speaker. The smaller battery provides the base bias voltage. This ensures that the transistor is switched on at all times. Both bias voltages may be obtained from a single power supply by the use of a potential divider as was done for the switch in the previous section.

In practice, a circuit as simple as this would give very poor quality reproduction. Very often more than one transistor is needed to produce sufficient amplification. In this case, the

collector current of the first transistor is fed into the base of the second, and so on.

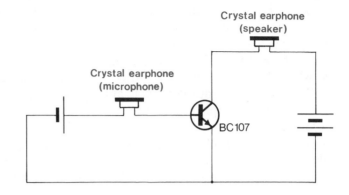

Fig. 28.14 Simple current amplifier

The Transistor as a Voltage Inverter

Consider the circuit shown in *Fig. 28.15*. If A is connected to the 6 V line a base current will flow and the transistor will be switched on. The collector current flows through the 1.2 kΩ resistor and so the there is a p.d. across it. Since the p.d.

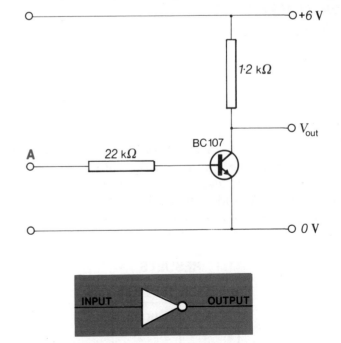

Fig. 28.15 NOT gate and symbol

between the collector and emitter is very small for all values of collector current (*see p.311*), this means that V_{out} is almost zero. In other words, when the input voltage is "high" the output voltage is "low". Conversely, if A is connected to the 0 V line no base current will flow, so no collector current will flow. Since there is now no current flowing through the 1.2 kΩ resistor there is no p.d. across it. Therefore, V_{out} is 6 V. That is, when the input voltage is "low" the output voltage is "high". Thus, the output voltage is always the inverse of the input voltage. For this reason the circuit is known as a **voltage inverter**.

The circuit of *Fig. 28.15* is another example of a logic gate. It is called a **NOT** gate — when the input is "high" the output is not "high". The truth table for a NOT gate is shown in *Fig. 28.16*.

A	Output
1	0
0	1

Fig. 28.16 Truth table for a NOT gate

EXPERIMENT 28.1
To Establish the Truth Table for a NOT Gate.

Apparatus: Transistor, 1.2 kΩ resistor, 22 kΩ resistor, voltmeter, battery or low voltage power supply.

Procedure:
1. Set up the circuit as shown in *Fig. 28.15*.
2. Connect point A to the positive of the battery/power supply. Note the reading on the voltmeter. Record the appropriate logic level, 1 or 0, in the truth table.
3. Disconnect A and connect it to the negative of the battery/power supply. Record the logic level.

RESULTS

Truth Table for NOT gate

A	Output

The Transistor as a Voltage Amplifier

We saw earlier how a bi-polar transistor could function as a current amplifier. A current amplifier may be converted to a voltage amplifier by placing a resistor, called a **load resistor**, in the collector circuit. As before, a small increase in the base current produces a large increase in the collector current. This now results in a large increase in the p.d. across the load resistor and therefore a decrease in the p.d. across the transistor.

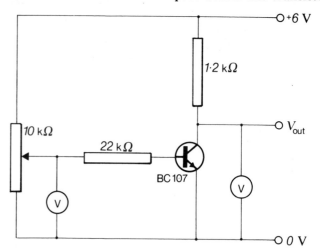

Fig. 28.17

The circuit shown in *Fig. 28.17* can be used to investigate the relationship between the input voltage and the output voltage. The output voltage is measured for a series of values of the input voltage and a graph is plotted of V_{out} against V_{in}, *Fig. 28.18*. Note from the graph that the value of V_{out} changes from almost 6 V to almost 0 V over a small range of values of V_{in}, *i.e.* for values of V_{in} near 1 V a small change in the value of V_{in} produces a large change in the value of V_{out}. In other words, the transistor is acting as a voltage amplifier. Note also that, over the range of values in question, the graph is a straight line, indicating that the change in V_{out} is proportional to the change in V_{in}. When the transistor is used as an amplifier the output voltage must lie on the linear part of the graph otherwise the output will be distorted, *i.e.* some parts will be more, or less, amplified than others. (The operation of the transistor as a voltage inverter may also be seen from this graph. When the input voltage is low the output voltage is high; when the input voltage is high the output voltage is low.)

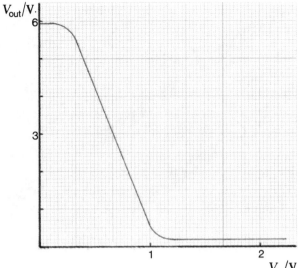

Fig. 28.18 Variation of output voltage with input voltage

In practice, a voltage amplifier is used to amplify a small varying voltage. The circuit shown in *Fig. 28.19* may be used to demonstrate the operation of a voltage amplifier. Firstly, the input voltage is set so that the output voltage is at the mid-point of the linear part of the graph of V_{out} against V_{in}. (This ensures the maximum possible range of values for V_{out}.) This is achieved by a suitable choice of values for the resistances of R_1 and R_2, which together make up a potential divider.

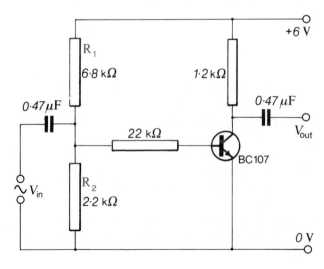

Fig. 28.19 Voltage amplifier

(The 10 kΩ variable resistor used in the previous circuit may also be used here.) A small varying p.d. is then applied between the base and the emitter. The magnitudes of the input and output voltages may then be compared using an oscilloscope, *Fig. 28.20*. Note that the output voltage is exactly out of phase with the input voltage, again showing that when the input voltage is high the output voltage is low and *vice versa*. Since the emitter is common to both the input and output circuits, this arrangement is known as a **common emitter** amplifier.

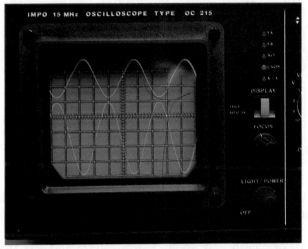

Fig. 28.20 Using an oscilloscope to compare input and output voltages

The capacitor connected in series with the input blocks any d.c. voltage from the a.c. source. Such a capacitor is necessary since d.c. voltage from the source might upset the base bias voltage, pushing the output voltage off the linear part of the graph. A similar capacitor in the output circuit blocks the d.c. voltage from the collector-emitter circuit, allowing through only the amplified varying voltage.

All of the circuits described in this section may be constructed with field effect transistors instead of bi-polar transistors. As an example, the circuit for an FET voltage amplifier is shown in *Fig. 28.21*. It is clear that the circuit is very similar to that of the amplifier based on the bi-polar transistor. It differs in the values of the components used, the bias voltages for FET's being generally larger than those for bi-polar transistors, and in the method of obtaining the appropriate bias voltages from a single battery/power supply. The main advantage of the FET amplifier is that it has a much higher input resistance and therefore draws a much smaller current from the source of the varying voltage.

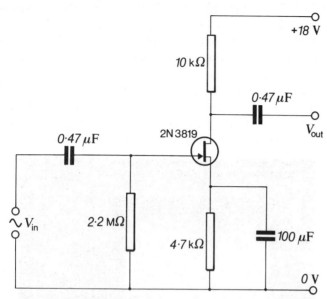

Fig. 28.21 *FET voltage amplifier*

Integrated Circuits (IC's)

An IC is a complete circuit, containing transistors, diodes, resistors and capacitors (not all IC's contain all these components) made of, and in, a single crystal of silicon. The crystal, or "chip", is about 5 mm square and less than 0.5 mm thick. IC's are packaged in plastic cases which have pins with which to connect the chip to other circuits, *Fig. 28.22*.

Fig. 28.22 *Integrated circuits*

Smaller metal cases are also sometimes used.

IC's were first made in the early 1960's with fewer than 100 components per chip. The possible number of components has increased gradually over the years and at present chips containing up to 1 million components are a possibility. Compared with circuits built using separate components, IC's are very much cheaper, smaller and more reliable. However, they cannot operate at high voltages (above about 30 V) and inductors and transformers cannot be built on them.

SUMMARY

A field effect transistor (FET) consists essentially of a piece of p-type material, called the gate, set in a channel of n-type material. Current flows through the channel of n-type material from the drain to the source. The width of the channel, and hence the current flowing, is controlled by the potential of the gate, which is reverse biased with respect to the channel.

A junction transistor consists of a piece of n-type material sandwiched between two pieces of p-type, or *vice versa*. The centre piece is called the base. The end pieces are called the collector and emitter, respectively. The emitter current is equal to the sum of the collector and base currents: $I_e = I_c + I_b$. The collector current is controlled by the base current. Small changes in the base current produce large changes in the collector current: $I_c = kI_b$. This property is used in the current amplifier. When a transistor is used as a switch, a change in the base current is used to switch on, or off, a device connected in the collector circuit.

A current amplifier may be converted to a voltage amplifier by connecting a load resistor in the collector circuit. A transistor may also be used as a voltage inverter or NOT gate.

An integrated circuit consists of a very large number of components constructed in a single small crystal of silicon.

Question 28

Section I

1. In an FET the current flows from.........................
   to..
   in a...................................... of n-type material.

2. In an FET the gate is .... with respect to the channel.

*3. What determines the size of the current flowing through an FET?...
   ....
   ....

4. Name the three sections of a bipolar transistor.........
   ....
   ....

*5. Name the currents flowing in a bipolar transistor and give the relationship between them...............................
   ....
   ....

*6. What is meant by saying that a bipolar transistor acts as a current amplifier?...
   ....

*7. How may a current amplifier be converted to a voltage amplifier? ....
   ....
   ....

*8. What is meant by saying that a transistor acts as a voltage inverter? ....
   ....

*9. Why is it necessary to have a capacitor in series with the output of a voltage amplifier?..............................
   ....

*10. What is an integrated circuit?...............................
   ....
   ....

Section II

1. *Fig. I* shows a light operated switch based on an l.d.r. (ORP 12). Given that the resistance of the l.d.r. is 1 MΩ in the dark and 100 Ω in bright light, explain how the switch works.

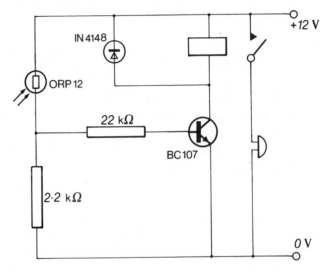

Fig. I

2. How could the circuit of *Fig. I* be modified so that the bell would ring in the dark. Suggest uses for this type of switch.

*3. The d.c. current gain of a transistor is 200. If the base current is 50 μA calculate (i) the collector current, (ii) the emitter current.

*4. The emitter current of a transistor is 45 mA. If 0.6% of the electrons entering the base combine there with holes calculate the value of (i) the base current, (ii) the collector current.

5. Draw a circuit diagram showing a switch based on two transistors. What is the advantage of using two transistors?
   Hint: The collector current from the first provides the base current for the second.

29 Radioactivity

29.1 Nuclear Structure

In 1911 Ernest Rutherford published results of a series of experiments which he had carried out over the previous five years. On the basis of these results, he proposed a new model of the atom. According to this model, which is basically the one used today, the atom is made up of a very small, dense, positive central part called the **nucleus**, surrounded by a cloud of negative electrons. Further experiments by a number of physicists, including the Irish physicist, E.T.S. Walton, over the following 20 years established that the nucleus is made up of particles, called **nucleons**. There are two types of nucleon, called **protons** and **neutrons**, respectively. Protons and neutrons have approximately the same mass ($\sim 10^{-27}$ kg) but differ in the fact that the protons are positively charged while neutrons are uncharged. The charge on the proton is equal in magnitude to the charge on the electron, so that an uncharged atom has equal numbers of protons and electrons.

All atoms of the same element have the same number of protons in their nuclei; atoms of different elements have different numbers of protons. The number of protons is called the proton number or the **atomic number**, Z, of the element. Thus, an atom of hydrogen which has one proton in its nucleus has an atomic number of 1; an atom of helium has two protons in its nucleus and so its atomic number is 2, and so on.

The total number of nucleons, *i.e.* protons and neutrons, in the nucleus of an atom is called the nucleon number or the **mass number**, A, of the atom. Thus, the mass number is equal to the atomic number plus the number of neutrons. Atoms which have the same atomic number but different mass numbers are called **isotopes**. For example, there are three known isotopes of hydrogen — one has no neutrons ($A = 1$), another (called deuterium) has one neutron ($A = 2$) and the third (called tritium) has two neutrons ($A = 3$); all of course have only one proton, *i.e.* $Z = 1$.

Radius of the Nucleus

The series of experiments which led Rutherford to propose his nuclear model of the atom involved firing alpha particles (particles made up of two protons and two neutrons, *see p. 321*) at very thin sheets of metal. When using a sheet of gold approximately 1 μm thick, he found that most of the particles passed straight through, while a small proportion were deflected through angles greater than 90°. This suggested that most of the volume of the atom was empty space with most of the mass concentrated in a very small fraction of the total volume and carrying a positive charge. Using these ideas

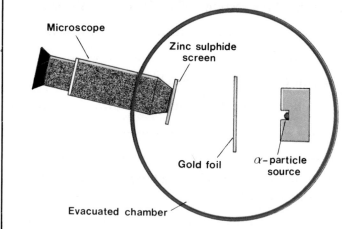

Fig. 29.1

Ernest Rutherford (1871 — 1937)

Rutherford was born in New Zealand, one of twelve children of a Scottish father who had migrated there thirty years previously. He excelled at school and, after graduation from New Zealand University, obtained a scholarship to Cambridge in 1895.

Rutherford investigated the newly discovered phenomenon of radioactivity. He named the three main types of radiation emitted and established the idea of half-life. In 1908, he showed that alpha particles are helium nuclei. Following a series of experiments between 1906 and 1911, he proposed the model of the atom which is accepted today. In 1919 he achieved the first artificial transmutation of one element to another when he bombarded nitrogen atoms with alpha particles, producing oxygen atoms and protons.

In 1908 Rutherford was awarded the Nobel Prize in Chemistry. He was appointed a professor of physics at Cambridge and became director of the Cavendish laboratory in 1919. He died in 1937 in London and is buried in Westminster Abbey with Newton and Kelvin.

Rutherford worked out the percentage of incident alpha particles which should be deflected through any given angle. In 1913 two of his assistants, Geiger and Marsden, using the apparatus shown in *Fig. 29.1*, carried out a series of experiments to test Rutherford's theory. The results of these experiments were exactly as predicted by Rutherford and so provided strong support for his model of the atom.

Using measurements from these experiments, Rutherford estimated that the radius of the nucleus of the atom was of the order of 10^{-15} m. Since the radius of the atom was known to be of the order of 10^{-10} m this means that most of the atom is empty. In fact, the total volume of the atom is about 10^{12} times the volume of all the particles which make it up.

Radioactivity

In 1896 the French physicist, Henri Becquerel was investigating the properties of uranium salts. On one occasion, he found that a photographic film, left beside a uranium salt, was exposed (blackened), even though it had been wrapped in black paper. Something coming from the uranium salt was able to penetrate the black paper and affect the film. The "something" is now called radiation and the uranium salt is said to be **radioactive**.

Further experiments have shown that radiation is emitted from an element when the nuclei of that element are unstable and break up to form more stable nuclei. The rate at which the nuclei of a particular element break up is found to be unaffected by physical or chemical changes; it is a purely random process. Radioactivity may thus be defined as follows.

> **Radioactivity is the spontaneous disintegration of unstable nuclei with the emission of one or more types of radiation.**

29.2 Detectors

After Becquerel's discovery of radioactivity in 1896 the radiation emitted by various substances was investigated by a number of physicists. Before considering the results of their experiments we shall look first at some of the devices which have been developed to detect the presence of radiation.

The Cloud Chamber

One of the first detectors was the cloud chamber, *Fig. 29.2*, developed by the Scottish physicist C.T.R. Wilson (1869-1959), in 1911. The device consists of a chamber containing dust-free air. At the bottom of the chamber is an absorbent pad soaked with a volatile liquid such as ethanol. The air in the chamber is thus saturated with ethanol vapour. The proportion of ethanol vapour in the air at a given time depends on the temperature. If the volume of the air is increased suddenly, by means of the pump, the temperature drops and, as a result, there is an excess of ethanol vapour in the air — it is said to be supersaturated. When radiation passes through the chamber while the air is in this condition, it ionises the air and the ethanol

Henri Becquerel (1852 — 1908)

Becquerel was born in Paris into a family of physicists. After a number of years as an engineer he succeeded, in 1892, to a post in the Museum of Natural History in Paris which had been held previously by both his father and his grandfather.

Becquerel studied fluorescence with a view to discovering if X-rays might be emitted as well as light. In 1896 he left photographic plates in a drawer with a uranium salt and discovered to his amazement a few days later that the plates were exposed, even though they were wrapped in black paper. (This phenomenon had been noticed almost 30 years earlier by another Frenchman, Abel Niépce. Niépce, however, failed to pursue his discovery.) Becquerel studied the radiation emitted from the uranium salt and, by 1900, was able to show that it consisted of negative electrons. In 1901 he established that the electrons were being emitted from the uranium atoms.

In 1898 Marie Curie named the phenomenon discovered by Becquerel "radioactivity" and in 1903 Becquerel shared the Nobel Prize in physics with Marie and Pierre Curie. The unit of activity of a radioactive source is named in his honour. One becquerel (Bq) is equivalent to one disintegration per second.

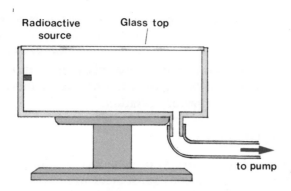

Fig. 29.2 Expansion cloud chamber

vapour condenses on the ions so formed. Thus a trail of tiny drops of ethanol is formed along the path of the radiation.

The Geiger-Muller (GM) Tube

The GM tube, *Fig. 29.3*, consists of a metal cylinder about 2 cm in diameter and containing a mixture of argon and bromine gas at low pressure ($\sim 10^4$ Pa). A metal wire runs up the centre of the tube and a potential difference of a few hundred volts is maintained between it and the case, the case being negative with respect to the wire. Radiation enters the tube through a very thin mica "window" at one end.

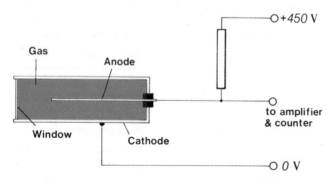

Fig. 29.3 Geiger-Muller tube

When radiation enters the tube it ionises some of the gas molecules. These ions are swept towards the electrodes, the positive ions going towards the case and the negative ions going towards the wire. The negative ions are accelerated rapidly in the strong electric field near the wire and cause further ionisation by collision. This process, known as gas amplification, ensures that, even if only a single pair of ions is produced initially by the radiation, a measurable charge is generated in the tube. When this charge reaches the electrodes

it passes around the external circuit as a pulse of current. This pulse produces a p.d. across the series resistor and this is used to operate a counting device *(see below)* and/or a loudspeaker. The p.d. across the resistor also reduces the p.d. across the tube and prevents a continuous discharge taking place. There is thus a "pause" before the tube is ready to detect more radiation. The duration of this pause, known as the "dead time", is typically of the order of a few nanoseconds.

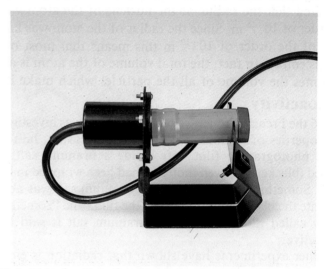

Fig. 29.4 GM tube. Note that the window is protected by a plastic cap

The Solid State Detector.

The solid state detector consists of a reverse biased p-n junction *(p.299)*. When the radiation falls on the depletion layer pairs of free electrons and holes are created *(cf. The Photodiode, p. 304)*. These electron-hole pairs constitute a pulse of current which passes around the external circuit, is amplified and used to operate a counter, as for the GM tube. Although silicon is used in solid state detectors, modern detectors are often made of indium antimonide (InSb) or gallium arsenide (GaAs).

The solid state detector is smaller, cheaper and less easily damaged than the GM tube. It also has the advantage that it operates at a low voltage and that it can be used directly to distinguish between the different types of radiation.

Counting Devices

Each "particle" of radiation entering a GM tube, or similar detector, causes a pulse of current. While this current could be detected with a sensitive meter, it is easier to use it to operate a counting device which records automatically the number of pulses received.

A **ratemeter** records the average number of "particles" of radiation entering the detector per minute or per second. It is most useful when the count rate is fairly high.

A **scaler** records the total number of "particles" detected in any given time. It is most useful when the count rate is low and for demonstrating the random nature of the radiation emitted from a radioactive source.

29.3 Types of Radiation

A large number of radioactive isotopes are known. One of these is radium-226, *i.e.* the isotope of radium whose mass number is 226. To investigate the nature of the radiation emitted from radium-226, a sample of the isotope is placed near a GM tube or a solid state detector which is connected to a scalar/ratemeter, *Fig. 29.5*, and the count rate is noted. A thin sheet of paper is now placed between the sample and the detector; the count rate is seen to drop considerably. However, the addition of further sheets of paper has no noticeable effect on the count rate. A thin sheet of aluminium is now placed between the sample and the detector; the count rate is further reduced. Again, the addition of further sheets has little effect. Finally, a block of lead is placed between the sample and the detector; the count rate is now reduced almost to zero.

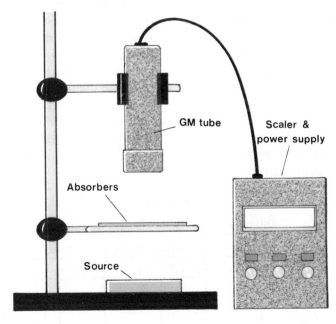

Fig. 29.5

This experiment, together with other similar experiments, leads to the conclusion that there are three distinct types of radiation which may be emitted from a radioactive source. (Not all radioactive sources emit all three types.) These three types of radiation are called α-**radiation**, β-**radiation** and γ-**radiation**. We shall now consider the properties of each of these three types of radiation in turn.

α — Radiation

This is the least penetrating of the three types of radiation, being stopped, as we have seen, by a thin sheet of paper. It is deflected in electric and magnetic fields, showing it to be charged. From the direction in which it is deflected it can be deduced that the charge is positive. Further experiments by Rutherford and his co-workers showed that α-radiation consists of particles having a charge equal in magnitude to twice the charge on an electron. They also showed that the mass of an α-particle was approximately equal to four times the mass of a hydrogen atom. In other words, α-particles are similar to helium atoms which have lost both their electrons. In 1909 Rutherford, together with Royds, carried out an experiment which showed conclusively that α-particles are identical to helium nuclei.

Since α-particles are relatively large, they cause a lot of ionisation as they pass through any material — in a cloud chamber they leave broad straight tracks. Consequently, they lose their energy quickly and so their penetrating ability is low. As we have seen, they cannot penetrate a thin sheet of paper. In air, they can penetrate about 6 cm. All the particles from a given source are emitted with the same speed ($\sim 10^7$ m s^{-1}). This is indicated by the fact that their tracks in a cloud chamber are all the same length.

Since α-particles have such a low penetrating ability, a GM tube used to detect them must have an extremely thin "window". They are also readily detected by a solid state detector, special photographic emulsions or a gold leaf electroscope. In a cloud chamber they leave short broad tracks.

β — Radiation

Like α-radiation, β-radiation is also deflected in electric and magnetic fields, only to a much greater extent and in the opposite direction. Experiments have shown that β-radiation consists of particles having the same charge and mass as electrons. The charge on a β-particle is therefore half that on an α-particle, while its mass is only approximately 1/7000 of

321

that of an α-particle. β-particles are much less likely to collide with the electrons of a material through which they pass, so, in a given distance, they produce fewer ions than an α-particle. Consequently, a β-particle will travel further, *i.e.* it is more penetrating, than an α-particle of similar energy. The most energetic β-particles penetrate about 5 mm of aluminium or 500 cm of air. The majority, however, are stopped by a thin sheet of aluminium. Again unlike α-particles, β-particles are emitted with a range of speeds from a particular source, the maximum speed being characteristic of the source. (β-particles may be emitted with speeds up to 2.7×10^8 m s^{-1}.)

β-particles may be detected efficiently with a GM tube, solid state detector or special photographic emulsions. In a cloud chamber β-particles produce tracks which are longer, but much thinner than those of α-particles.

γ — Radiation

γ-radiation is electromagnetic radiation of very short wavelength ($\sim 10^{-11}$ m). As such it cannot be deflected in electric or magnetic fields and, like all electromagnetic radiation it travels at the speed of light. γ-rays cause very little ionisation per cm of their path and are therefore very penetrating — they may penetrate several cm of lead. Since they cause so little ionisation in a given distance, they are very difficult to detect. A GM tube detects — indirectly — about 1% of the γ-rays passing through it. Solid state detectors, special photographic emulsions and cloud chambers may also be used.

The properties of α-, β- and γ-radiations may be summarised under a number of headings as shown in *Table 29.1*.

	Mass/u	Charge/e	Ionising Ability	Penetrating Ability	Deflected in Electric and Magnetic Fields
α-particles	4	+2	good	poor	yes
β-particles	0.0005	−1	medium	medium	yes
γ-rays	0	0	poor	good	no

Table 29.1

EXPERIMENT 29.1

To Investigate the Range of the Three Types of Radiation in Air

Apparatus: α-particle source, β-particle source, γ-ray source, detector, counter.

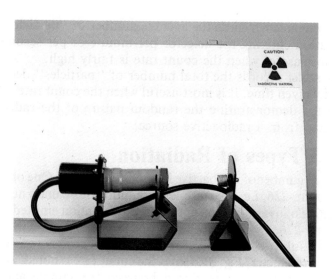

Fig. 29.6

Procedure:

1. Set up the detector and counter. Set the counter to zero and record the number of counts over a five minute period. Calculate the average number of counts per second. This is the background count rate.
2. Place the α-particle source in front of the detector. Measure the distance, d, from the source and record the count rate.
3. Move the counter back 0.5 cm and record the count rate. Repeat until the count rate reaches background level. Plot a graph of count rate, A_d, against distance.
4. Replace the α-particle source with the β-particle source and repeat the procedure. In this case the detector should be moved in steps of 10 cm and it may not be possible for the count rate to reach background level. Plot a graph of count rate against distance, using the same axes as before.
5. Replace the β-particle source with the γ-ray source and repeat Step 4.

RESULTS
Background count rate =   s^{-1}.

α-particles		β-particles		γ-rays	
d/cm	A_d/s^{-1}	d/cm	A_d/s^{-1}	d/cm	A_d/s^{-1}

Marie Curie
(1867 — 1934)

Marie Curie was born Marie Sklodowska in Warsaw, Poland. Both of her parents were teachers, her father being a physics teacher. In 1891, Marie followed her brother and sister in search of education to Paris where she entered the Sorbonne. In 1895 she married a French chemist, Pierre Curie.

Marie Curie studied the radiation emitted from uranium. She discovered that some minerals were more radioactive than could be accounted for by uranium alone. By July 1898, with the help of her husband, she had isolated a new element from the uranium ore. This she called polonium in honour of her native land. However, this still did not account for all the activity of the ore. By the end of the year she and her husband had detected a second new element which they called radium. In order to obtain the new element in measurable quantities, the Curies spent the next four years, and most of their money, in painstakingly refining large quantities of the ore. Eventually, 8 t of the ore yielded just 1 g of radium.

In 1903, Marie and Pierre shared the Nobel Prize in physics with Henri Becquerel who had discovered radioactivity seven years previously. In 1906 Pierre Curie was killed in a traffic accident. Five years later Marie was awarded the Nobel Prize in chemistry for her discovery of the two new elements. She is the only person ever to have received a second Nobel Prize in science.

An idealist, Marie Curie refused to profit from her discoveries. She died of leukaemia, induced by overexposure to radiation, on July 4, 1934.

Nuclear Reactions

An unstable nucleus achieves stability by the emission of radiation, usually in a series of steps *(see Table 29.2)*. The resulting stable nucleus is usually a different element (different atomic number from the original nucleus). The series of steps ending in a stable nucleus is called a **decay chain**. Each step in the decay chain is a nuclear reaction which may be represented by an equation in much the same way as chemical reactions may be represented by equations.

An α-particle is the same as a helium nucleus, *i.e.* it consists of two protons and two neutrons and so has an atomic number of two and a mass number of four. Thus, when a nucleus emits an α-particle it loses two protons and two neutrons. Therefore, its atomic number decreases by two while its mass number decreases by four. When writing equations to represent nuclear reactions it is usual to represent the nucleus with the chemical symbol for the atom with the atomic number as a subscript and the mass number as a superscript. Thus, $^{226}_{88}\text{Ra}$ represents a nucleus of the radium isotope which has an atomic number of 88 and a mass number of 226, *i.e.* the nucleus has 88 protons and 138 neutrons.

Radium-226 is radioactive and decays with the emission of an α-particle. This reaction may be represented as follows.

$$^{226}_{88}\text{Ra} \rightarrow \, ^{222}_{86}\text{X} + \, ^{4}_{2}\text{He}$$

Reference to the Periodic Table of the elements shows that the daughter nucleus, X, is in fact an isotope of radon. The reaction may therefore be written as

$$^{226}_{88}\text{Ra} \rightarrow \, ^{222}_{86}\text{Rn} + \, ^{4}_{2}\text{He}$$

Note that the sum of the atomic numbers on both sides of the equation must be the same and the sum of the mass numbers on both sides must also be the same.

The emission of a β-particle cannot change the mass number of the parent nucleus since the mass of an electron is much less than ($\sim 1/2000$) the mass of a proton or a neutron. However, the electron carries one unit of negative charge, therefore the nucleus must gain one unit of positive charge, *i.e.* it must gain a proton. The only way in which a nucleus can gain a proton without a change in its mass number is for a neutron to be converted to a proton. Thus,

$$^{1}_{0}\text{n} \rightarrow \, ^{1}_{1}\text{p} + \, ^{0}_{-1}\text{e}$$

(Note: An electron cannot have a mass number or an atomic number. The respective values of 0 and —1 are given to the electron simply to balance the equation.)

Thus, when a nucleus emits a β-particle its atomic number increases by 1 while its mass number remains unchanged. For example, the decay of lead-214 may be represented by the equation

$$^{214}_{82}\text{Pb} \rightarrow \ ^{214}_{83}\text{Bi} + \ ^{0}_{-1}\text{e}$$

The emission of a γ-ray does not change either the atomic number or the mass number of the nucleus. It simply results from a rearrangement of the protons and the neutrons, which lowers the internal energy of the nucleus and so forms a more stable arrangement.

Recent research has indicated that some nuclei may decay with the emission of particles other than α-particles and β-particles. In particular it has been shown that radium-223 may decay to lead-209 with the emission of a carbon-14 nucleus. However, only about 1 in 10^{10} radium nuclei decay by this mechanism. The rest decay in a series of steps with the emission of α-particles and β-particles.

It should be noted that all types of nuclear reaction result in a release of energy from the nucleus involved. This energy is released in the form of a photon of electromagnetic radiation and/or kinetic energy of the nucleus and the emitted particles. Since energy is emitted as a result of the reaction the total mass of the particles involved must decrease, according to Einstein's equation, $E = mc^2$. Indeed, as we noted in *Chapter 8*, it is only in nuclear reactions that the change in mass is an appreciable percentage of the original mass. It was from such a reaction that Einstein's equation was first verified experimentally. The experiment was carried out by Cockroft and Walton in 1932 and involved directing a beam of protons onto a lithium target. The resulting reaction may be represented by

$$^{7}_{3}\text{Li} + \ ^{1}_{1}\text{H} \rightarrow \ ^{4}_{2}\text{He} + \ ^{4}_{2}\text{He}$$

Cockroft and Walton found that the difference between the mass of the lithium and hydrogen nuclei and the mass of the two helium nuclei (or α-particles) was equivalent to the kinetic energy of the two helium nuclei.

Element	Atomic Number	Mass Number	Type of Radiation	Half-life
Uranium (U)	92	238	α	4.5×10^9 y
Thorium (Th)	90	234	β, γ	24 d
Protactinium (Pa)	91	234	β, γ	71 s
Uranium (U)	92	234	α	2.5×10^5 y
Thorium (Th)	90	230	α, γ	8.0×10^4 y
Radium (Ra)	88	226	α, γ	1.6×10^3 y
Radon (Rn)	86	222	α	3.8 d
Polonium (Po)	84	218	α	3.1 min
Lead (Pb)	82	214	β, γ	27 min
Bismuth (Bi)	83	214	β, γ	20 min
Polonium (Po)	84	214	α	1.6×10^{-4} s
Lead (Pb)	82	210	β, γ	22 y
Bismuth (Bi)	83	210	β	5 d
Polonium (Po)	84	210	α, γ	138 d
Lead (Pb)	82	206	Stable	

Table 29.2 The Uranium decay series. Only the principal modes of decay are given.

E. T. S. Walton (b. 1903)

Walton was born in Dungarvan, Co. Waterford, and was educated at Trinity College, Dublin, graduating in 1926. In 1927 he went to Cambridge where he worked with John Cockroft in the Cavendish laboratory under the direction of Lord Rutherford.

Walton, with Cockroft, worked to invent a device for accelerating charged particles. In 1929 they succeeded in devising a linear accelerator in which particles could be accelerated through a p.d. of 700 kV. In 1932 they directed a beam of protons from the accelerator onto a lithium target and found that alpha particles were emitted. This was the first time an accelerator had been used to produce a nuclear reaction — a technique which has been fundamental to the development of nuclear physics. Equally importantly, they were able to verify, for the first time, Einstein's mass/energy equation, $E = mc^2$.

In 1947 Walton was appointed professor of physics at Trinity College, Dublin. In 1951 he and Cockroft were awarded the Nobel Prize in physics. He is the only Irish person to have been awarded a Nobel Prize in science.

29.4 The Law of Radioactive Decay

As mentioned earlier, the process of radioactive decay, or disintegration, is a purely random one. Given a sample of a radioactive material, it would be impossible to predict when a particular nucleus would disintegrate. However, since even a very small sample contains a very large number of nuclei, it is possible to predict the average number of nuclei which will decay in a given time. This is similar to tossing a coin. If you toss a penny once you cannot predict whether it will come up "heads" or "tails". If you toss it 1000 times you may be reasonably certain that you will get approximately 500 "heads" and 500 "tails".

Since radioactivity is a random process, the more nuclei there are in a sample the more disintegrations are likely to occur in a given time. The **Law of Radioactive Decay** states that:

> **The number of disintegrations per second is proportional to the number of nuclei present.**

The number of disintegrations per second is called the **activity**, A, of the sample. The unit of activity is the **becquerel**, Bq. 1 Bq is equivalent to one disintegration per second.

In symbols, the law of radioactive decay may be written as

$$\frac{dN}{dt} \propto N$$

where N is the number of nuclei present at time t.

The law may also be written as

$$\frac{dN}{dt} = -\lambda N$$

where λ is a constant, called the **decay constant**.

Example

The decay constant for plutonium-239 is 9.2×10^{-13} s^{-1}. Calculate the activity of a sample which contains 5×10^{14} nuclei (approximately 0.2 μg).

$$\left| \frac{dN}{dt} \right| = \lambda N$$

$$= 9.2 \times 10^{-13} \times 5 \times 10^{14}$$

$$= 460 \text{ Bq}$$

Ans. 460 Bq

Half-life

Since the rate of decay decreases with the number present, the time taken for any given percentage of a sample to decay is always the same, no matter how many nuclei are in the sample to start with. An example, using small numbers, illustrates this. Suppose there are 1000 nuclei in a sample of a certain radioactive isotope and the rate of decay is 10 per second. Then the time taken for 10% (100) of the nuclei to decay is 10 s. Now if there are 500 nuclei in the sample the rate of decay is 5 per second (half the number, half the rate of decay) and the time taken for 10% (50) of the nuclei to decay is still 10 s. (Note: This is only an illustration. Since the rate of decay decreases as the number of nuclei decreases, the time taken for 10% to decay is longer than 10 s; it is, in fact, 10.54 s. Since the rate is changing continuously the time must, in a practical case, be calculated using the Law of Radioactive Decay.) Rather than the time taken for 10% of a sample to decay it is usual to consider the time taken for 50% to decay. This time is called the **half-life** of an element.

> **The half-life, $T_{1/2}$, of an element is the time taken for half of the nuclei in any given sample to decay.**

Since the activity of a given sample is proportional to the number of nuclei in it, the half-life may also be defined in terms of the activity rather than the number of nuclei. Thus,

> **The half-life, $T_{1/2}$, of an element is the time taken for the activity of any given sample to decrease to half of its original value.**

Example
The half-life of Actinium-228 is 6.1 h. A sample has a mass of 16 μg at noon on Monday. What would be the mass of the actinium at 12.24 p.m. on Tuesday?

From noon on Monday to 12.24 p.m. on Tuesday is 24 h 24 min.

$$24 \text{ h } 24 \text{ min} = 4 \text{ half-lives}$$

Therefore, the fraction of the mass remaining will be

$$\tfrac{1}{2} \times \tfrac{1}{2} \times \tfrac{1}{2} \times \tfrac{1}{2} = (\tfrac{1}{2})^4 = \tfrac{1}{16}$$

That is, the mass remaining is 1.0 μg.

Ans. 1.0 μg

Note: The remaining 15 μg does not just disappear. Most of it is accounted for by the masses of the daughter product and the emitted particles. The remainder is accounted for by the kinetic energy of particles and nuclei.

It may be shown, using integral calculus *(see Appendix B),* that the half-life of an element is related to the decay constant by

$$T_{\frac{1}{2}} = \frac{\ln 2}{\lambda}$$

where ln 2 is the natural log of 2, *i.e.* the log of 2 to the base e.

Example
Calculate the half-life of plutonium-239. The decay constant for plutonium-239 is $9.2 \times 10^{-13} \text{ s}^{-1}$.

$$T_{\frac{1}{2}} = \frac{\ln 2}{\lambda}$$

$$= \frac{0.693}{9.2 \times 10^{-13}}$$

$$= 7.5 \times 10^{11} \text{ s}$$

$$= 24\,000 \text{ y}$$

Ans. 24 000 y

The half-life of an element is unaffected by all known physical and chemical changes and is thus an important characteristic

of the element. A knowledge of half-lives can therefore be used in identifying unknown elements. Half-lives of known radioactive isotopes range from 10^{-6} s to 10^9 y.

EXPERIMENT 29.2

To Determine the Half-life of Radon

Apparatus: Sample of thorium in a plastic bottle, plastic cell, detector, counter, stop clock.

Procedure:
1. Connect the detector to the counter. Determine the background count rate as in the previous experiment.
2. Arrange the bottle containing the thorium as shown in *Fig. 29.7.* Open the clip and squeeze the bottle to fill the cell with radon. Close the clip and start the clock and the counter.

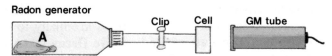

Radon generator · Clip · Cell · GM tube · A

Fig. 29.7

3. Record the total count at 10 s intervals and hence determine the average count rate, A_d, during each interval. Continue until the count rate has reached background level. Plot a graph of count rate against time and from the graph determine the half-life of radon. Take at least two readings from the graph and find the average.

RESULTS

Background count rate = ___ s^{-1}.

t/s	Total Count	A_d/s

Half-life of radon = ___ s.

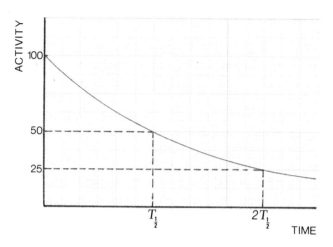

Fig. 29.8

29.5 Uses and Hazards of Radiation

The radiation from radioactive isotopes, or radioisotopes, is now widely used in many different areas. Here we can consider only briefly a few of these applications.

Fig. 29.9   Radioactive sources must be stored in a secure place and be suitably shielded

Medicine

Over 100 radioisotopes are in use in medicine, although a few, *e.g.* cobalt-60 and iodine-131, are much more commonly used than others. γ-rays from cobalt-60 are used in the treatment of cancer. The rapidly dividing cancerous cells are particularly easily killed by γ-rays. To minimise damage to the surrounding healthy tissue the beam of γ-rays is directed at the cancer from a number of different angles so that the healthy cells receive only a small dose. Radiation treatment may also be administered internally by implanting the radioisotope near to the cancer. If the radioisotope has a sufficiently short half-life it may not be necessary to remove it.

Radioisotopes are also used in the diagnosis of several diseases. Since radioactive isotopes are chemically identical to non-radioactive isotopes of the same element, they may be absorbed in the body in the normal manner. For example, all iodine in the body gathers in the thyroid gland. Injecting a sample of iodine-131 into the body makes possible the diagnosis of several diseases of that gland.

Bacteria are killed by γ-radiation and thus it is possible to sterilise medical instruments, *etc.*, by irradiating them with γ-rays. This method of sterilisation has the advantage that, since no heating is involved, plastic instruments which would otherwise melt may be sterilised. Furthermore, such instruments may be sterilised after being sealed in plastic bags.

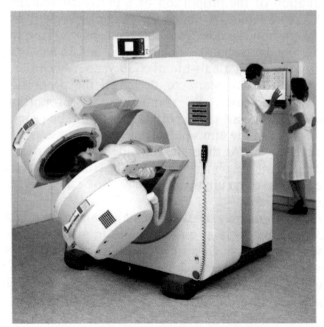

Fig.29.10   Gamma ray camera (Courtesy of Siemens)

327

Industry

Radioisotopes are used in industry to trace the flow of materials, *e.g.* the flow of lubricating oil through a machine. Blockages and leaks in underground pipes may be found by passing radioactive materials along the pipe and monitoring their progress above ground with a GM tube or other detector.

The rate of wear of an engine may be determined by including radioisotopes in the pistons, *etc.*, and examining the lubricating oil for radiation. Much smaller quantities of material can be detected in this way than could be detected by any chemical or physical means.

The thicknesses of various products, *e.g.* sheet steel, may be monitored by measuring the intensity of radiation passed through them. This measurement can then be used to control automatically the device producing the product.

Other Uses

In agriculture, radioisotopes are used to monitor fertiliser uptake in plants, in the evaluation of pesticides, in insect eradication and in numerous other ways.

The age of rocks and fossils may be established from an examination of the proportion of a particular radioisotope which they contain. A plant, for example, while it is alive absorbs carbon dioxide from the atmosphere. This carbon dioxide contains carbon-12 which is stable and carbon-14 which is radioactive with a half-life of 5730 years. When the plant dies the carbon-14 starts to decay so the proportion of carbon-14 decreases. By comparing the proportion of carbon-14 in a fossil sample of the plant with the proportion of carbon-14 in the atmosphere is is possible to establish the age of the fossil. At the same time the age of the rock from which the fossil was obtained may also be established. A similar procedure, based on uranium-238 may also be used to establish the age of rocks.

Radioisotopes may be used as energy sources in remote or inaccessible places, *e.g.* on the moon, in a spacecraft or in remote lighthouses. One form of heart pacemaker uses a plutonium-238 source which has a power output of 0.16 mW and which is capable of giving the heart 70 shocks per minute for a minimum of 10 years.

Hazards

All forms of radiation, including X-rays, are potentially harmful to the human body. The effects can range from radiation burns to the initiation of some forms of cancer (mainly leukaemia) and genetic changes. While the short term effects of high levels of radiation are fairly well understood, the long term effects of low levels of radiation are, of their nature, more difficult to establish.

α-particles, because of their great ionising ability, cause the most damage. However, since they cannot penetrate skin, they are only dangerous if ingested through the lungs or stomach.

β-particles can penetrate skin and so some form of shielding must be used with β-particle sources. One source of β-particles is strontium-90 which is absorbed into the bones where it produces bone-marrow disease.

Because of its great penetrating ability, γ-radiation is the most dangerous and the most difficult to shield against. Several cm of lead are required for adequate protection. Because of the cost of lead, shielding on a large scale is obtained from 1-2 m of concrete.

The greatest care should always be taken with radioactive materials. They should never be handled directly; they should be stored in a special, secure place, shielded as necessary; one should never eat nor drink in their vicinity.

Fig. 29.11   Marie Curie

SUMMARY

The nucleus of an atom is made up of protons and neutrons. The number of protons is the atomic number, A, and the total number of protons and neutrons is the mass number, Z.

Radioactivity is the spontaneous disintegration of unstable nuclei with the emission of radiation. The radiation emitted from radioactive sources may be detected directly using photographic emulsions or a cloud chamber. It may also be detected using a GM tube or solid state detector in conjunction with a loudspeaker or a scaler/ratemeter.

The three main types of radiation emitted by radioactive substances are α-, β- and γ-radiation. These are helium nuclei, electrons and electromagnetic radiation, respectively. All cause ionisation and can penetrate solid substances. α-particles cause the most ionisation and are the least penetrating while γ-rays cause the least ionisation and are the most penetrating. α-particles and β-particles, being charged, are deflected in electric and magnetic fields; γ-rays are not. All affect photographic emulsions and cause fluorescence.

The law of radioactive decay states that the activity of a sample, *i.e.* the number of nuclei decaying per second, is proportional to the number of nuclei in the sample. The constant of proportionality is called the decay constant, λ. The half-life, $T_{1/2}$, of an element is the time taken for half of the nuclei in any given sample to decay.

$$T_{1/2} = \frac{\ln 2}{\lambda}$$

Radioisotopes are widely used in medicine, industry and agriculture. Examples of their use are: the treatment of cancer; examining an engine for wear; tracing the flow of fluids in a body, plant or machine; evaluating insecticides and pesticides; energy sources in remote or inaccessible locations.

Questions 29

DATA: Avogadro's Number, $N_A = 6.0 \times 10^{23}$ mol^{-1}.

Section I

1. Protons and neutrons have approximately the same....

....

but ....

2. Explain: Atomic number; Mass number....................

....

....

3. Define the term radioactivity...................................

....

....

4. Radioactivity is unaffected by................................

and................ changes — it is a ....process.

5. Name two devices used to detect the radiations emitted from radioactive substances...................................

....

6. What is the difference between a ratemeter and a scaler?

....

....

7. Name three types of radiation emitted from radioactive substances..

....

....

8. State the law of radioactive decay...........................

....

....

9. What is meant by the half-life of a substance?...........

....

10. Would a sample of a radioactive substance ever decay completely? Explain...

....

11. Give three uses of radioisotopes...........................

....

....

12. Give two precautions which should be taken when dealing with radioactive substances...................................

....

....

Section II

1. Complete the following nuclear reactions.

   $^{14}_{7}N + ^{4}_{2}He \rightarrow$

   $^{8}_{4}Be \rightarrow \quad + ^{4}_{2}He$

   $^{12}_{5}B \rightarrow ^{12}_{6}C +$

2. Calculate the number of α-particles and the number of β-particles emitted in the decay of $^{235}_{92}U$ to $^{215}_{84}Po$.

*3. The mass of a certain radioisotope decreases from 10 mg to 2.5 mg in 20.4 days. What is the half-life of the isotope? What happens to the remaining 7.5 mg?

*4. A radioactive isotope has a half-life of 10 days. Calculate the time taken for seven eighths of the isotope to decay.

*5. The half-life of an isotope is 60 s. What percentage of the original isotope will remain after 5 minutes?

*6. Protactinium-231 decays to actinium-227. Given that the masses of these two nuclei are 231.0350 u and 227.0269 u, respectively, and that the mass of the alpha particle is 4.0026 u calculate the kinetic energy of the α-particle, given that 97% of the energy released in the reaction appears in this form. (Take 1 u = 1.66×10^{-27} kg.)

7. $^{238}_{92}U$ decays to $^{222}_{86}Rn$ in the following steps. Name the particle emitted at each stage.

   $^{238}_{92}U \rightarrow ^{234}_{90}Th \rightarrow ^{234}_{91}Pa \rightarrow ^{234}_{92}$
   $\rightarrow ^{230}_{90}Th \rightarrow ^{226}_{88}Ra \rightarrow ^{222}_{86}Rn.$

*8. A sample of air containing radon-222 is found to emit 630 α-particles per second. Given that the half-life of radon-222 is 3.8 d and assuming that 1 mol of radon-222 has a mass of 222 g, calculate the mass of radon in the sample.

*9. Uranium-234 decays to thorium-230 with the emission of an α-particle. The decay constant for the reaction is 8.8×10^{-14} s^{-1}. Calculate the number of α-particles emitted per second from 2.8 μg of uranium-234. Assume that the mass of 1 mol of uranium-234 is 234 g. Is this assumption strictly correct? Explain.

*10. A radioactive source contains 3.8×10^{-17} g of bismuth-211 which decays by beta emission. The source is found to emit 570 β-particles per second. Calculate (i) the decay constant for the reaction, (ii) the half-life of bismuth-211. What is the daughter product of the reaction? (Assume mass of 1 mol of bismuth-211 is 211 g.)

*11. Uranium-235 decays by α-emission with a half-life of 7.1×10^{8} y. Calculate the number of α-particles emitted in one minute from 1.0 kg of uranium-235, assuming that one mole has a mass of 235 g.

30 Nuclear Energy

30.1 Fission

We saw in the previous chapter that the nucleus of an atom is made up of positively charged protons and uncharged neutrons. Since like charges repel each other, we would expect the protons in the nucleus to fly apart. Obviously this does not happen, so we must conclude that there exists a force of attraction between the protons which is stronger than the electrostatic repulsion between them. Since this force is apparent only in the nucleus it must act only over very small distances ($\sim 10^{-15}$ m). This nuclear force acts equally between protons, between protons and neutrons and between neutrons. It is very much greater ($\sim 10^6$ times) than the electrostatic force of attraction between a proton and an electron.

As we know, changes involving the electrons of an atom, *e.g.* chemical changes, can release considerable quantities of energy, *e.g.* the burning of coal or oil. From what we learned in the previous paragraph, we might expect that it would be possible to obtain much greater quantities of energy from reactions involving the nucleus of the atom. This is indeed so and one of the methods of obtaining this energy is the process known as **fission.**

Fission is the breaking up of a large nucleus into two smaller nuclei of similar size with the release of energy.

Nuclear fission was discovered by the German scientists Otto Hahn (1879 — 1968) and Fritz Strassmann (b. 1902) in Berlin in 1938. The Germans failed to capitalise on the discovery and the idea was exported to the U.S.A. There, in Chicago in 1942, the Italian physicist Enrico Fermi (1901 — 1954) built the first atomic pile or reactor. The reactor first produced energy at 3.45 p.m. on December 2, 1942.

The Fission Reaction

The pairs of nuclei produced when a uranium atom undergoes fission vary from one nucleus to the next. The most commonly occurring pairs have mass numbers around 95 and 140, respectively. The following reaction is a typical example.

$$^{235}_{92}\text{U} + ^{1}_{0}\text{n} \rightarrow ^{141}_{56}\text{Ba} + ^{92}_{36}\text{Kr} + 3\,^{1}_{0}\text{n}$$

There are two things to note about this reaction. Firstly, as suggested earlier, it results in the release of large amounts of energy — the fission of all the nuclei in 1 kg of uranium-235 would release as much energy as the burning of about 3000 t of coal. Secondly, for each neutron causing the reaction three more are produced. (Not all reactions result in three neutrons being produced; sometimes two, and sometimes only one, are produced. On average, 2.5 neutrons are released by each uranium nucleus undergoing fission.) Therefore, under certain conditions, a chain reaction, *Fig. 30.1*, may take place.

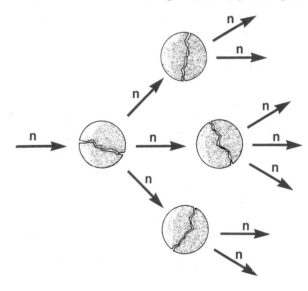

Fig. 30.1 Chain reaction

For a chain reaction to proceed, at least one of the neutrons from each fission reaction must cause fission of another nucleus. In a small mass of uranium, most of the neutrons escape before they collide with another nucleus and so the reaction dies out. Above a certain mass a smaller fraction of the neutrons escape and a **chain reaction** takes place. For uranium-235, the so-called critical mass is about 10 kg and is approximately the size of a hurling ball.

The Atomic Bomb

In the atomic bomb, more properly called a nuclear or fission bomb, two pieces of fissile material of less than the critical mass are brought suddenly together at the appropriate moment by means of a chemical explosion. The first atomic bomb was exploded on July 16, 1945, at Alamagordo, New Mexico. It contained plutonium-239 and the energy released was equivalent to that released by the explosion of 20 000 t of TNT. Three weeks later, on August 6, a uranium bomb, also equivalent to 20 kt of TNT, was exploded over Hiroshima. Two thirds of the city were totally devastated and 75 000 people were killed. Three days later a second bomb, this time containing plutonium, was dropped on Nagasaki. Between 35 000 and 40 000 people were killed and half the city was destroyed.

The Fission Reactor

In a nuclear reactor the energy released in the fission process is used to generate electricity. The rate at which energy is released is controlled by ensuring that, on average, only one of the neutrons produced in each fission reaction causes another reaction. The arrangement of a typical reactor is shown in *Fig. 30.2*.

Natural uranium consists mainly of two isotopes, U-235 (0.7%) and U-238 (99.3%). Unlike U-235, U-238 is not fissionable. Furthermore, it absorbs the fast neutrons produced in the fission of U-235, thus preventing a chain reaction from taking place in natural uranium. The fission of U-235 is effected most easily by slow, or thermal, neutrons. So natural uranium can be used as fuel in a reactor provided that the fast neutrons produced in the fission of U-235 can be slowed down before they are absorbed by the U-238. The material used to slow down the neutrons is called the moderator and the substance most commonly used for this purpose is graphite. "Heavy water" (deuterium oxide, D_2O) is also used.

In a typical reactor, the fuel (natural uranium or uranium slightly enriched in U-235) is contained in long cylindrical rods. A large number of these rods are inserted into the reactor core where they are surrounded by the moderator and the coolant, usually gas (carbon dioxide) or water. The rate of the reaction is controlled by control rods, usually made of steel containing cadmium or boron, which can be lowered to any desired depth into the reactor core. The rods absorb excess neutrons, so that

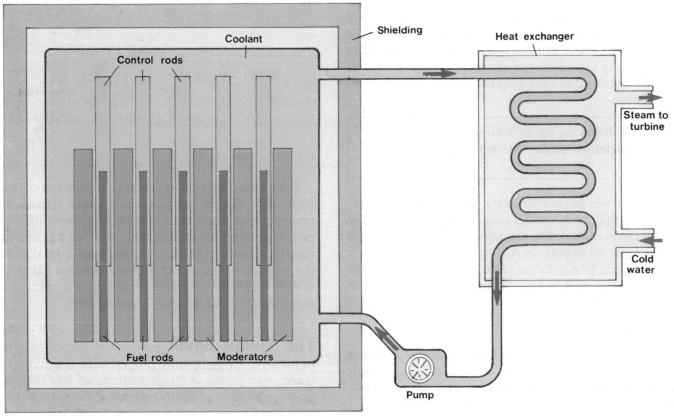

Fig. 30.2 Fission reactor

the reaction proceeds at a constant rate. The reaction can be stopped completely by dropping the control rods fully into the reactor core.

When the fission reaction is in progress, most ($\sim 85\%$) of the nuclear energy released appears first as kinetic energy of the various particles produced. These particles are stopped by the materials in and around the fuel rods and their kinetic energy is converted into internal energy in the core. This energy is carried by the coolant to the heat exchanger, where it is used to produce steam which drives a turbogenerator, just as in a power station using coal or oil as fuel.

The Breeder Reactor

As we have seen U-238 is not fissionable. However, when U-238 is bombarded with neutrons the following reaction takes place.

$$^{238}_{92}U + ^{1}_{0}n \rightarrow ^{239}_{94}Pu + 2 \, ^{0}_{-1}e$$

The plutonium-239 produced in this reaction is, like U-235, fissonable by slow neutrons. In a breeder reactor, a core of highly enriched uranium (uranium which contains a high percentage of U-235) is surrounded by a blanket of natural uranium. Some of the neutrons produced in the fission of U-235 are absorbed by U-238, producing plutonium-239 which in turn undergoes fission. The reactor thus "breeds" its own fuel. In fact, the plutonium is produced faster than it is used. The excess, or indeed all, of the plutonium can be removed and used for various purposes, including making bombs.

Fig. 30.3 Nuclear reactor (Loire, France)

Plutonium-239 is radioactive with a half-life of 24 360 years. It is also one of the most toxic substances known; less than one tenth of a microgram can be fatal.

Most of the radioisotopes used in medicine and industry *(see p. 327)* are artificial isotopes, *i.e.* they do not occur naturally. They are produced by inserting stable isotopes into the core of a reactor where they are bombarded by neutrons. For example, the cobalt-60 used in the treatment of cancer is produced from cobalt-59 in the reaction

$$^{59}_{27}Co + ^{1}_{0}n \rightarrow ^{60}_{27}Co.$$

30.2 Fusion

In certain circumstances two light nuclei may be made to join together to form one heavier nucleus. This process is known as nuclear **fusion**.

> **Fusion is the joining together of two small nuclei to form one larger nucleus with the release of energy.**

The following is a typical example of a fusion reaction.

$$^{2}_{1}H + ^{2}_{1}H \rightarrow ^{3}_{2}He + ^{1}_{0}n$$

Since the deuterium nuclei both have the same charge there is a repulsive electrostatic force between them. To overcome this force, and so get close enough for the attractive nuclear force to become effective, the two nuclei must approach each other at very high speeds. To attain these speeds the gas must be heated to very high temperatures ($\sim 10^8$ K). While such temperatures are difficult to maintain on earth, they are common in the sun and other similar stars. It is believed that the energy emitted by such stars is produced in fusion reactions similar to the one given above.

The Hydrogen Bomb

The first hydrogen, or H-bomb (more properly called a fusion bomb) was constructed in 1952-53. In an H-bomb, a substance containing a mixture of deuterium ($^{2}_{1}H$), tritium ($^{3}_{1}H$) and lithium-6 ($^{6}_{3}Li$) is arranged around a uranium fission bomb. When the fission bomb is detonated, the temperature rises very rapidly to the level required to start fusion reactions in the mixture, resulting in the release of enormous amounts of energy in a very short time. The amount of energy released can be further increased by encasing the bomb in a shell of natural uranium. Neutrons produced in the fusion reactions then promote fission reactions in the uranium. H-bombs are much more powerful than atomic bombs, having an effect equivalent to that of several million tonnes of TNT.

Fusion Reactors

An H-bomb is essentially an uncontrolled fusion reaction. Since the early 1950's research into the production of a controlled fusion reaction has been going on — so far with only limited success. While some such reactions have been produced on a small scale, the energy required to produce them has always far exceeded the energy yield. The problem is to heat the ionised gas, called a plasma, to a temperture of the order of 10^8 K and to keep it hot long enough for the fusion reaction to take place. One method being investigated involves trapping the plasma in specially shaped magnetic fields while passing a very large current through it to heat it. Another is to fire a large number of very powerful lasers simultaneously at a small sphere of the gas.

The European Atomic Energy Commission (Euratom) is currently developing one of the largest magnetically confined fusion plasmas in the world (JET — Joint European Torus) at Culham, England. This machine will ultimately, it is hoped, use the reaction

$$^2_1H + {}^3_1H \rightarrow {}^4_2He + {}^1_0n + 17\ MeV$$

As an EEC member, Ireland is contributing to this development of fusion technology, notably in the area of electronic and laser techniques for monitoring the plasma properties.

A fusion reactor would have a number of important advantages over the fission reactors presently in operation. Firstly, it would be safer in operation since the process gives rise to much lower levels of radiation. Secondly, the serious problem of radioactive waste associated with fission reactors would not exist. Thirdly, the fuel — deuterium — would be very cheap, widely available and, most importantly, virtually limitless. It has been estimated that there is enough deuterium in the oceans of the world to provide energy at the current rate of usage for 1000 million years!

SUMMARY

When certain nuclei, *e.g.* U-235 and Pu-239, are struck by neutrons they split in two with the release of large amounts of energy and of more neutrons. This process is called nuclear fission. In a large enough piece of uranium or plutonium a chain reaction can take place, *i.e.* the neutrons released in the fission of one nucleus cause fission of further nuclei.

In a nuclear reactor the fission reaction is controlled so that energy is released at a steady rate. This energy is carried away by the coolant (gas or water) and used to produce steam which drives a turbogenerator. In a breeder reactor U-238 (which is not fissionable) absorbs neutrons produced in the fission of U-235. The resulting isotope of uranium, U-239, decays with the emission of two β-particles to plutonium-239, which is fissionable.

At very high temperatures two or more small nuclei can be made to join together to form a larger nucleus, again with the release of large amounts of energy. This process is called nuclear fusion and is the source of the energy produced in stars and in the hydrogen bomb. Fusion reactors are not yet technically feasible but are very attractive, especially because of the ready availability of the fuel, deuterium.

Questions 30

DATA: $1 \text{ u} = 1.66 \times 10^{-27} \text{ kg}$

$\qquad c = 3.00 \times 10^8 \text{ m s}^{-1}$

$\qquad e = 1.60 \times 10^{-19} \text{ C.}$

Section I

1. What is meant by nuclear fission?...........................
....
....

2. Name the Italian physicist who built the first atomic reactor ....

3. What is a chain reaction?......................................
....
....

4. Why does a sample of natural uranium not explode?..
....
....

5. What is the function of the moderator in a nuclear reactor?
....
....

6. The control rods in a reactor control......................
....

7. What is a breeder reactor?....................................
....
....

8. What is meant by nuclear fusion?...........................
....
....

9. Why is it difficult to develop a commercial fusion reactor?
....
....

10. Give two advantages which a fusion reactor would have compared with a fission reactor..............................
....

Section II

1. The fission of one uranium-235 atom results in the release of 200 MeV of energy, 85% of which appears as kinetic energy of the particles produced. Calculate (i) the difference in the total mass of the particles before and after the reaction, (ii) the total kinetic energy of the particles in joules.

2. In a fission reactor the fission of one atom of uranium-235 results in the release of 200 MeV of energy, 40% of which ultimately appears as electrical energy. Calculate the number of uranium atoms undergoing fission per second for each kilowatt of electrical power output. If the total power output of the reactor is 100 MW calculate the mass of U-235 "used" in one hour. (Assume one mole of U-235 has a mass of 235 g. Avogadro's number $= 6.0 \times 10^{23} \text{ mol}^{-1}$.)

3. Given that the masses of the proton and neutron are 1.0078 u and 1.0087 u, respectively, and that the mass of the helium-4 nucleus is 4.0026 u calculate the total energy released in the formation of the helium nucleus.

4. Calculate the energy released in the reaction

$$\,^2_1\text{H} + \,^2_1\text{H} \;\rightarrow\; \,^3_2\text{He} + \,^1_0\text{n}$$

In what form is this energy initially released?
(Mass of H-2 = 2.0141 u; mass of He-3 = 3.0160 u; mass of neutron = 1.0087 u.)

REVISION EXERCISES E

DATA:
$$c = 3 \times 10^8 \text{ m s}^{-1}$$
$$e = 1.6 \times 10^{-19} \text{ C}$$
$$h = 6.6 \times 10^{-34} \text{ J s}$$
$$m_e = 9.1 \times 10^{-31} \text{ kg.}$$

Section I

1. The following values were obtained for the current flowing through a diode for a series of values of the p.d. across it.

V/V	0.40	0.50	0.60	0.64	0.68	0.72	0.76	0.80
I/mA	0.5	1.0	2.5	5.0	9.5	20	35	100

Use these data to plot the characteristic curve of the diode. Draw a diagram of the circuit you would use for this experiment. Why might it be desirable to have a resistor in series with the diode?

2. In an experiment to determine the half-life of an isotope the total count recorded on a scaler was noted at 10 s intervals as shown in the following table.

Time/s	0	10	20	30	40	50	60	70	80	90	100
Count	0	562	960	1211	1369	1410	1510	1566	1601	1629	1654

From these data plot a suitable graph and hence determine the half-life of the isotope, given that the background count rate was 28 min^{-1}.
Describe the precautions which should be taken when handling radioactive materials.

Section II

*1. Outline the principle of Milikan's oil drop experiment, explaining the object of the experiment and the results obtained.

*2. What is meant by thermionic emission?
Explain, with the aid of a labelled diagram, how a beam of electrons is produced and controlled in a cathode ray tube.

3. What is the photoelectric effect? In what respect(s) is it similar to thermionic emission?

4. Describe an experiment to show how the photocurrent depends on the intensity of the incident radiation.

*5. Describe an experiment to investigate the relationship between the frequency of the incident radiation and the kinetic energy of the photoelectrons.

*6. What is the minimum frequency of the radiation required to liberate electrons from a metal whose work function is 3.5 eV?

*7. The maximum wavelength of radiation which will liberate electrons from sodium is 5.4×10^{-7} m. What is the work function of sodium?

*8. The work function of gold is 4.9 eV. Calculate the maximum kinetic energy of the electrons emitted when radiation of wavelength 2.0×10^{-7} m falls on a piece of gold. What is the maximum speed of the electrons?

9. Explain, with particular reference to line spectra, how emission spectra and absorption spectra are produced.

10. Draw a labelled diagram of an X-ray tube and explain how the X-rays are produced. What determines the frequency of the X-rays produced?

11. Describe an experiment to plot the characteristic curve of a semiconductor diode.

*12. Describe experiments, one in each case, to establish truth tables for AND, OR and NOT gates.

13. Use circuit diagrams to show how a transistor may be used as (i) a switch, (ii) an amplifier.

14. Give the properties of α-particles, β-particles and γ-rays. How may the relative penetrating abilities of these radiations be determined experimentally?

15. Describe two different types of nuclear radiation detector and explain how each is used.

16. What is meant by the half-life of an isotope? Describe an experiment to determine the half-life of a relatively short-lived isotope.

17. A certain isotope has a half-life of 2 hours. At 9.00 a.m., a container held 2.56×10^{-6} kg of the isotope. What mass of the original isotope will remain in the container at (i) 11.00 a.m., (ii) 1.00 p.m., (iii) 7.00 p.m, (iv) 11.00 p.m.?

18. Explain how nuclear energy is converted to electrical energy in a nuclear power station.

Appendix A

Equations in Mechanics Using Calculus

Calculus provides a more rigorous, and often simpler, approach to many topics in physics. Here we shall use it to derive some of the equations met in the Mechanics section of the course. Each section in this appendix should be studied in conjunction with the appropriate chapter in the text. All symbols representing vector quantities in this appendix are printed in **bold type**.

A.1 Velocity and Acceleration

(Chapter 5)

The average velocity of a body is given by

$$v = \frac{\Delta s}{\Delta t}$$

where Δs is the change in displacement in time Δt.

The velocity at any instant is therefore

$$v = \text{Lt.} \frac{\Delta s}{t}$$
$$t \rightarrow 0$$
$$= \frac{ds}{dt}$$

That is,

$$\boxed{v = \frac{ds}{dt}}$$

or,

> **Velocity is the rate of change of displacement with respect to time.**

Similarly, from the definition of acceleration, the acceleration at a given instant is

$$\boxed{a = \frac{dv}{dt} = \frac{d^2s}{dt^2}}$$

or,

> **Acceleration is the rate of change of velocity with respect to time.**

A.2 Equations of Motion

(Chapter 5)

From the definition of acceleration given above

$$dv = a\,dt$$
$$\Rightarrow \quad v = \int a\,dt$$

If a is constant, *i.e.* does not vary with time, then

$$v = at + C \quad \text{(i)}$$

where C is a constant.
When $t = 0$, $v = u$, where u is the initial velocity.
Substituting in *Eqn. (i)* gives

$$u = 0 + C$$
$$\Rightarrow \quad C = u$$

Therefore, *Eqn. (i)* becomes

$$\boxed{v = u + at}$$

From the definition of velocity given in the previous section

$$ds = v\,dt$$
$$\Rightarrow \quad s = \int v\,dt$$
$$= \int (u + at)\,dt$$
$$= ut + \tfrac{1}{2}at^2 + C \quad \text{(ii)}$$

again assuming a to be constant.
If $s = 0$ when $t = 0$ then

$$0 = 0 + 0 + C$$
$$\Rightarrow \quad C = 0$$

Eqn. (ii) then becomes

$$\boxed{s = ut + \tfrac{1}{2}at^2}$$

A.3 Circular Motion

(Chapter 10)

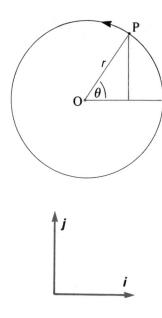

Fig. I

Consider a particle travelling in a circular path of radius r, *Fig. I*. At a particular time, t, the displacement of the particle, relative to O, is

$$s = r \cos \theta \; \mathbf{i} + r \sin \theta \; \mathbf{j} \quad (i)$$

where $\mathbf{i}$ and $\mathbf{j}$ are unit vectors in the directions shown. If the angular speed, ω, of the particle is constant, then

$$\theta = \omega t$$

and *Eqn. (i)* becomes

$$s = r \cos \omega t \; \mathbf{i} + r \sin \omega t \; \mathbf{j} \quad (ii)$$

But,

$$v = \frac{ds}{dt}$$

Therefore,

$$v = - r\omega \sin \omega t \; \mathbf{i} + r\omega \cos \omega t \; \mathbf{j} \quad (iii)$$

But,

$$a = \frac{dv}{dt}$$

Therefore,

$$a = - r\omega^2 \cos \omega t \; \mathbf{i} - r\omega^2 \sin \omega t \; \mathbf{j}$$
$$= - \omega^2 (r \cos \omega t \; \mathbf{i} + r \sin \omega t \; \mathbf{j})$$

That is,

$$a = - \omega^2 s$$

Thus, the acceleration of the particle is in the opposite direction to its displacement, *i.e.* it is towards the centre, and is of magnitude

$$\boxed{a = \omega^2 r}$$

From *Eqn. (iii)* the magnitude of the velocity, *i.e.* the speed, is given by

$$v^2 = r^2\omega^2 \sin^2 \omega t + r^2\omega^2 \cos^2 \omega t$$
$$= r^2\omega^2 (\sin^2 \omega t + \cos^2 \omega t)$$
$$= r^2\omega^2$$
$$\Rightarrow v = r\omega$$

From *Eqn. (ii)* the displacement is along a line whose slope (relative to axes parallel to $\mathbf{i}$ and $\mathbf{j}$) is

$$m_1 = \tan \omega t$$

From *Eqn. (iii)* the velocity is along a line whose slope is

$$m_2 = - \cot \omega t$$

The product of the slopes is

$$m_1 m_2 = (\tan \omega t) \times (- \cot \omega t)$$
$$= - 1$$

Therefore, the velocity is at right angles to the displacement. Since the displacement is along a radius the velocity at any point is along a tangent to the circle at that point.

A.4 Simple Harmonic Motion

(Chapter 11)

A simple harmonic motion is defined by the equation

$$a = -\omega^2 s$$

In calculus notation this becomes

$$\frac{d^2s}{dt^2} = -\omega^2 s$$

This is a second order differential equation, a particular solution of which is

$$s = A \sin \omega t \quad \text{(i)}$$

where A is a constant. Since the maximum value of a sine function is 1, A is the maximum value of the displacement, *i.e.* the amplitude. Differentiating *Eqn. (i)* with respect to t gives the velocity. Thus,

$$v = A\omega \cos \omega t$$

The Periodic Time

Consider two instants, t_1 and t_2, at which the displacement is the same.

$$s_1 = s_2$$
$$\Rightarrow \quad \sin \omega t_1 = \sin \omega t_2$$
$$\Rightarrow \quad \omega t_1 = \omega t_2 + n(2\pi), \text{ where n} = 0, 1, 2, \textit{etc.}$$
$$\Rightarrow \omega(t_1 - t_2) = n(2\pi)$$
$$\Rightarrow \quad t_1 - t_2 = \frac{n(2\pi)}{\omega}$$

The periodic time, T, is the time interval from when s has a particular value until it next has the same value. Therefore, putting n = 1 in *Eqn. (ii)*,

$$\boxed{T = \frac{2\pi}{\omega}}$$

Appendix B

B.1 Torque on a Coil in Magnetic Field

(Chapter 24)

Consider a rectangular coil, of length a and width b, *Fig. I*. The coil carries a current, I, flowing in the direction shown.

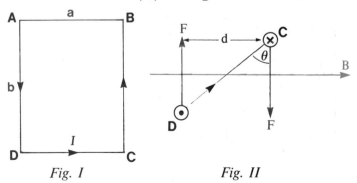

Fig. I   *Fig. II*

The coil is now placed in a uniform magnetic field of flux density, B, with the length of the coil being perpendicular to the page, *Fig. II*. The current now flows into the page at C and out of the page at D. Each of the sides AD and BC now experiences a force, F, in the directions shown. These forces are equal in magnitude and opposite in direction and so constitute a couple. The torque of the couple is

$$T = Fd$$
$$= Fa \cos \theta$$

In a radial field the plane of the coil is always parallel to the field, *i.e.*

$$\theta = 0$$
$$\Rightarrow \quad \cos \theta = 0$$
$$\Rightarrow \quad T = Fa$$

But,

$$F = IbB$$

$$\Rightarrow \quad T = BIab$$

But,

$$ab = A$$

where A is the area of the coil. Therefore,

$$T = BIA$$

This is the torque on a single loop of wire. If the coil has a total of N turns the total torque is given by

$$\boxed{T = BIAN}$$

B.2 The Law of Radioactive Decay

(Chapter 29)

The law of radioactive decay, in mathematical form, is

$$\frac{dN}{dt} = -\lambda N$$

Rearranging gives

$$\frac{1}{N} \, dN = -\lambda \, dt$$

$$\Rightarrow \int \frac{1}{N} \, dN = \int -\lambda \, dt$$

$$\Rightarrow \quad \ln N = -\lambda t + C \quad \text{(i)}$$

If the number of nuclei present when $t = 0$ is N_0, then *Eqn. (i)* becomes

$$\ln N_0 = 0 + C$$

$$\Rightarrow \quad C = \ln N_0$$

Substituting in *Eqn. (i)* for C gives

$$\ln N = -\lambda t + \ln N_0$$

$$\Rightarrow \quad \ln N - \ln N_0 = -\lambda t$$

$$\Rightarrow \quad \ln \frac{N}{N_0} = -\lambda t \quad \text{(ii)}$$

When t is equal to the half-life, $T_{\frac{1}{2}}$

$$N = \tfrac{1}{2} N_0$$

Substituting for N in *Eqn. (ii)* gives

$$\ln \tfrac{1}{2} = -\lambda T_{\frac{1}{2}}$$

$$\Rightarrow \quad \ln 2 = \lambda T_{\frac{1}{2}}$$

$$\Rightarrow \quad \boxed{T_{\frac{1}{2}} = \frac{\ln 2}{\lambda}}$$

Appendix C

Physical Quantities, Units and Symbols

Quantity	Symbol	Unit	Symbol
Acceleration	a	metre per second squared	m s^{-2}
Activity	A	becquerel	Bq
Amount of substance	n	mole	mol
Angular speed	ω	radian per second	rad s^{-1}
Area	A	metre squared	m^2
Atomic number	Z	—	—
Capacitance	C	farad	F
Charge	Q	coulomb	C
Charge density	σ	coulomb per metre squared	C m^{-2}
Coefficient of friction	μ	—	—
Density	ρ	kilogram per metre cubed	kg m^{-3}
Detector count rate	A_d	per second	s^{-1}
Electric current	I	ampere	A
Electric field intensity	E	newton per coulomb	N C^{-1}
		volt per metre	V m^{-1}
Electromotive force	E	volt	V
Energy	E	joule	J
— electrical	W		
— internal	U		
— kinetic	E_k		
— potential	E_p		
Force	F	newton	N
Frequency	f	hertz	Hz
Grating spacing	s	metre	m
Half—life	$T_{1/2}$	second	s
Inductance			
— mutual	M	henry	H
— self	L	henry	H
Length	l	metre	m
Magnetic flux	ϕ	weber	Wb
Magnetic flux density	B	tesla	T
Magnification	m	—	—
Mass	m	kilogram	kg
Mass number	A	—	—
Momentum	p	kilogram metre per second	kg m s^{-1}
Neutron number	N	—	—
Nucleon number	A	—	—
Order of interference	n	—	—
Periodic time	T	second	s
Permittivity	ε	farad per metre	F m^{-1}
Potential difference	V	volt	V
Power	P	watt	W
Pressure	p	Pascal	Pa
Proton number	Z	—	—
Radioactive decay constant	λ	per second	s^{-1}
Refractive index	n	—	—
Resistance	R	ohm	Ω
Resistivity	ρ	ohm metre	Ω m
Sound intensity	I	watt per metre squared	W m^{-2}
Specific heat capacity	c	joule per kilogram kelvin	J kg^{-1} K^{-1}
Specific latent heat	l	joule per kilogram	J kg^{-1}
Speed	u,v	metre per second	m s^{-1}
Speed of waves	c	metre per second	m s^{-1}
Temperature	θ	degree Celsius	°C
— absolute	T	kelvin	K
Time	t	second	s
Torque	T	newton metre	N m
Velocity	u,v	metre per second	m s^{-1}
Volume	V	metre cubed	m^3
Work	W	joule	J

Appendix D

Values of Physical Constants and other Useful Data

Quantity	Symbol	Value
Speed of electromagnetic radiation (in vacuum)	c	3.0×10^8 m s^{-1}
Gravitational constant	G	6.7×10^{-11} N m^2 kg^{-2}
Permittivity of vacuum	ε_0	8.85×10^{-12} F m^{-1}
Electronic charge	e	1.6×10^{-19} C
Specific charge of electron	e/m_e	1.76×10^{11} C kg^{-1}
Rest mass of electron	m_e	9.1×10^{-31} kg
Rest mass of proton	m_p	1.67×10^{-27} kg
Planck's constant	h	6.6×10^{-34} J s
Avogadro's constant	N_A	6.0×10^{23} mol^{-1}
Acceleration due to gravity	g	9.8 m s^{-2}
Speed of sound in air		340 m s^{-1} (approx.)
Wavelength of sodium light		590 nm
S.h.c. of water		4.18 kJ kg^{-1} K^{-1}
S.h.c. of copper		390 J kg^{-1} K^{-1}
S.h.c. of aluminium		910 J kg^{-1} K^{-1}
S.l.h. of vaporisation of water		2.3×10^6 J kg^{-1}
S.l.h. of fusion of ice		3.3×10^5 J kg^{-1}
Refractive index of water		1.33
Refractive index of glass		1.5 (approx.)
Normal atmospheric pressure		1.01×10^5 Pa (76 cm of Hg)
Half-life of radon		53.5 s
E.c.e. of copper		3.3×10^{-7} kg C^{-1}

Appendix E

Useful Equations

The following list includes most of the equations which are
"boxed" in the text. The page number refers to the page on
which the equation first appears.

			page
R	$=$	$2f$	13
m	$=$	$\dfrac{v}{u}$	14
$\dfrac{1}{f}$	$=$	$\dfrac{1}{u} + \dfrac{1}{v}$	15
n	$=$	$\dfrac{\sin i}{\sin r}$	23
n	$=$	$\dfrac{\text{Real Depth}}{\text{Apparent Depth}}$	26
$\dfrac{1}{n}$	$=$	$\sin c$	27
M	$=$	$\dfrac{f_o}{f_e}$	43
v	$=$	$u + at$	52
s	$=$	$ut + \frac{1}{2}at^2$	53
v^2	$=$	$u^2 + 2as$	54
p	$=$	mv	56
ω	$=$	$\dfrac{\theta}{t}$	100
v	$=$	$r\omega$	100
F	$=$	ma	63
F	$=$	$mr\omega^2$	102

			page
F	$=$	$\dfrac{GMm}{r^2}$	66
g	$=$	$\dfrac{GM}{r^2}$	67
T^2	$=$	$\dfrac{4\pi^2 d^3}{GM}$	103
F	$=$	μR	72
T	$=$	Fr	74
T	$=$	Fd	77
p	$=$	$\dfrac{F}{A}$	77
p	$=$	$\rho g h$	79
a	$=$	$-\omega^2 s$	108
T	$=$	$\dfrac{2\pi}{\omega}$	109
T	$=$	$2\pi \sqrt{\dfrac{l}{g}}$	110
W	$=$	Fs	87
E_k	$=$	$\frac{1}{2}mv^2$	88
E_p	$=$	mgh	89
E	$=$	mc^2	91
c	$=$	$f\lambda$	119
f_o	$=$	$\dfrac{f_s}{(1 + v/c)}$	128
$n\lambda$	$=$	$s \sin \theta$	133
f	$=$	$\dfrac{n}{2l} \sqrt{\dfrac{T}{\mu}}$	148
Intensity level	$=$	$\log_{10} \left(\dfrac{I}{10^{-12}} \right)$	154

$$pV = k \qquad 162$$

$$\frac{T}{273.16} = \frac{pV_T}{pV_{t.p.}} \qquad 164$$

$$p = \frac{1}{3}\frac{mN\overline{c^2}}{V} \qquad 172$$

$$p = \frac{1}{3}\rho\overline{c^2} \qquad 172$$

$$T \propto \overline{E}_k \qquad 173$$

$$c = \frac{\Delta E}{m\,\Delta\theta} \qquad 176$$

$$l = \frac{\Delta E}{m} \qquad 179$$

$$F = \frac{1}{4\pi\varepsilon}\frac{Q_1 Q_2}{r^2} \qquad 195$$

$$E = \frac{1}{4\pi\varepsilon}\frac{Q}{r^2} \qquad 203$$

$$E = \frac{F}{Q} \qquad 202$$

$$\psi = EA\cos\theta \qquad 203$$

$$\psi = \frac{Q}{\varepsilon} \qquad 204$$

$$E = \frac{\sigma}{\varepsilon} \qquad 206$$

$$C = \frac{Q}{V} \qquad 209$$

$$C = \frac{\varepsilon A}{d} \qquad 210$$

$$W = \frac{1}{2}CV^2 = \frac{Q^2}{2C} \qquad 211$$

$$R = \frac{V}{I} \qquad 217$$

$$R = \rho\frac{l}{A} \qquad 218$$

$$R = R_1 + R_2 \quad \text{(Series)} \qquad 219$$

$$\frac{1}{R} = \frac{1}{R_1} + \frac{1}{R_2} \quad \text{(Parallel)} \qquad 220$$

$$P = VI = RI^2 \qquad 234$$

$$m = zIt \qquad 243$$

$$F = IlB \qquad 251$$

$$F = qvB \qquad 257$$

$$\Phi = BA \qquad 264$$

$$E = \frac{d\Phi}{dt} \qquad 266$$

$$V_{r.m.s.} = \frac{V_0}{\sqrt{2}} \qquad 269$$

$$\frac{V_{in}}{V_{out}} = \frac{N_p}{N_s} \qquad 273$$

$$E = hf \qquad 290$$

$$hf = hf_0 + \frac{1}{2}mv^2 \qquad 290$$

$$T_{\frac{1}{2}} = \frac{\ln 2}{\lambda} \qquad 326$$

APPENDIX F

Experiments

2.1 To Verify that the Angle of Incidence Equals the Angle of Reflection 10

2.2 To Verify that the Image Distance Equals the Object Distance for a Plane Mirror 11

*2.3 To Find the Focal Length of a Concave Mirror 16

2.4 To Find the Focal Length of a Convex Mirror 19

*3.1 To Verify Snell's Law and hence Determine the Refractive Index of Glass 24

3.2 To Determine the Refractive Index of Water 26

3.3 To Investigate the relationship between the Angle of Incidence and the Angle of Deviation 28

*3.4 To Determine the Focal Length of a Converging Lens 33

3.5 To Determine the Focal Length of a Diverging Lens 35

*5.1 To Measure Velocity and Acceleration 55

*5.2 To Verify the Principle of Conservation of Momentum 58

*6.1 To Verify that $a \propto F/m$ 64

*6.2 To Measure the Acceleration due to Gravity, g (Free fall method) 68

*7.1 To Measure the Coefficient of Dynamic Friction 73

*7.2 To Verify the Principle of Moments 76

*9.1 To Verify the Parallelogram Law 97

11.1 To Determine the Value of g using a Simple Pendulum 111

*13.1 To Measure the Wavelength of Light 134

*14.1 To Investigate the Factors which Determine the Natural Frequencies of a Stretched String 146

*14.2 To Measure the Speed of Sound in Air 150

14.3 To Measure the Speed of Sound in a Gas/Solid 152

*15.1 To Calibrate and Use a Termocouple Thermometer 161

*15.2 To Verify Boyle's Law 162

*17.1 To Measure the s.h.c. of Water 177

*17.2 To Compare the s.h.c.'s of Copper and Water 178

*17.3 To Measure the s.l.h. of Fusion of Ice 180

*17.4 To Measure the s.l.h. of Vaporisation of Water 181

21.1 To Investigate the Variation of Current with p.d. for (i) a length of wire, (ii) a filament lamp 217

21.2 To Determine the e.m.f. of a cell using a Potentiometer 223

*21.3 To Determine the Resistivity of Nichrome 226

*21.4 To Investigate the Variation of Resistance of a Copper Wire with Temperature 228

*21.5 To Investigate the Variation of Resistance of a Thermistor with Temperature 229

21.6 To Calibrate and Use a Thermistor Thermometer 229

*22.1 To Verify Joule's Law 234

23.1 To Investigate the Relationship between p.d. and Current for an Electrolyte 242

*23.2 To Measure the Electrochemical Equivalent of Copper 243

*23.3 To Measure the Internal Resistance of a Cell 247

*27.1 To Plot the Characteristic Curve of a Diode 301

*27.2 To Establish Truth Tables for AND and OR Gates 306

*28.1 To Establish the Truth Table for a NOT Gate 314

29.1 To Investigate the Range of the Three Types of Radiation in Air 322

*29.2 To Determine the Half-life of Radon 326

*Indicates mandatory student experiment.

Answers to Numerical Questions

Chapter 2 (p.21)

Section II

1. 30°
2. (i) 24 cm in front of mirror; 0.6. (ii) 60 cm in front of mirror; 3. (iii) 30 cm behind mirror; 3.
4. 13.3 cm in front of mirror; 3.75 cm.
5. 60 cm in front of mirror; 20 cm.
6. 15 cm.
7. 7.5 cm behind mirror; 0.25.
9. 60 cm in front of mirror; 12 cm behind mirror.
10. 20 cm.
11. 20 cm.

Chapter 3 (p.37)

Section I

9. 30 cm on opposite side of the lens; 1.

Section II

1. 22° 37'.
2. 1.48.
3. 1.4; 2.9 cm
4. 1.6.
5. 1.33.
6. 2.3 m^2
7. 1.41.
8. (i) 30 cm on opposite side of lens; 2. (ii) 10 cm on same side of lens; 2. (iii) 16.7 cm on opposite side of lens; 0.67.
9. 100 cm on opposite side of lens; 0.25.
10. 20 cm; 60 cm on opposite side; 0.33.
11. 53.3 cm; 26.7 cm.
12. (i) 8.6 cm on same side of lens; 0.43. (ii) 10.9 cm on same side of lens; 0.27. (iii) 6.0 cm on same side of lens; 0.6.
13. 10.1 cm.
14. 100 cm; 16.7 cm.

Chapter 4 (p.47)

Section II

3. 8 cm.
6. (i) 65 cm. (ii) 12.
7. 97.6 cm; 2.4 cm.
8. 5 cm.
9. 32'; 10° 33'.
10. 42° 25'.

Revision Exercises A (p.49)

Section I

1. 10.0 cm.
2. 15.6 cm.
3. 1.53.
4. 1.34.
5. 5.0 cm.
6. 19.8 cm.

Section II

5. (i) 27 cm in front. (ii) 3.3 cm.
6. 12 cm behind; 1.6.
7. 15 cm.
10. 5.5 cm behind; 0.45.
11. 8 cm.
15. 43° 36'.
19. 5 cm.
21. 8.2 cm on same side; 0.59.
24. 80 cm.

Chapter 5 (p.60)

Section II

1. 1.3 m s^{-2}.
2. (i) 22.5 m s^{-1}. (ii) 169 m.
3. 6.7 s; 4.5 m s^{-2} N.
4. 3 m s^{-1}.
5. 5.9 m; 29.4 m s^{-1}.
6. (i) 7.3 m. (ii) 0.53 s; 1.92 s. (iii) 4.2 m s^{-1}.
7. 17.8 m
8. 3.3 ms^{-1}

9.   2.0 kg m s^{-1}.
10.   2.5 m s^{-1}.
11.   5.0 m s^{-1}.
12.   (i) 23.6 m s^{-1} S. (ii) 1.4 m s^{-1}. (iii) 53.6 m s^{-1}.
13.   0.113 kg m s^{-1}.
15.   (i) 0.533 m s^{-2}; 1.6 ms^{-2} (ii) 2480 m.

Chapter 6 (p.69)

Section I

2.   25 N upwards.
7.   1.6 m s^{-2}.
15.   25 N; 40.8 kg.

Section II

1.   2.2 kN forward horizontally; 2.75 m s^{-2}.
2.   1.5 m s^{-2}; 750 N.
3.   2.5 kN.
4.   11.8 N; 3.67 m s^{-1}; 16.9 N.
5.   8 m s^{-1} N; 2.8 kN S; 16 m.
6.   39.8 m.
7.   29 s; 1.06 km.
8.   180 kg m s^{-1}.
9.   (a) 549 N; (b) 627 N; (c) 549 N; (d) 470 N.
10.   (i) 2.0 × 10^{20} N. (ii) 3.6 × 10^{22} N.
11.   9.7 m s^{-2}; 2.7 × 10^{6} m
12.   1.7 m s^{-2}.
13.   1708 N.
14.   3.4 × 10^{8} m.

Chapter 7 (p.84)

Section I

10.   1.0 × 10^{3} kg.

Section II

1.   78.4 N.
3.   29.2 m.
4.   (i) 3.4 kN; (ii) 4.3 kN.
5.   0.34; 0.34 m s^{-2}.
6.   39 cm.

7.   3.8 cm; 7.6 cm; 11.4 cm.
8.   20.4 cm.
9.   9.8 × 10^{-3} m^2.
10.   5.5 × 10^{3} kg m^{-3}
11.   101.3 kPa.
12.   183 N.
13.   (i) 3 N. (ii) 3.1 × 10^{-4} m^3. (iii) 1.32 × 10^{3} kg m^{-3}.
14.   2.04 × 10^{6} kg.
15.   200 t.
16.   720 kg m^{-3}.

Chapter 8 (p.93)

Section I

3.   176 J.
9.   30 kJ.

Section II

1.   (i) 200 J. (ii) 14 m s^{-1}.
2.   (i) 500 J. (ii) 14 m s^{-1}. (iii) 4 m s^{-2}. (iv) 3.5 s.
3.   80 N.
4.   245 J; 14 m s^{-1}.
5.   (a) 49 J. (b) (i) 10 J; 39 J. (ii) 30 J; 20 J.
6.   6.4 m s^{-1}; 5.1 m.
7.   8 kg m s^{-1}.
8.   58.7 J; 0.6 J.
9.   4.6 m s^{-1}.
10.   (i) 196 kJ; (ii) 19.8 m s^{-1}. (i) 148 kJ; (ii) 17.2 m s^{-1}.
11.   69 W.
12.   360 kW.
13.   97 kJ; 2.4 kW.
14.   1.8 kJ.
15.   1.4 MW.

Chapter 9 (p.98)

Section I

2.   25 km N.
3.   13 m S 67° 23′ W.
4.   17.3 N at 60° to horizontal.
6.   12 N.

Section II

1. 350 km; 250 km S 53° 8′ E.
2. 233 N at 30° 58′ to horizontal.
3. 112 m E 63° 26′ N.
4. 1.7 m s^{-1} S at 28° 4′ to horizontal.
5. 10.3 m s^{-1} S 14° 2′ W; 12 s.
6. 1.1 m s^{-2} S; 1.1 kN S.
7. (i) −3 m s^{-2}. (ii) 4 m s^{-2}. (iii) 5 m s^{-2} W 53° 8′ S.
8. 153 N; 129 N.
9. 195 km; 195 km.
10. 0.08 kg m s^{-1}; 0.139 kg m s^{-1}.
11. 136 N. (i) 3536 J. (ii) 118 W.
12. 21.7 N at 5° 25′ to the horizontal.

Chapter 10 (p.106)

Section I

10. 2.5 to 1.

Section II

1. 314 rad s^{-1}; 62.8 m s^{-1}.
2. 3.49 rad s^{-1}; 34.9 cm s^{-1}.
3. 9 m s^{-1}.
4. 11.25 m s^{-2}.
5. 0.5 rad s^{-1}.
6. (i) 10 rad s^{-1}. (ii) 40 m s^{-2}. (iii) 100 N.
7. (i) 2.5 m s^{-1}. (ii) 15 N. (iii) 27 N; 3 N.
8. 45 N; 35 N.
9. 4.4 m s^{-1}.
11. 7.3 × 10^{-5} rad s^{-1}; 3.1 km s^{-1}.
12. 2.3 × 10^{6} s.
13. 4.3:1.

Chapter 11 (p.111)

Section I

8. 2 to 1.

Section II

1. 1.97 m s^{-2}.
2. (i) 8 s. (ii) 6.5 m.
3. 1.27 s; 2.54 s.
5. 1.6 m s^{-2}.
7. (i) 1.55 s; (ii) 1.55 s; (iii) 1.47 s; (iv) 1.66 s.

Revision Exercises B (p.113)

Section I

1. 58 cm s^{-2}.
5. 0.40.

Section II

2. 2.5 m s^{-2}; 80 m.
3. (i) 20.4 m. (ii) 10.3 m s^{-1}. (iii) 0.73 s; 3.35 s.
4. 3.1 m s^{-2}; 2.2 kN.
5. 26.5 m; 3.86 s.
6. 7.8 m s^{-1}.
7. 3.0 m s^{-1}.
9. 2.4 kN.
10. (i) 1.5 m s^{-2}. (ii) 15 m s^{-1}. (iii) 75 m.
12. 7.3 × 10^{22} kg.
14. 0.51.
15. 0.58.
16. 0.34.
17. 0.52
18. 0.8 kN.
20. 25 cm.
21. 150 N m.
22. 220 N and 180 N vertically upwards.
23. 47 kPa.
26. 213 N.
28. (i) 160 kJ. (ii) 1.6 kN.
29. 100 W.
30. 18 N at 56° 19′ to the horizontal.
31. 260 N.
32. 4.5 m s^{-1} at 26° 34′ to the bank.
33. 10.2 m s^{-1} N 11° 19′ W.
34. 13 kN.
35. (i) 15 m s^{-1}. (ii) 37.5 N.
36. 40 N; 35 N.
37. (i) 3.9 × 10^{8} m. (ii) 1.0 km s^{-1}.

Chapter 12 (p.129)

Section II

4. 2.0×10^8 m s^{-1}.
5. 4.3×10^{14} Hz to 8.1×10^{14} Hz.
6. 0.3 m s^{-1}.
7. 1.28 MHz; 3.21 m.
8. 340 m s^{-1}.
9. (i) 550 Hz. (ii) 458 Hz.
10. 403 Hz; 324 m s^{-1}.

Chapter 13 (p.141)

Section I

4. 3.3×10^{-6} m.

Section II

3. 5.2×10^{-7} m.
4. $16° 36'$; $343° 24'$; $34° 50'$; $325° 10'$; $58° 58'$.
6. 5.86×10^{-7} m.
7. 3.
8. $2° 58'$.
9. 2.3 cm; 97.7 cm.
11. 4.6×10^{-7} m.
12. 71 km.
13. 3125.

Chapter 14 (p.156)

Section II

3. 16.8 cm; 336 m s^{-1}.
4. 0.227.
5. 425 Hz; 1275 Hz.
6. 850 Hz; 1700 Hz.
7. 18 m.
8. 1 to 15.
9. (i) 120 Hz; (ii) 339 Hz.
10. (i) 7.8×10^{-4} kg m^{-1}; (ii) 110 Hz.
11. 4.7×10^4 N m kg^{-1}.
13. 30 dB; 3 dB.
14. (i) 239 μW m^{-2}; 84 dB. (ii) 60 μW m^{-2}; 78 dB.
15. 3.2×10^{-4} W m^{-2}; 78 dB.

Chapter 15 (p.167)

Section II

2. 44 °C.
3. (i) 62.5 °C. (ii) 335.5 K.
4. 9.71×10^4 Pa.
5. 12 Pa m^3.
6. 328 K.
7. 1.36.
8. 375 K.

Chapter 16 (p.174))

Section II

1. 1.5×10^{24}.
2. 2.68×10^{19}.
3. (i) 0.08 mol. (ii) 2.6×10^{-27} kg.
4. 2.1×10^{17}.
5. 5.5; 5.7.
6. 1.4×10^3 m s^{-1}.
7. 0.15 kg m^{-3}.
8. 15:1.

Chapter 17 (p.185)

Section II

1. 50.2 kJ.
2. 2100 J kg^{-1} K^{-1}.
3. 32.1 °C.
4. 3.3 kJ kg^{-1} K^{-1}.
5. 7.5 K.
6. 15:8.
7. 0.3 K.
8. 23 °C.
9. 19.8 kJ.
10. 9.5 g.
11. 6.18×10^6 J.
12. 1.3 K.

Revision Exercises C (p.187)
Section I
1. 5.90×10^{-7} m.
2. 5.8×10^{-5} kg m^{-1}.
3. 360 m s^{-1}.
5. 2100 J kg^{-1} K^{-1}.
6. 10.8 : 1
7. 3.3×10^5 J kg^{-1}.
8. 2.4×10^6 J kg^{-1}.

Section II
4. 5.80×10^{-7} m.
8. 0.17 m.
13. 25 °C.
15. 75 °C.
16. 2.7×10^{22}.
19. 4.7×10^2 J kg^{-1} K^{-1}.
21. 3.3×10^5 J kg^{-1}.
22. 2.3×10^6 J kg^{-1}.
23. 51 °C.
24. 21 °C.

Chapter 18 (p.197)
Section II
2. 7.2×10^{-5} N; 1.1×10^{-9} N.
3. 0.19 mN.
4. 21 mN
5. 17 mN
6. 95 mN.
7. 2.5×10^{26} m s^{-2}.
8. 7.7 nC.

Chapter 19 (p.207)
Section I
3. 80 μJ.

Section II
1. 2.0 kV.
2. 6000 J.

3. 5.9×10^7 m s^{-1}.
4. 2.6 kV.
5. 1.3×10^5 V m^{-1}.
6. (i) 1.0 μC; (ii) 3.6 μC m^{-2}.
7. 9.9×10^4 N m^2 C^{-1}.
8. (i) 0.92 MV m^{-1} from A.
   (ii) 0.67 MV m^{-1} from A. (iii) 0.12 MV m^{-1} towards A.
9. 1.5×10^4 V m^{-1}.
10. (i) 5.4×10^{16} m s^{-2}; (ii) 3.1×10^5 V m^{-1}.
11. 3.0 mm.
12. 6.7 mm at 26° 34′ to initial velocity.

Chapter 20 (p.213)
Section II
1. 1.4 nC.
2. 0.18 μF; 3.6 μC.
3. 1.1×10^7 V m^{-1}.
4. (i) 2.3 μm. (ii) 26 V.
5. 0.94 J; 94 mC.
6. 20 V.
7. 9.6×10^{-2} J.
8. 90 000.

Chapter 21 (p.231)
Section II
1. 2×10^{-7} N.
2. 72 μC.
3. 0.28 A.
4. (i) 0.5 A. (ii) 3 V.
5. (i) 1.5 A. (ii) 0.9 A. (iii) 0.6 A.
7. (i) 2.1 A. (ii) 0.38 A. (iii) 5.6 V.
8. 0.74 mm.
9. 46 m.
10. 7.2 V; 0 V.
11. 1.3 V; 2.1 V.
12. 1.25 A; 0.75 A; 0.5 A.
13. 0.59 A; 0.32 A; 0.27 A.
14. 6 V; 0.30 A.
15. 7.5 Ω.

Chapter 22 (p.240)
Section II
1. 21.6 MJ.
2. 75 kJ.
3. 12.7 A; 0.45 A; 5.7 A.
4. 807 Ω; 24.2 Ω; 194 Ω.
5. 3.2 Ω; 3.8 A.
6. 200 J.
7. £1.78.
8. 13 A; 1 A; 3 A.
9. 26 K.
10. 9.9 MJ; 18p.

Chapter 23 (p.249)
Section I
11. 15.

Section II
1. 0.99 g.
2. 1.1×10^{-6} kg C^{-1}.
3. 1.2×10^3 s.
4. (i) 4.56 g. (ii) 2.8 μm.
5. 3.8×10^{19}; 0.1 A.
6. (i) 1.8 A. (ii) 1.6 kC. (iii) 5.1×10^{21}.
7. $8.9 \times 10^5 C kg^{-1}$
8. (i) 1.5 A. (ii) 40 A.
9. (i) 8.82 V. (ii) 2%.
10. 40 A.
11. 0.5 A; 0.33 A; 1.0 Ω.

Chapter 24 (p.262)
Section II
1. 2.0 N vertically.
2. 0.33 T.
3. 33° 38′.
4. 0.63 pN.
5. 0.98 m.
6. 1.2 m.
7. (i) 1.01 Ω. (ii) 0.02 Ω.
8. (i) 2425 Ω. (ii) 49.9 kΩ.
9. 0.37 A; 7.5%.
10. 150 Ω.

Chapter 25 (p.277)
Section II
1. 5.1×10^{-7} Wb; 2.55 μV.
2. 0.15 Wb; 0.15 V.
3. 15 mT.
4. 9.2 μT; 79° 24′.
5. 9.8 mJ.
6. 5.5 V.
7. 40.
8. 16 500.
9. 111 W; 0.5 A; 30 mA.
10. 14 V.

Revision Exercises D (p.280)
Section I
1. 1.43 V.
2. 1.29×10^{-6} Ω m.
3. 3.7 Ω.
4. 740 Ω.
6. 3.32×10^{-7} kg C^{-1}.
7. 1.07 Ω.

Section II
1. 27 N.
2. 0.33 V m^{-1} from A to BC and at right angles to BC.
3. (i) 45 μC. (ii) 3.0 μA.
4. (i) 0.5 A. (ii) 3 V.
5. (i) 30 Ω. (ii) 0.4 A. (iii) 4.0 V. (iv) 0.24 A.
6. (i) 15 Ω. (ii) 0.4 A. (iii) 2.0 V.
9. 37.5 kJ.
10. 18 kJ; 750 C.
11. (a) (i) 5 A; (ii) 1 kW. (b) (i) 1 A; (ii) 40 W.
12. 3.2 Ω; 7.5 A.
16. 0.99 g.
17. 1.1×10^{-8} kg C^{-1}.
20. 7.9×10^{-2} N.
24. (i) 0.1001 Ω. (ii) 2.0004×10^{-2} Ω.
25. (i) 4.9 kΩ. (ii) 49.9 kΩ.
30. 18 000.

Chapter 26 (p.294)
Section II

2. 6.6×10^{-19} J; 2.6×10^{-19} J.
3. 6.2×10^{-7} m.
4. (i) 1.5×10^{-19} J. (ii) 5.8×10^5 m s^{-1}.
5. (i) 4.5 eV. (ii) 1.1×10^{15} Hz.
6. (i) 6.6×10^{-34} J s. (ii) 4.2 V.
7. 1.6×10^{-14} J; 1.9×10^8 m s^{-1}.
8. 1.2×10^{19} Hz; 2.5×10^{-11} m.
9. 9.1×10^{15}

Chapter 27 (p.307)
Section II

5. (i) 5 V. (ii) 0.02 V.

Chapter 28 (p.317)
Section II

3. (i) 10 mA. (ii) 10.05 mA.
4. (i) 0.27 mA. (ii) 44.7 mA.

Chapter 29 (p.329)
Section II

2. 5; 2.
3. 10.2 days.
4. 30 d.
5. 3.1%.
6. 8.0×10^{-13} J.
8. 1.1×10^{-16} kg.
9. 632.
10. (i) 5.3×10^{-3} s^{-1}. (ii) 2.2 min.
11. 4.7×10^9.

Chapter 30 (p.335)
Section II

1. (i) 0.214 u. (ii) 2.72×10^{-11} J.
2. 7.8×10^{13}; 11 g.
3. 28.4 MeV.
4. 3.3 MeV.

Revision Exercises E (p.336)
Section I

2. 20 s.

Section II

6. 8.5×10^{14} Hz.
7. 2.3 eV.
8. 1.3 eV; 6.7×10^5 m s^{-1}.
17. (i) 1.28 mg. (ii) 0.64 mg. (iii) 0.08 mg. (iv) 0.02 mg.

INDEX

α — radiation 321
Absolute zero 164
Absorption spectra 140, 292
A.c. and capacitors 269
A.c. and inductors 270
A.c., effective value of 269
Acceleration 52
Acceleration due to gravity 67
Acceleration, centripetal 101
Accommodation, power of 39
Accumulators 248
Activity (of radioactive source) 325
Air track 62
Alternating current 267, 268
Alternator 267
Ammeter 260
Ampere 215
Ampère, André Marie 216
Amplitude 118
AND gate 305
Aneroid barometer 81
Angle of dip 255
Angular displacement 100
Angular speed 100
Antinode 121
Apparent depth 25
Arago's disc 275
Archimedes' principle 81
Artificial radioactivity 333
Astigmatism 134
Atmospheric pressure 80
Atomic bomb 331
Atomic mass unit 169
Atomic number 318
Atomic theory 193
Avogadro constant (number) 169
Avogadro's law 173
β - radiation 321
Back e.m.f. 268
Band spectra 140
Barometer 80
Beats 123
Becquerel 325
Becquerel, Henri 319
Bel 154
Bi-polar transistor 309, 310
Binoculars 45
Blind spot 40
Bourdon gauge 78
Boyle's law 161
Boyle, Robert 163
Breeder reactor 333
Brownian motion 169

Callan, Nicholas 275
Camera 40
Capacitance 209
Capacitor 209
Capacitor, electrolytic 212
Capacitor, energy in 211
Capacitor, variable 212
Capacitors, uses of 212
Cathode ray tube 286
Cathode rays 284
CdS cell 298
Cell, internal resistance of 246
Cells 245
Celsius scale 158, 159
Celsius, Anders 159
Centre of curvature 12
Centripetal acceleration 101
Centripetal force 102
Chain reaction 331
Charge 190
Charging by induction 193
Circuit breakers 238
Cloud chamber 319
Coefficient of friction 72
Coherent 121
Colour thermometer 160
Complementary colours 137
Composition of vectors 96
Compression 119
Conduction 182
Conductors 192
Conservation of flux 203
Constant volume gas thermometer 165
Constructive interference 121
Control rods 332
Convection 183
Copernicus 59
Corpuscular theory 131
Cosmic rays 284
Coulomb 195, 216
Coulomb's law 195
Coulomb, Charles 197
Couple 77
Critical angle 26
Curie, Marie 323
Current amplifier 311, 313
D.c. motor 258
Decay chain 323
Decay constant 325
Decibel 154
Density 78
Depletion layer 299
Destructive interference 121
Deviation, angle of 28
Dielectric 210
Diffraction 119

Diffraction grating 132
Diode 300
Dip circle 255
Discharge 293
Dispersion 136
Displacement 51
Displacement, angular 100
Doping 297
Doppler effect 126
Dry cell 245
Dynamic friction 72
Earth's magnetic field 255
Earthing 192, 239
Einstein, Albert 91
Electric charge 190
Electric circuits 238
Electric current 215
Electric field flux 203
Electric field intensity 202
Electric field lines 194
Electric fields 194
Electric flux 203
Electrochemical equivalent 243
Electrolysis 241
Electrolysis, Faraday's law of 243
Electrolyte 241
Electromagnet 253
Electromagnetic induction 264
Electromagnetic relay 254, 313
Electromotive force 200
Electron 193, 283
Electron, specific charge of 285
Electronvolt 199
Electroplating 244
Electroscope 191
Electrostatic forces 190
Elliptical orbit 104
E.m.f. 200
E.m.f. of cell 247
E.m.f., back in motor 268
E.m.f., induced 264
Energy 87
Energy and mass 91
Energy levels 291
Energy, conservation of 88
Energy, forms of 87
Energy, kinetic 88
Energy, potential 89
Equations of motion 52
Equilibrium 75
Experimental error 7
Extrinsic conduction 298
Eye 39
Farad 209
Faraday cage 205

Faraday, Michael 206
FET 309
Field effect transistor 309
Field intensity and potential 206
Fission 331
Fission reactor 332
Fixed points 158
Fletcher's trolley 55
Flotation, law of 83
Fluorescence 139
Flux linkage 266
Focal length 12
Focus 12
Force 62
Force on current 256
Force on moving charge 256
Force, centripetal 102
Forward bias 300
Fraunhofer lines 140
Frequency 119
Friction 72
Friction, limiting 72
Fuel rods 332
Fuses 237
Fusion 333
γ - radiation 322
Galileo 59
Galvanometer 259
Gases, conduction in 292
Gates 304
Gauss's law 204
Geiger-Muller (GM) tube 320
Generators 266
Geostationary orbit 104
Glow discharge 293
Gold leaf electroscope 191
Gravitation, Newton's law of 66
Gravity 66
Greenhouse effect 185
Half-life 325
Harmonics 145
Heat capacity 178
Heat pump 179
Hertz 119
Hertz, Heinrich 139
Holes 296
Horsepower 92
Huygens, Christiaan 132
Hydrogen bomb 333
Hydrometer 83
Ideal gas 164
Ideal gas scale 164
Impurity element 297
Induced current 264
Induced e.m.f. 264
Inductance 270

Induction 193
Induction coil 270
Induction motor 274, 276
Inertia 56
Infra-red radiation 138
Initial velocity 53
Insulation 183
Insulators 192
Integrated circuits 316
Intensity level 154
Intensity of sound 153
Interference 120
Interference pattern 121
Internal energy 176
Internal reflection 26
Internal resistance of cell 246
Intrinsic conduction 297
Inverse square law 195
Inverted image 13
Ion 241
Isotopes 318
JET 334
Joule 87
Joule's law 234
Joule, James 235
Junction diode 300
Junction voltage 299
Kelvin 164
Kinetic theory 169
Kinetic theory and temperature 172
Kinetic theory equation 171, 172
Kinetic theory, assumptions 170
Kirchhoff's laws 224
Kundt's tube 152
Laminated (core of transformer) 274
Latent heat 179
Lateral inversion 11
Lead-acid cell 248
L.e.d. 303
Left hand rule 256
Lens formula 32
Lens, converging 31
Lens, diverging 35
Lenz's law 265
Lever 76
Light dependent resistor 298
Light, speed of 135
Light, wavelength of 134
Light-emitting diode 303
Lightning conductor 202
Load resistor 314
Logic gates 304
Long-sighted 39
Longitudinal wave 119

Loudness 149
Loudspeaker 257
Luminous 9
Magnetic declination 255
Magnetic dipole 251, 253
Magnetic domains 253
Magnetic field 251
Magnetic field due to current 253
Magnetic field lines 252
Magnetic field, force on current in 256
Magnetic flux 264
Magnetic flux density 251
Magnetic forces 251
Magnetic variation 255
Majority charge carriers 298
Mass 56
Mass number 318
Metre 51
Metre bridge 226
Michelson's experiment 135
Microscope, compound 41
Microscope, simple 41
Millikan's oil drop experiment 283
Millikan, Robert 284
Minority charge carriers 298
Mirage 30
Mirror formula 14
Mirror, concave 12
Mirror, convex 18
Moderator 332
Mole 170
Moment of force 74
Moments, principle of 75
Momentum 56
Momentum, conservation of 56
Monochromatic light 132
Motor, d.c. 258
Moving coil loudspeaker 257
Multimeter 261
Mutual induction 270
N-type semiconductor 298
Neutrons 193, 318
Newton 62, 63
Newton's law of Gravitation 66
Newton's laws of motion 62
Newton, Isaac 65
No-parallax technique 11
Node 121
Noise 149
Normal adjustment 42
NOT gate 313
Notes 149
Nuclear reactions 323
Nucleus 318

Nucleus, radius of 318
Ohm 217
Ohm's law 217
Ohm, Georg Simon 218
Ohmmeter 261
Oil drop experiment 283
Optical fibres 30
OR gate 305
Overtones 145
P-n junction 299
P-type semiconductor 298
Parallelogram law 96
Parsons, William 46
Pascal 77
Pendulum 110
Pendulum, period of 110
Period (of wave) 118
Periscope 11, 29
Permittivity 195
Phase 118
Photodiode 304
Photoelectric emission 288
Photoelectric law 290
Photons 290
Pitch 149
Planck's constant 290
Plasma 334
Platinum resistance thermometer 160
Plugs 239
Point discharge 202
Point effect 201
Polarisation 125
Polarisation (cell) 245
Polarisation of light 135
Positive holes 296
Potential difference 199
Potential divider 221
Potential energy 89
Potential, zero of 199
Potentiometer 222
Powder track timer 54
Power 92
Pressure 77
Pressure, atmospheric 80
Primary cells 245
Primary colours 137
Principle of moments 75
Prism 28
Proof plane 201
Protons 193, 318
Quality 149
Quanta 290
Quantum theory 289
R.m.s. voltage/current 269
Radial magnetic field 259

Radiation 184
Radiation, hazards of 328
Radiation, uses of 327
Radioactive decay, law of 325
Radioactivity 319
Rarefaction 119
Ratemeter 321
Real image 13
Rectifier 302
Reflection 9
Reflection, laws of 9
Reflection, total internal 26
Refraction 23
Refraction, laws of 23
Refractive index 23
Relative atomic mass 169
Resistance 217
Resistance and temperature 228
Resistance, measuring 225
Resistivity 218
Resistors 219
Resistors in parallel 220
Resistors in series 219
Resolution of vectors 97
Resonance 125
Resultant force 62
Reverse bias 300
Rheostat 221
Ring main circuit 238
Röntgen, Wilhelm 288
Root-mean-square current 269
Root-mean-square speed 172
Root-mean-square voltage 269
Rosse, Earl of 46
Rutherford, Ernest 318
S.h.m., period of 108
Satellite, period of 103
Scalar quantity 51
Scaler 321
Secondary cells 248
Secondary colours 137
Self-induction 270
Semiconductors 296
Seven segment display 304
Shock, electric 237
Short-sighted 40
Sign convention (mirrors) 15
Sign convention (Doppler effect) 128
Significant figures 7
Silicon 296
Simple harmonic motion 108
Slip rings 267
Smoothing circuit 303
Snell's law 23
Solar constant 184

Solenoid 253
Solid state detector 320
Sonometer 146
Sound, intensity of 153
Sound, interference of 143
Sound, refraction of 144
Sound, speed of 149
Sound, transmission of 143
Spark plug 272
Specific charge of electron 285
Specific heat capacity 176
Specific latent heat 179
Spectra 139, 291
Spectrometer 45
Spectrum, absorption 140
spectrum, band 140
Spectrum, continuous 139
Spectrum, electromagnetic 138
Spectrum, emission 139
Spectrum, line 139
Spectrum, visible 137
Speed of light 135
Speed, angular 100
Split-ring commutator 259
Standard temperature/pressure 165
Standing wave 121
Stationary waves 121, 145
Superconductivity 228
Superposition, principle of 120
Symbols 6
System, closed 56
SI system 6
Telescope, astronomical 42
Telescope, magnifying power of 42
Telescope, reflecting 44
Telescope, terrestrial 43
Temperature 158
Tesla 251
Thermionic emission 284
Thermistor 229, 299, 312
Thermocouple 160
Thermometers 159
Thermometric property 158
Threshold frequency 289
Threshold of audibility (hearing) 153
Ticker tape timer 55
Torque 74
Torque on coil 259
Transformer 272

Transformer, energy losses in 274
Transformer, uses 274
Transistor 309
Transistor as a switch 312
Transverse wave 119
Triangle law 95
Triple point 164
Truth table 305
Two-way switch 239
U-value 183
Ultra-violet radiation 139
Ultrasonics 152
Unified atomic mass unit 169
Van de Graaff generator 190
Vector quantity 51
Vectors, addition of 95
Vectors, composition of 96
Vectors, resolution of 97
Velocity 51
Virtual image 10
Volt 199
Volta, Alessandro 245
Voltage amplifier 314
Voltage inverter 313
Voltameter 241
Voltmeter 261
Walton, E.T.S. 324
Watt 92
Watt, James 92
Wave pulse 117
Wave theory 131
Wavelength 118
Waves 117
Waves, longitudinal 119
Waves, reflection of 124
Waves, refraction of 124
Waves, stationary 121
Waves, transverse 119
Weber 264
Weight 67
Wheatstone bridge 225
Work 87
Work function 290
X-rays 287
X-rays, uses of 288
Young's experiment 131
Young, Thomas 134

12.—PHYSICS

This syllabus has been designed as a complete course in physics for second level pupils. It will also provide a basic course in physics for pupils who intend to progress to further studies/training in science.

The general aims of the syllabus are to contribute to the general education of second level pupils by:—

1. Developing the ability to observe, to think logically, to understand and use scientific method and to communicate effectively;

2. Providing a reasonably broad perspective of physics, and so developing an understanding of the physical environment and Man's interactions with it;

3. Giving an understanding of the fundamental principles of physics and their application to everyday life;

4. Developing an appreciation of physics as a human endeavour thereby enriching their experience of life.

Physics is an experimental subject and practical work by the pupils is to be regarded as an integral part of the course. A list of suitable experiments to be undertaken by pupils is included in each section of the syllabus. Such experiments are essential for a proper understanding of the syllabus content. Teachers are encouraged to extend the list of experiments where feasible.

As the syllabus has been drawn up on the basis that pupils will devote an appropriate amount of time to laboratory work, a candidate will not be admitted to the Leaving Certificate Examination in this subject in any case where the Department considers that an adequate course of laboratory work has not been followed by such candidate. For this purpose records of practical work done should be kept and be available for inspection.

Teachers should also use practical demonstrations where appropriate: they may find the computer a useful aid in such work. They are recommended to arrange liaison with local industry so that the applications of physics in technology may be observed. Teachers, in addition, are encouraged to refer to the historical development of the subject where appropriate. In this regard reference should be made to the lives and work of great physicists.

The solution of problems, based as far as possible on everyday examples, forms an integral part of this course. Pupils should be familiar with the SI system of units, including the use of standard symbols for units. They should be aware of the possible sources of error in experimental work and should have an understanding of the importance of significant figures. It is recommended that teachers encourage the use of calculators in the classroom/laboratory.

The syllabus content given in *italics* will not be examined at Ordinary Level.

SYLLABUS

Ordinary and Higher Level Courses

MECHANICS

MOTION

1. Linear motion.	Units of mass, length and time – definition of units not required; displacement, velocity and acceleration – definitions and units. Equations of motion – derivation and simple applications, including vertical motion under gravity.
2. Vectors and scalars.	Distinction between vector and scalar quantities. *Composition and resolution of coplanar vectors. Reference should be made, where appropriate throughout the rest of the course, to the vector nature of the various physcial quantities, e.g: velocity, acceleration.*
3. *Circular motion.*	*Uniform motion in a circle. Definition of angular velocity. Derivation of formulae for tangential velocity and centripetal acceleration.*
4. *Simple harmonic motion.*	*Definition:* $a = -\omega^2 x$. *Derivation of equation for period. Motion of simple pendulum as simple harmonic motion.*

FORCES

1. Newton's laws of motion.	Demonstration of laws using air track, CO_2 pucks, or other suitable method. Force and momentum – definitions and units. $F = ma$ as special case of Newton's second law.
2. Conservation of momentum.	Principle of conservation of momentum—demonstration using air track or other suitable method; application to acceleration of spacecraft, jet aircraft (reference should be made to the fact that these phenomena may also be explained by Newton's third law). Problems involving change of mass need not be considered.
3. Pressure.	Definition and unit, i.e. the pascal. Reference to the fact that fluid pressure depends on depth and density. Archimedes' principle.
4. Moments.	Definition. Levers. Only problems involving coplanar, parallel forces need be considered.
5. Gravity.	Newton's Universal Law of Gravitation. Weight = mg. Variation of g, and hence W, with distance from centre of earth (effect of centripetal acceleration not required). Value of acceleration due to gravity on other bodies in space, e.g. moon, etc. *Circular satellite orbits – derivation of the relationship between the period, the mass of the central body and the radius of the orbit. Reference to the fact that orbits are frequently elliptical rather than circular.*
6. Friction.	Limiting friction—definition and demonstration. Reference to importance of friction in everyday experience, e.g. walking, cycling, brakes, etc. Reference to use of lubricants, including air, e.g. hovercraft. Definition of coefficients of static and dynamic friction. Problems involving static friction on a horizontal plane only need be considered.

ENERGY

1. Work.

Definition and unit. Problems involving force and displacement in same direction only need be considered.

2. Energy.

Energy as ability to do work. Different forms of energy. Mass as a form of energy: $E = mc^2$. Conversions from one form to another. Demonstrations of different energy conversions. Principle of conservation of energy. P.E. $= mgh$ and K.E. $= \frac{1}{2}mv^2$. *Derivation of formulae. Proof that P.E. + K.E. is constant for a freely falling body.*

3. Power.

Power as rate of doing work or rate of energy conversion. Unit. Estimation of average power developed by person running upstairs, repeatedly lifting a " weight ", etc.

EXPERIMENTS

1. Measurement of velocity and acceleration.
2. *Verification of the parallelogram law.*
3. To show that $a \propto F/m$.
4. Verification of the principle of conservation of momentum.
5. Verification of the principle of moments.
6. Measurement of g.
7. Measurement of the coefficient of dynamic friction.

HEAT

HEAT AND TEMPERATURE

1. Concept of temperature.

Heat as the energy transferred between two places as a result of the temperature difference between them.

2. Thermometric properties.

Demonstration of the variation of length, pV for a fixed mass of gas, e.m.f., resistance and colour with temperature.

3. Thermometers.

Reproducible temperatures. Choice of interval: a Celsius scale defined by $\theta = 100 \, (Y_\theta - Y ice)/(Y steam - Y ice)$ where Y is the thermometric property and the ice and steam points are assigned the numbers zero and 100 respectively; the Kelvin scale as expressed by the relationship $T = k(pV)$ for the ideal gas, and by assigning the number 273.16 to the highly reproducible triple point of water. (Ideal gas behaviour achieved with real gases at successively lower pressures.) Relationship between Celsius and Kelvin scales. *Demonstration of the constant volume gas thermometer. The gas thermometer as the standard thermometer.*

4. *The kinetic theory.*

Demonstration of Brownian movement. Assumptions of the kinetic theory. Derivation of the kinetic theory equation and the relationship of this to Boyle's law and the concept of temperature. Proof of Avogadro's law on the assumption that the average kinetic energy of all gas molecules is the same at a given temperature. Avogadro's constant.

QUANTITY OF HEAT

1. Specific heat capacity.

Definition and unit. Applications, e.g. storage heaters.

2. Specific latent heat.

Definition and unit. Reference to practical applications, e.g. heat pump.

HEAT TRANSFER

1. Conduction.

Qualitative comparison of rates of conduction through various solids. Reference to U-values in domestic situations.

2. Convection.

Reference to everyday examples of convection.

3. Radiation.

Radiation from the sun. Solar constant. Reference to solar heating.

EXPERIMENTS

1. Verification of Boyle's law.
2. Calibration and use of a thermometer.
3. Measurement of specific. heat capacity, e.g. of water or of a metal by a mechanical or electrical method.
4. Comparison of specific heat capacities by method of mixtures.
5. Measurement of specific latent heat of fusion and of vaporisation of water.

WAVE MOTION

WAVES

1. Properties of waves.

Longitudinal and transverse waves. Frequency, amplitude, wavelength, velocity. The relationship $v = f\lambda$.

2. Wave phenomena.

Polarisation, reflection, refraction, interference and diffraction – simple demonstrations using slinky, ripple tank, microwaves or other suitable method. Stationary waves – relationship between internode distance and wavelength. Diffraction effects at an obstacle and at a slit with reference to the significance of wavelength.

3. The Doppler effect.

Qualitative treatment as illustrated by sound from a moving source, e.g. a train. *Simple quantitative treatment for moving source and stationary observer.*

SOUND

1. Wave nature of sound.

Demonstration of interference, e.g. using two speakers and a signal generator. Demonstration that sound requires a medium. Speed of sound in various media.

2. Characteristics of notes.

Amplitude and loudness, frequency and pitch. Factors which affect quality. Frequency limits of audibility.

3. Resonance.

Demonstration using tuning forks or other suitable method.

4. *Intensity of sound.*

Necessity for a log scale. Threshold of hearing. Definition of the bel. The decibel scale.

5. Vibrations in strings.

Factors which determine the natural frequency of a string.

EXPERIMENTS

1. Measurement of the speed of sound in air.
2. Investigation of the variation of the frequency of a string with length and tension.

LIGHT

REFLECTION

1. Laws of reflection.

Demonstration using ray box, laser or other suitable method.

2. Mirrors.

Images formed by plane and spherical mirrors. Simple exercises on mirrors by ray tracing or use of formula. Magnification.

REFRACTION

1. Laws of refraction.

Demonstration using ray box, laser or other suitable method. Refractive index. Deviation by a prism using ray tracing. *Refractive index in terms of relative speeds.*

2. Lenses.

Images formed by single thin lenses. Simple exercises on lenses by ray tracing or use of formula. Magnification.

3. Total internal reflection.

Critical angle. Relationship between critical angle and refractive index. Demonstration and explanation of transmission of light through optical fibres.

OPTICAL INSTRUMENTS

1. The eye.

Optical aspects only need be considered.

2. **The microscope.**

Single lens microscope (simple microscope) – ray diagram to show formation of image. *Optical system of the compound microscope – image formation.*

3. The astronomical telescope.

Ray diagram to show formation of image in a refracting telescope in infinite (normal) adjustment. *Magnifying power of the telescope in normal adjustment.*

4. *The spectrometer.*

The spectrometer and the function of its parts.

WAVE NATURE OF LIGHT

1. Diffraction and interference.

Young's slits to demonstrate the wave nature. Use of the diffraction grating formula, $n\lambda = d\sin\theta$. *Derivation of formula.*

2. Light as a transverse wave motion.

Demonstration of polarisation using pieces of polaroid or other suitable method.

SPECTRA

1. Dispersion.

Dispersion by a prism and a diffraction grating. Recombination by a prism or Newton's disc.

2. Colours.

Primary, secondary and complementary colours. Addition of colours, e.g. stage lighting and television. Pigment colours need not be considered.

3. Electromagnetic spectrum.

Relative positions of the radiations in terms of wavelength, frequency and energy. Detection and uses of ultra-violet and infra-red radiation. *Description of a terrestrial method for measuring the speed of light.*

4. Emission and absorption spectra.

Demonstration of various types of spectra. Reference to energy levels.

EXPERIMENTS

1. Measurement of the focal length of a concave mirror.
2. Verification of Snell's law.
3. Measurement of the focal length of a converging lens.
4. Measurement of refractive index of a liquid or a solid.
5. Measurement of the wavelength of monochromatic light.

ELECTRICITY

CHARGES

1. Electrification by contact.

Charging by the rubbing together of dissimilar materials. Types of charge. Demonstration of forces between charges. Conductors and insulators.

2. Electrification by induction.

Separation of charges by induction – demonstration using an insulated conductor and a nearby charged object.

3. Electroscope.

Structure and uses.

4. The Van de Graaff generator.

Principle of the generator – total charge resides on the outside of a metal object; charges tend to accumulate at points. Point discharge.

ELECTRIC FIELD

1. Coulomb's law.

Coulomb's law as an example of the inverse square law. *Forces between coplanar charges.*

2. Electric Fields.

Demonstration of field patterns using oil and semolina or other method. Idea of lines of force. *Electric field intensity, electric flux – vector nature of electric field to be stressed. Conservation of total electric flux – simple applications. Simple calculations of field intensity.*

3. Potential difference.

Definition of potential difference – work per unit charge needed to transfer a charge from one point to another. Definition of volt. Earth at zero potential. The electroscope as an indicator of potential difference.

CAPACITANCE

1. Capacitance.

Definition: $C = Q/V$. Unit of capacitance.

2. Practical capacitors.

Parallel-plate capacitor - demonstration that capacitance depends on the common area, the distance between the plates and the nature of the dielectric. *Derivation of* $C = \epsilon_0 A/d$. *Derivation of formula for energy stored in a capacitor:* $E = \frac{1}{2}CV^2$. Common uses of capacitors.

ELECTRIC CURRENT

1. Sources of electric current.

Van de Graaff generator. Description of a simple cell and a lead-acid accumulator – details of chemical reactions not required. Definition of electromotive force.

2. The ampere.

Demonstration of forces between current-carrying conductors – definition of the ampere. Definition of the coulomb.

3. Heating effect of electric current.

Demonstration of heating effect. Effect of current and time – Joule's law ($P \propto I^2$). Advantage of use of E.H.T. in transmission of electrical energy.

4. Resistance.

Relationship between potential difference and current for various types of conductor (solid, liquid and gas). Demonstration of Ohm's law. Resistance – definition and unit. The resistance of a metallic conductor varies with length, cross-sectional area and temperature. Resistivity. Resistors in series and parallel – *derivation of formulae*. Simple problems. *Kirchhoff's laws and simple networks containing not more than two loops. The metre bridge. The potentiometer. Internal resistance of a source of e.m.f.*

5. Domestic circuits.

Use of fuses, plugs, general household circuitry, including the ring circuit. Earthing and general safety precautions. The kilowatt-hour.

ELECTROMAGNETISM

1. Magnetism.

Demonstration of the magnetic effect of an electric current. Magnetic poles. *Current loops regarded as magnetic dipoles.*

2. Magnetic fields.

Magnetic field due to magnets and to the current in a long straight wire, a loop and a solenoid – demonstration and description without mathematical details. Magnetic flux.

Magnetic flux density. The earth's magnetic field – concept of dip and declination. *Vector nature of magnetic field to be stressed.*

3. Current in a magnetic field.

Force on a current-carrying conductor in a magnetic field depends on the current, the length of wire and the magnetic flux density: $F = IlB$. *Derivation of $F = qvB$. (Students should be made aware that it is the component of B that is perpendicular to the current which is instrumental in producing the force on the current.)* Demonstration of force on a coil in a magnetic field. Simple d.c. motor. Principle of operation of a moving-coil loudspeaker and electromagnetic relay. Principle of a moving-coil galvanometer. Conversion of a galvanometer to an ammeter, to a voltmeter and to an ohmmeter.

ELECTROMAGNETIC INDUCTION

1. Laws of electromagnetic induction.

Faraday's law. Lenz's law. Simple calculations. *Concept of mutual inductance and self-inductance.*

2. Practical applications.

Structure and principle of operation of simple a.c. and d.c. generators. Variation of voltage and current with time, i.e. alternating voltages and currents. *Peak and r.m.s. values of alternating currents and voltages.* Structure and principle of operation of induction coil and transformer. Uses of generators, induction coil and transformer. Simple problems on transformers. *Effect of inductors and capacitors on a.c. – without mathematics or phase relationships. Principle of the induction motor.*

CHEMICAL EFFECT

1. Electrolysis.

Conduction due to ions. Simple examples of electrolysis, e.g. in molten sodium chloride or in copper sulphate solution.

2. Faraday's first law.

Electrochemical equivalent. Electroplating.

EXPERIMENTS

1. Verification of Joule's law.
*2. Measurement of the resistivity of the material of a wire.
*3. To investigate the variation of the resistance of a metallic conductor with temperature.
4. *Measurement of the internal resistance of a source of e.m.f.*
*5. *To investigate the variation of the resistance of a thermistor with temperature.*
6. Measurement of electrochemical equivalent.

* An ohmmeter or a metre bridge may be used in these experiments.

ATOMIC PHYSICS

THE ELECTRON

1. The electron.

The electron as the indivisible quantity of charge. Reference to mass and location in the atom. *The principle of each experiment: (i) to determine the electron charge; (ii) to determine the charge to mass ratio for the electron. These experiments may be simulated on a large scale model or on a computer.*

2. Thermionic emission.

Principle of thermionic emission and its application to the production of a beam of electrons (cathode rays). Use of cathode ray tube to demonstrate the production of a beam of electrons and their deflection in electric and magnetic fields. *The action of the grid and the need for the focusing anode.*

3. Photoelectric emission.

Effect of intensity of light on the photocurrent. *Relation between frequency of light and kinetic energy of electrons – Einstein's photoelectric law. Threshold frequency. The photon.*

4. X-rays.

Principles of the hot cathode X-ray tube. Detection and uses.

SEMICONDUCTORS

1. Conduction in semiconductors.

The distinction between intrinsic and extrinsic conduction. P-type and n-type semiconductors.

2. The p-n junction.

Basic principles underlying current flow across a p-n junction.

3. Applications of the diode.

P-n diode used as a half wave rectifier. *Light emitting diode (LED): Principle of operation. Practical applications. AND and OR gates: Construction of AND and OR gates using diodes and a resistor.*

4. The transistor.

Basic structure of a bi-polar or a uni-polar (field effect) transistor. *Basic structure and simple descriptive account of the basic principles of (i) bi-polar (npn) transistor, (ii) uni-polar (n-channel) transistor. Operation of a bi-polar transistor in a common emitter circuit.*

5. Applications of the transistor.

Demonstration of the switching action of a transistor. Applications of the transistor as a switch should be indicated, e.g. to switch a relay. *Demonstration of the transistor as a voltage amplifier – the purpose of bias and load resistors. Demonstration of the transistor as a voltage inverter and in the implementation of the NOT function.* *Reference to integrated circuits.*

6. Effect of light and temperature on semiconductors.

Light dependent resistor, e.g. cadmium sulphide (CdS) cell. The thermistor and its use as a thermometer. *Reference should be made to the non-linear behaviour of resistance with temperature. The photo-diode.*

THE NUCLEUS

1. Structure of the nucleus.

Atomic nucleus as protons plus neutrons. Mass number, atomic number, isotopes. *Reference to the order of magnitude of the radius of the nucleus – principle of Rutherford's experiment. This experiment may be simulated on a large scale model or on a computer.*

2. Radioactivity.

Nature and properties of alpha, beta and gamma emissions. Demonstration of the ionising effect of the radiations by means of a cloud chamber or other suitable method. Experimental evidence for the relative ranges of the radiations in air. Principle of operation of ionisation chamber, G-M tube or solid state detector. Law of radioactive decay. Concept of half-life. *Decay constant. Relationship between decay constant and half-life. Numerical problems (not requiring calculus).* Artificial radioactivity induced by neutron capture. Uses of radio-isotopes.

3. Nuclear energy.

Principles of fission and fusion. Mass-energy conservation in nuclear reactions. The nuclear reactor (fuel, moderator, control rods, shielding and heat-exchanger). Outline of fission reactions in uranium 235; breeding of plutonium in uranium.

4. Radiation and health hazards.

General health hazards and precautions in the use of ionising radiations, including X-rays.

EXPERIMENTS

1. Plotting the characteristic curve of a semiconductor diode.
2. *Establishing truth tables for AND, OR and NOT circuits.*
3. Measurement of the half-life of a short-lived radioactive isotope.